Electric Power Plant Engineering

Joshua Weingreen

MAXWELL PRESS

Chennai Trichy New Delhi

MAXWELL PRESS

This edition has been published in india by arrangement with Carson Books, UK

ISBN 978-93-5528-180-7

Maxwell Press
No. 44, Nallathambi Street,
Triplicane, Chennai 600 005

Printed and bound in India

MJP 1273

Publisher : C. Janarthanan

Electric Power Plant Engineering

MAXWELL PRESS

Publisher's Note

The legacy of a country is in its varied cultural heritage, historical literature, developments in the field of economy and science. The top nations in the world are competing in the field of science, economy and literature. This vast legacy has to be conserved and documented so that it can be bestowed to the future generation. The knowledge of this legacy is slowly getting perished in the present generation due to lack of documentation.

Keeping this in mind, the concern with retrospective acquiring of rare books has been accented recently by the burgeoning reprint industry. Maxwell Press is gratified to retrieve the rare collections with a view to bring back those books that were landmarks in their time.

In this effort, a series of rare books would be republished under the banner, "Maxwell Press". The books in the reprint series have been carefully selected for their contemporary usefulness as well as their historical importance within the intellectual. We reconstruct the book with slight enhancements made for better presentation, without affecting the contents of the original edition.

Most of the works selected for republishing covers a huge range of subjects, from history to anthropology. We believe this reprint edition will be a service to the numerous researchers and practitioners active in this fascinating field. We allow readers to experience the wonder of peering into a scholarly work of the highest order and seminal significance.

MAXWELL PRESS

PREFACE TO THIRD EDITION

A thorough revision of the text has been made in this edition in order to record the progress of the electrotechnical industries since the second edition was printed. There are new chapters and paragraphs on oil switches, open-air switches and lightning arresters, as well as on sub-stations with special regard to their use in out-door service. In addition, the author treats in greater detail the use of reactive coils and synchronous condensers.

The chapter on wall outlets has been supplemented and brought up to date by a description of the latest types of construction, and the chapters on central stations and sub-stations have been revised in accordance with present practice.

The third edition thus becomes virtually a new and enlarged treatise. It is the hope of the author that it may receive from students and practicing engineers an even more favorable reception than was accorded to the previous editions.

J. WEINGREEN.

CRACOW, POLAND,
January, 1922.

PREFACE TO SECOND EDITION

The present revision has been made to include recent progress and development in the art of energy control as applied in the generation, transmission and distribution of electrical energy.

The chapters on storage batteries, potential regulators and lightning arresters have been supplemented by descriptions of European apparatus and practice, as it is believed that knowledge of such practice will be of interest and value to American engineers.

The entire book has been corrected and revised to accord with present practice and has been brought fully up to date.

The author desires to express his appreciation of the kind reception accorded the first edition. He hopes that in its revised form it will prove still more useful, and that it will continue to serve as a practical handbook to students and practicing engineers.

J. WEINGREEN.

CRACOW, AUSTRIA,
June, 1913.

PREFACE TO FIRST EDITION

In the last decade the electrical industry has developed to such an exceptional extent that it has become desirable to formulate some sort of rules and regulations which may be used as guides in the various problems of construction arising in every-day practice. Up to the present there has been published but little suitable literature dealing with the control of the generation and the distribution of electrical energy. The author therefore feels justified in offering this treatise to the technical public, with the hope that it may at least partially fill the void. Its object is to offer to the contractor and engineer, as well as to the student, material which will help them to understand the methods of handling electrical energy. It is assumed that the reader is familiar with the basic principles of electrical engineering, as well as with electrical machinery and the ordinary instruments. The aim has been throughout to restrict theoretical discussions as much as possible and to eliminate higher mathematics. The book is intended as a useful handbook for those concerned with practical problems.

I have limited the work to American power station engineering. The material represents exclusively present-day practice, and the lines which future development may be expected to follow are pointed out.

All conclusions set forth herein are based on personal experience and a careful study of recognized standard constructions and the opinions of such authorities as Professor Charles P. Steinmetz, Dr. Louis Bell, Frank J. Sprague, W. C. Gotshall, Paul M. Lincoln, Professor Dr. F. Niethammer, Stephen G. Hayes, and others.

I am particularly indebted, for assistance and information, to the following organizations: American Institute of Electrical Engineers; Electric Storage Battery Company; Ford, Bacon and Davis; General Electric Company; Gould Storage Battery Company; Hartman Circuit Breaker Company; Indianapolis and Louisville Traction Company; Interborough Rapid Transit Company; Locke Insulator Manufacturing Company; New York

Edison Company; Westinghouse, Church, Kerr and Company; and Westinghouse Electric and Manufacturing Company.

I desire also to recognize the assistance rendered by the publishers of the *Electric Railway Journal*, the *Electrical World*, the *Electric Journal*, the *Electrical Review and Western Electrician* and the *Elektrotechnische Zeitschrift.*

The author further desires to express special gratitude to Mr. Trifon von Schrenk, C. E., for assistance in editing and proof-reading.

J. WEINGREEN.

NEW YORK,
July, 1909.

CONTENTS

ELECTRIC POWER PLANT ENGINEERING

CHAPTER I

INTRODUCTION

With the development of modern systems of electric traction, power and light distribution, and with the increase in size of the electrical machines, it has become necessary to introduce a prompt and reliable means of control which will serve at any time as an indicator of the quantity and efficiency of the power generation and distribution. The subdivision of systems into central and sub-stations, often far apart, has further made it imperative to protect expensive apparatus and machinery, inside and outside the stations, against any harmful effects of disturbances in the system. These functions are performed by the "switchgear." Under this collective term we include all apparatus, instruments, cells, and compartments with their connections, as well as accessories and place of installation, as distinguished from the term "switchboard." Although this formerly referred to the then simple switching system in the station, it has come to apply to that part of the "switchgear" which is assembled on a row of slate or marble panels. The functions of the switchgear may be summed up as follows:

1. To start the machines, to maintain them in service, or to cut them out of the system.

2. To gather and distribute the electrical energy, to control its consumption and output, and to record its characteristic fluctuations.

3. To afford protection against disturbances due to short-circuits, lightning, overload, or other causes, either in the entire system or in any individual portion of it.

4. To afford protection for operators.

The switchgear in general is the flexible link between the source of supply and the consumer. In designing such a linkage, one must take into consideration the best economy, reliability, adaptability, and efficiency of the service in the near future, as well as the unknown conditions of the more remote future.

The main part of the switchgear is the switchboard. It constitutes the nerve center of the entire system, whence the four above-mentioned functions are performed. We distinguish between two kinds of switchboards:

A. Direct-control switchboards, which have all instruments and apparatus fastened on them (direct-current and low-tension alternating-current switchboards).

B. Distant-controlled switchboards, which have the busbars, oil switches, circuit breakers, etc., located away from the panel, and hence operated by means of cranks, levers, toggles, chains or gears, or by means of motors or solenoids (high-tension and extra-high-tension switchboards).

In order to be able to perform its functions the switchgear must respond to the following conditions:[1]

1. The apparatus and their manner of installation must be fireproof (installed on slate or marble).

2. The conduits and connections which carry current must be so chosen as not to become overheated.

3. All parts must be easily accessible.

4. All live parts with the exception of those of low tension must be kept away from the front of the board. In general, high tension should not be carried to the board.

5. The arrangement of the wires should be symmetrical and as simple as possible.

6. As far as practicable and without unnecessary complication of the arrangement of the apparatus, it should be rendered impossible to make a wrong connection which would result in serious consequences.

7. The possible enlargement of the switchboard should be taken into consideration.

8. A disturbance in one part of the board should not affect the entire system.

9. A sufficient number of safety devices should be provided.

10. All necessary instruments and apparatus for the operation and control of the output of the generators and feeders should be installed.

The apparatus and instruments used must naturally be suited to special cases; low-tension d.c. requires different apparatus from a high-tension a.c. The apparatus must be adapted to the kind

[1] "The Standard Handbook for Electrical Engineers." McGraw-Hill Book Company, N. Y.

of current, d.c. or a.c., to the voltage, to the kw. rating of the units and to the local conditions of the plant. We accordingly divide switchgears into two main groups, i.e.:

d.c. switchgear.

a.c. switchgear.

The first group includes mainly direct-controlled, and the second, both direct- and distant-controlled switchboards.

It is characteristic of American practice that products manufactured on a larger scale are standardized, a method which allows of better, quicker, and cheaper construction of products, together with more accurate and economical design. This also has reference to the production of switchboard panels. The installation of such standard panels in a station requires a minimum amount of time, money, and intelligence, which is of especial advantage in the delivery of such material to foreign lands. The panels are carefully tested before shipment. (The General Electric Company's test voltage for instruments rated up to 1,000 volts is 2,500 volts, and double the service voltage for instruments of higher rating.) The panels most often standardized are those used in railway service and, to a certain extent, those for lighting purposes. Low- and medium-voltage boards are distinguished by the fact that a separate panel is supplied for each generating or feeding unit. The entire board consists of a number of unit panels or groups. Separate panels are required for each generator, d.c., synchronous, or induction motor, for the d.c. and a.c. sides of synchronous converters, for transformer sets, storage batteries, groups of arc or incandescent lamps, for every feeder or group of feeders, and in large stations for the power-house instruments.

Another division is sometimes made according to kind of current and system, as, for example, single, double, three, four, and six-phase a.c. system, or two, three, or five-wire d.c. system. A large number of combinations is therefore possible, covering all cases arising in practice.

Although American switchboard arrangements are less decorative and simpler than those in European use, they nevertheless afford easier orientation and safer operation. A single glance at the board will at once reveal the number of the different units in the plant and the ways and means for their control. The symmetrical arrangement of connections on the back of the board makes possible an easy and safe access, and facilitates tracing of connections.

CHAPTER II

DIRECT-CURRENT GENERATORS

In the last decade electrical energy in the form of direct current has predominated for traction, as well as for light and power distribution. Due to the increase in variety and size of the applications of electrical energy and the advance in inventions and improvements in the field of a.c. apparatus, especially motors, the d.c. is being steadily replaced by the a.c. This does not mean that the a.c. is in all cases a substitute for the d.c., for, on account of its characteristics, the d.c. is in some instances indispensable, while under other special conditions it may be more advantageous.

We will classify the d.c. sources as follows:

1. Generators.
2. Converters (synchronous converter, mercury rectifier).
3. Storage batteries.

Figure 1 shows a wiring diagram of a generator equalized on the negative. The generator has a compound field winding consisting of a shunt and series coils. The shunt field winding has great inductance on account of a large number of turns. It is connected on one side to the negative and on the other side through a field-discharge resistance switch and a variable resistance to the positive. When the generator is started the residual field is sufficient to generate a low voltage which then builds up rapidly, the field current and machine voltage mutually reacting to increase each other. The machine voltage does not increase in proportion to the field current. After a short time it reaches a certain value which is approximately the normal voltage of the machine. The coils of the series winding contain few turns of low resistance and are wound on the same pole as the shunt coil. The shunt winding is predominant in its effect, and the series winding may either intensify or oppose the magnetism produced by the shunt winding. To run compound machines successfully in parallel it is necessary to connect their series field windings in parallel through a low-resistance connection, so that if the load on one machine increases, the additional current will divide through the series coils of the other machines and raise their voltage correspondingly. This connection between the machine sides of the series coils is called the equalizer. We therefore have in the case of several generators one positive, one negative, and one equalizing bus, each of which connects the respective poles

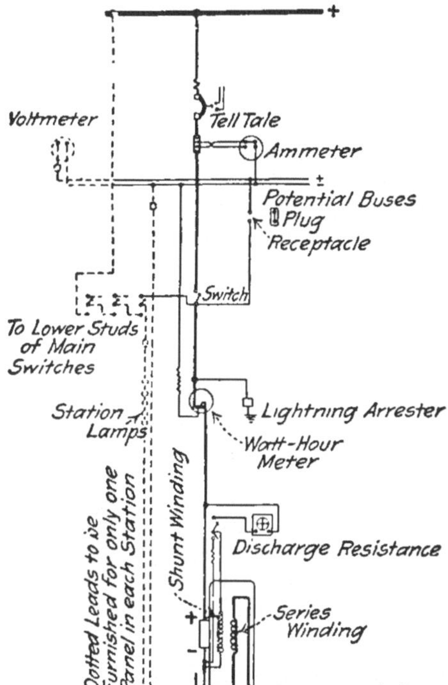

FIG. 1.—Wiring diagram of a single-pole, direct-current generator, panel generator equalized on the negative side.

of the machine in parallel. The method of connection shown is used in the case of large central stations with several generators, and affords a considerable saving of copper in cables. The positive bus is mounted on the back of the switchboard, while the negative and equalizing buses are installed in the foundations of the machine. The apparatus for putting the machine in and out of service and for regulating and recording the current and e.m.f. consists of a quick-break lever switch, which is one of the main connections between the positive brush of the generator and the positive busbar. In case it is necessary to repair one of the machines it must be disconnected from the live bus, for although the machine is not supplying energy, there is, nevertheless, a current from the bus to the machine if the connections are not open, a condition which would render handling of the machine parts unsafe. The lever switch in the connection to the equalizer bus and the automatic circuit breaker in the series-winding circuit serve the same purpose. In addition to the above, a further means of disconnecting the machine is provided in the form of an automatic circuit breaker which is connected to the bus on the positive side of the machine. This apparatus operates in case the positive side of the machine is accidentally grounded. Since the negative is previously grounded (in case of railways, being connected with the rail), an excessive flow of energy through the earth would take place, which could seriously damage the machine. The rheostat for adjusting the field current of the machine is generally operated directly from the switchboard by means of handwheels, chains, and sprockets. In case it becomes impossible to operate from the switchboard, on account of the large size of the rheostat or lack of space for setting it up near the switchboard, it is operated by means of a pilot wheel mounted on a tripod in front of the panel corresponding to the respective generator. This facilitates reading the instruments which indicate the necessary adjustment.

In general, a rheostat is a device for dissipating electrical energy which has been converted into heat. Therefore, it should not be boxed up or installed in confined, out-of-the-way places but should stand in large open areas to which cooling air currents will have ready access. (See Fig. 2.)

For generators, ammeters are always necessary. They indicate whether or not the load is properly divided between them. The connection is made by means of a shunt. It is desirable to locate the shunts as near the instruments as possible, thus avoiding the use of long leads. One voltmeter is sufficient to indicate the voltage of all the machines or busbars, the connection being made through plug switches. The voltmeter is protected against abnormal voltage by a renewable fuse. For large boards the voltmeter is placed on a swinging bracket on the side of the generator panel. Watt-hour meters are desirable in so far as they indicate the energy output of the individual generators.

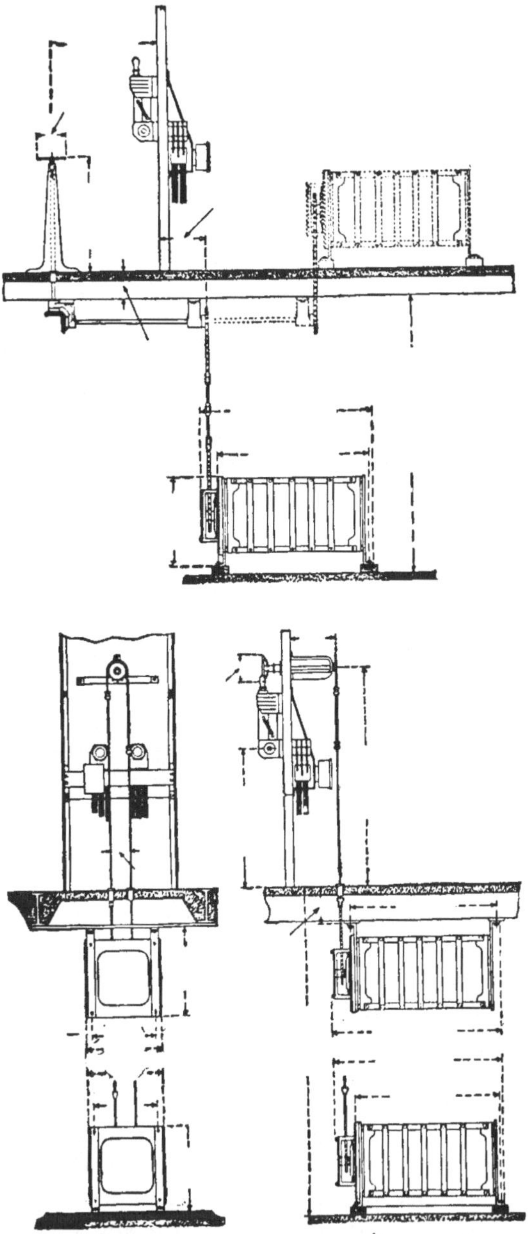

Fig. 2.—Methods of mounting field rheostats.

Potential wires across the backs of all the panels serve to ease the connections for the respective instruments. They are connected to the positive and negative sides of the machine at any convenient points. A lightning arrester is inserted between the quick-break switch and the watt-hour meter in order to protect the machine and instruments against lightning. A num-

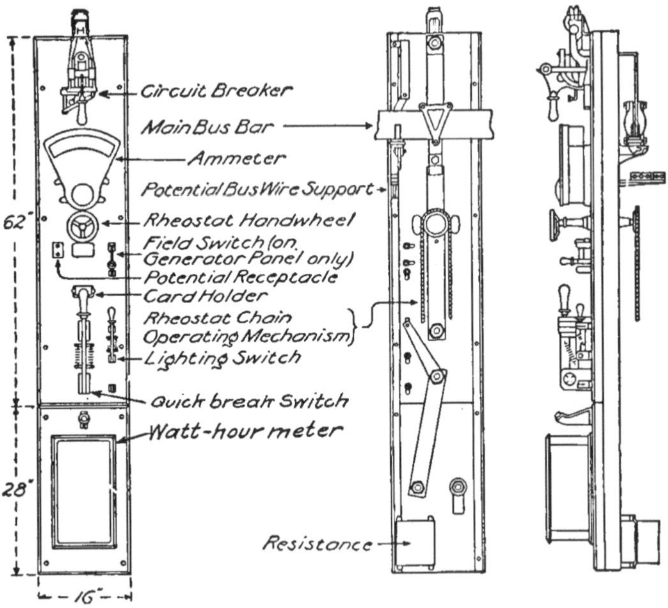

Fig. 3.—Generator or synchronous converter panel for ratings not exceeding 1,200 kw.

ber of incandescent lamps may be inserted across the main voltage as shown by the dotted line on the diagram. In order to call the attention of the operator to the fact that one of the circuit breakers on the positive side has opened, an electric tell-tale is connected with it, which rings when the circuit breaker opens. Instead of tell-tales, signal lamps are sometimes employed. Figures 3 and 4 show the arrangement and dimensions of generator panels for various kw. ratings as manufactured by the General Electric Company. Such panels can be used when either the positive or negative bus is mounted on the board. In the latter case a change in the connections between the ammeter and the shunt and a different location of the lightning arrester is necessi-

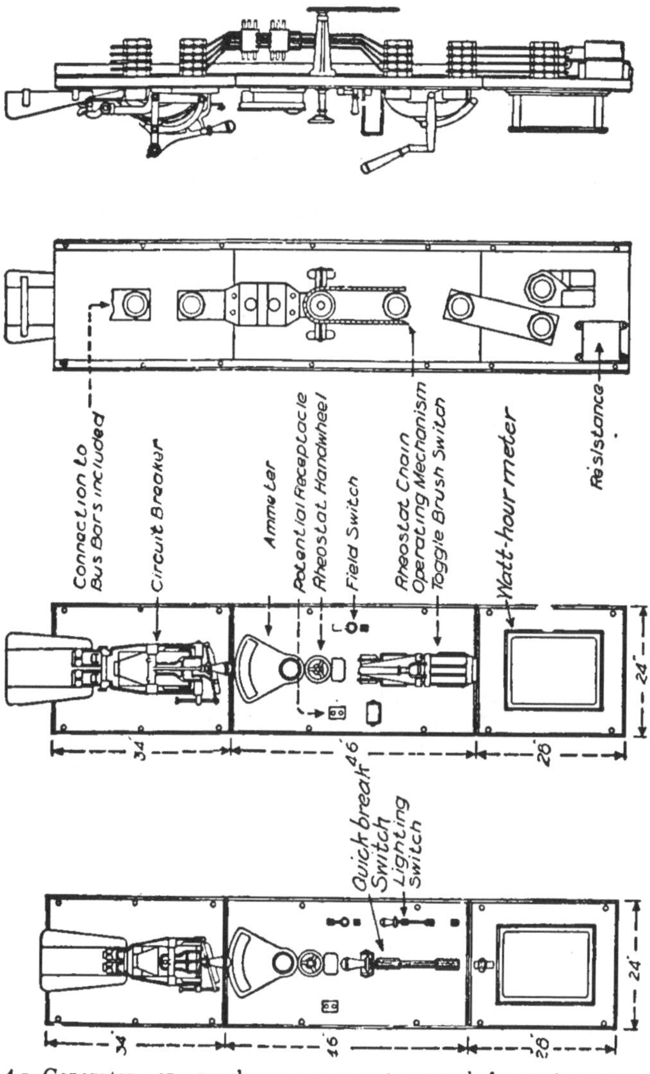

FIG. 4.—Generator or synchronous converter panel for ratings exceeding 1,200 kw.

tated. (See Fig. 5.) This condition often obtains in practice in small central stations. The negative bus on the board is grounded, while the generator is equalized on the positive side.

The panels are made of slate or of white Italian or blue Vermont marble, 2 inches thick and from 16 to 24 inches in width,

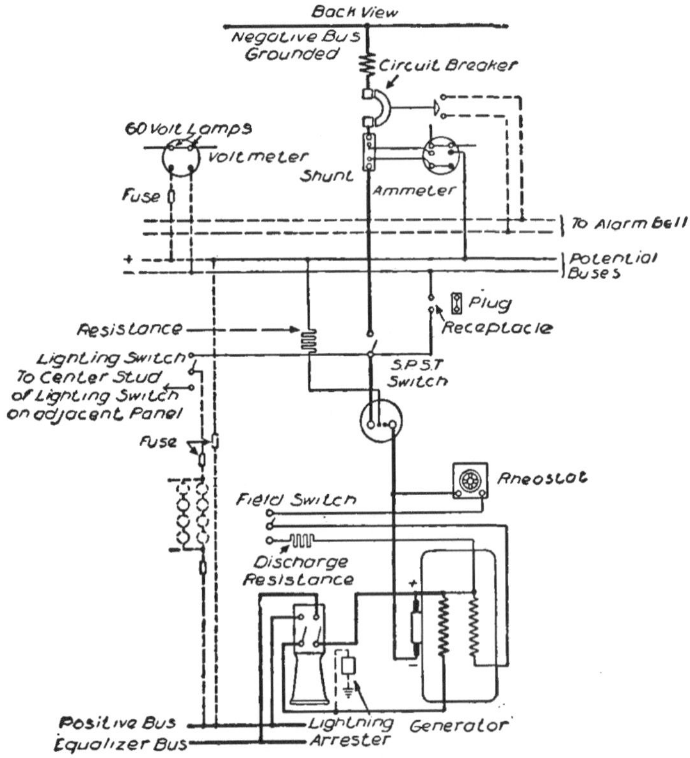

FIG. 5.—Wiring diagram of a single-pole, direct-current generator panel, generator equalized on the positive side.

and are built in two or three sections, giving a total height of from 90 to 108 inches (G. E. make). The supporting framework is composed of angles and tees. Gas pipes connected to the panels with movable cast-iron clamps have recently come into use. For connections between different instruments, apparatus, and busbars for higher current values, copper strips are employed, allowing one square inch per 1,000 amp. Standard thicknesses and widths are used, sometimes combining several strips for one connection, with a ventilating space between them.

Aluminum may be used in place of copper, in which case the current density may be taken as 750 amp. per sq. in. These strips must be carefully bent in order to bridge other strips, or contact studs of equal or different polarity. For voltages

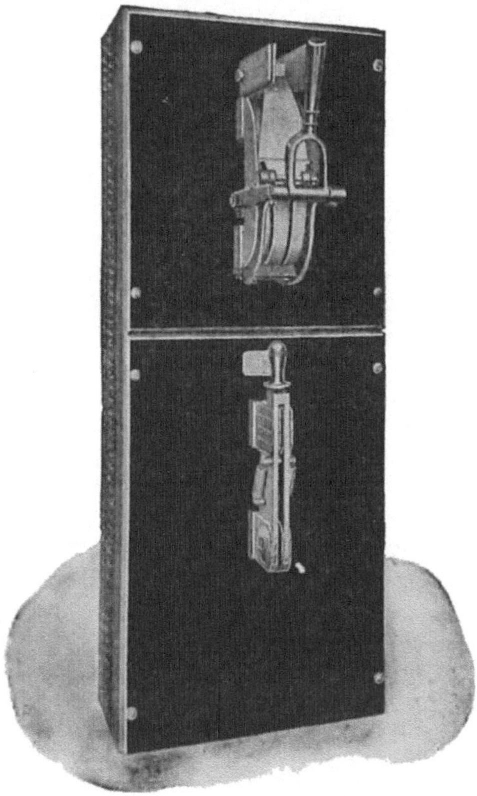

FIG. 6a.—Panel for main and equalizer switches. (Front view.)

up to 600 it is customary to allow a minimum space of one inch between connections or between live parts and ground. The Westinghouse Company employs elongated contact bolts in place of bent copper strips. These bolts are copper rods of convenient lengths with brass castings similar to a union and with both right- and left-hand threads. A straight connection between one set of bolts bridges a similar connection between another set. The standard sizes of busbars are 2 inches by 0.25

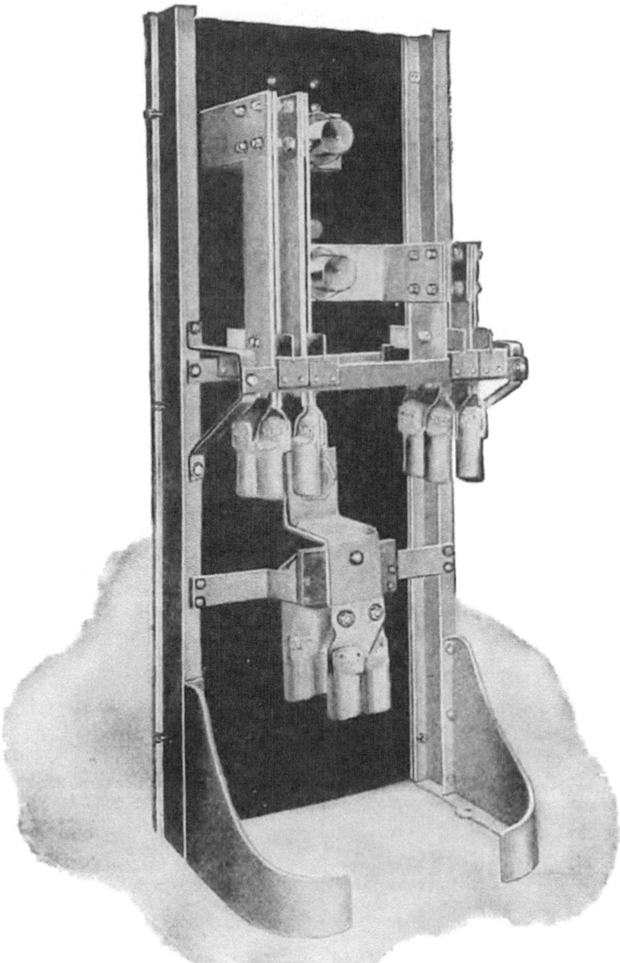

FIG. 6b.—Panel for main and equalizer switches. (Back view.)

inch, 3 inch by 0.25 inch, 5 inch by 0.25 inch, and 10 inches by 0.25 inch. These and other sizes are used for connections. All live parts are designed for a rise in temperature not exceeding 20° C. at normal load. The apparatus should be designed to perform its rated functions accurately and safely, to meet successfully the most severe conditions that may be imposed upon it, and to present a handsome appearance.

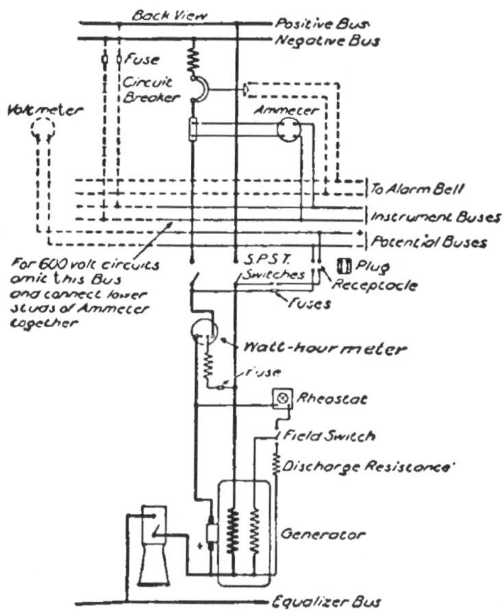

FIG. 7.—Wiring diagram of a double-pole, direct-current generator pane generator equalized on the positive side.

The circuit breaker for the negative side and the equalizer switch in Fig. 1, and the equalizer switch and positive switch in Fig. 5, are mounted on separate pedestals or panels near the machine so as to facilitate handling. (Fig. 6.)

Up to this point we have treated switchboards having one of the buses either positive or negative mounted on the back of the board. This construction has the advantage of greater insurance against short-circuiting, since with the uninsulated live parts only one polarity prevails. These switchboards are mainly employed for railway service up to 600 volts. For 125, 250, and 600 volts, and from 25 to 1,000 kw., double-pole switchboards are often used, having both buses mounted on the back of the board,

They must, therefore, have two lever switches on the board for both current directions. Hence, the pedestal near the machine holds only the equalizer switch. In all other respects connections, instruments, etc., are identical with those of the first-mentioned case. Figure 7 shows a wiring diagram, and Figs. 8 and 9 show a switchboard for this class of machine. The mounting of the

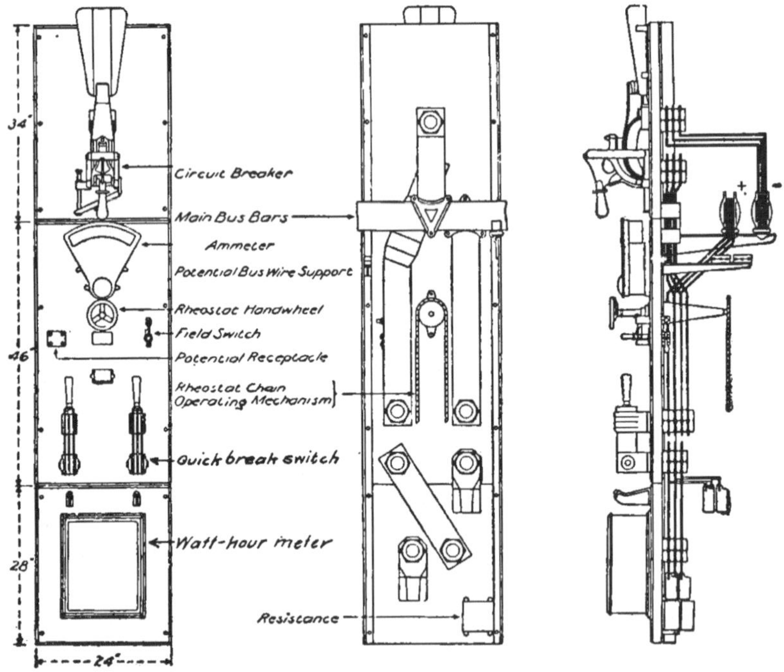

FIG. 8.—Double-pole generator panel for 550 kw., 250 volts.

positive and negative buses should be noted. They are of the same standard width and thickness as noted above. The buses are designed to carry 1,000 amp. per sq. in. for the normal load, plus a certain guaranteed percentage for overload. They are mounted on insulators attached to brackets, giving for the positive bus a distance of 5 inches from the board for normal sizes of apparatus and 11 inches for larger sizes. The negative bus is 6 inches from the positive one. (Fig. 10.) Since we have a fixed distance between the busbars and the upper edge of the panel and assumed distances between instruments, we can easily lay out with great accuracy the bending and drilling of copper

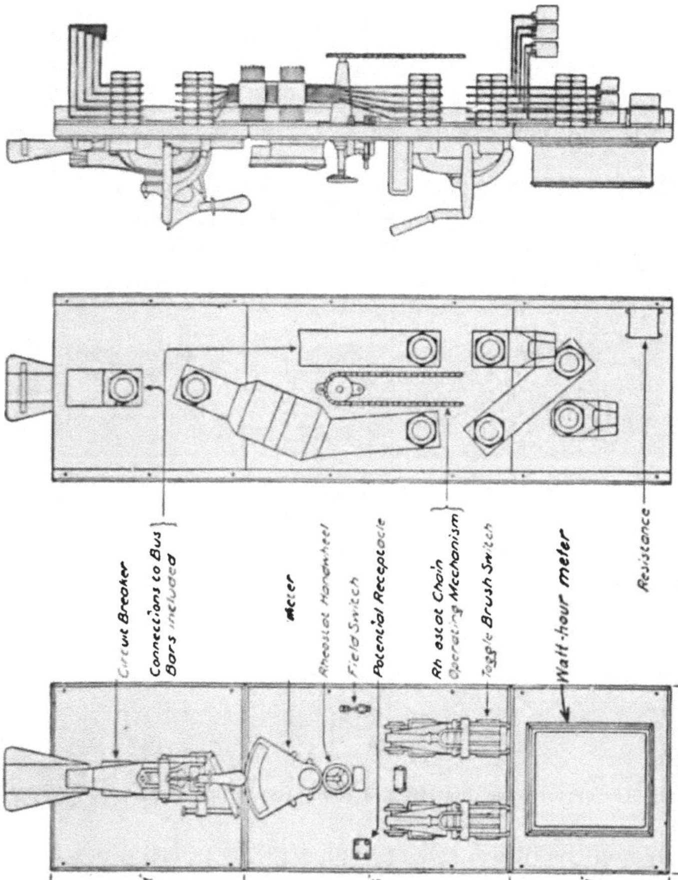

FIG. 9.—Double-pole generator panel for ratings exceeding 550 kw.

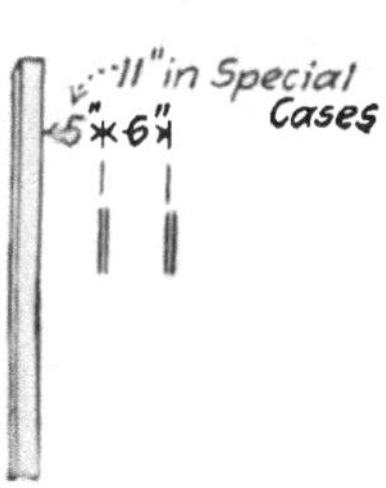

FIG. 10.—Method of mounting of direct-current busbars.

connections. When strips of over 2 inches in width are to be connected with busbars of 3 inches, 5 inches, or 10 inches, we may use special cast-iron clamps instead of resorting to drilling, riveting, and soldering. (Fig. 8.)

For generators of smaller rating and voltage, as for instance 25 to 100 kw., and 125 to 250 volts, the equalizer bus is also

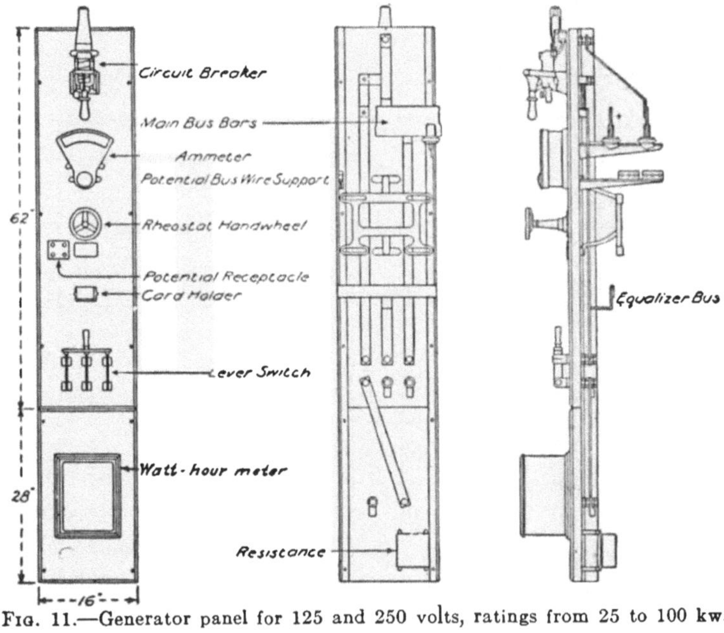

FIG. 11.—Generator panel for 125 and 250 volts, ratings from 25 to 100 kw.

mounted on the board. This eliminates the pedestal in front of the machine, and in place of it a three-pole lever switch on the board is used, which takes care of the positive, negative, and equalizer sides of the machine. (Fig. 11.) The rheostat is much smaller and is mounted on a tripod or on pipe supports on the back of the board.

In order to throw one generator into parallel with others in service, proceed as follows:

1. Close main and equalizer switches (on pedestal or panel near machine).[1]

[1] If one or both lever switches are on the board, it is obvious that they perform the same functions as stated above. Hence the same procedure should be followed as here indicated.

2. Close field switch (on panel).
3. Close circuit breaker.
4. Insert potential plug in receptacle and adjust voltage by means of rheostat.
5. When the proper voltage is obtained, close the other main switch.

CHAPTER III

SYNCHRONOUS CONVERTERS

A synchronous converter is essentially a d.c. generator, possessing, besides the ordinary commutator, several collector rings which are connected with the armature winding at corresponding points. When this machine is driven through the application of mechanical power it delivers either direct or alternating current. If, on the other hand, electrical energy is employed to drive the machine, available mechanical energy will be the result. It is thus evident that the machine can be run either as a synchronous or d.c. motor or as a converter of a.c. into d.c. or *vice versa*. In this chapter we will treat only the d.c. side of the converter.

In the first case, when the converter acts as a d.c. generator, the switching arrangements are almost identical with those of the ordinary d.c. machine. Figure 12 differs from Fig. 1 only in that the negative side of the machine is grounded directly or is directly connected with a grounded negative busbar, so that we have only one lever switch for the equalizing bus. This switch is located near the machine or on the frame of the converter itself. The automatic safety device which guards against grounding is found on the a.c. side of the generator. The automatic circuit breaker on the positive side is supplied with a special coil which opens the circuit breaker at a given low voltage. This winding is connected with a speed-limit device mounted on the shaft of the converter, which closes the circuit of the winding at a given speed limit and operates the circuit breaker. Figures 3 and 4 may also represent converters for which the positive bus is mounted on the board. The series field winding of the converter is connected to the negative side. The above considerations apply to a converter started from the a.c. side, which is the most frequent case in service, because the converter does not require to be synchronized. (See Chap. XXI.) After the converter has been started on the a.c. side, and it is desired to throw it in parallel with other machines on the d.c. side, proceed as follows:

1. Close the equalizing lever switch on the machine.

2. Close the automatic circuit breaker on the board.

3. Ascertain the voltage, using plug switch and voltmeter on movable arm.

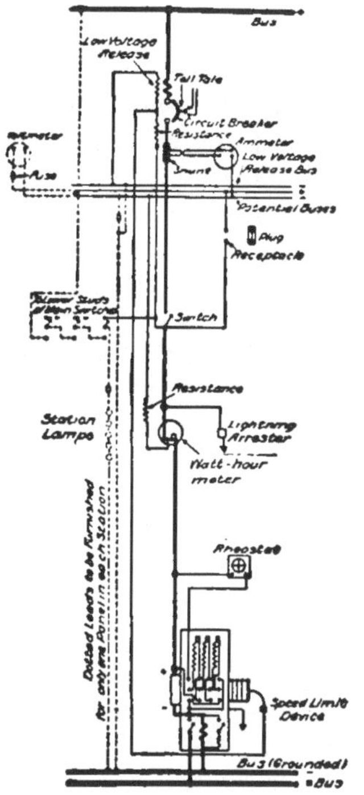

FIG. 12.—Wiring diagram of a continuous-current synchronous converter panel.

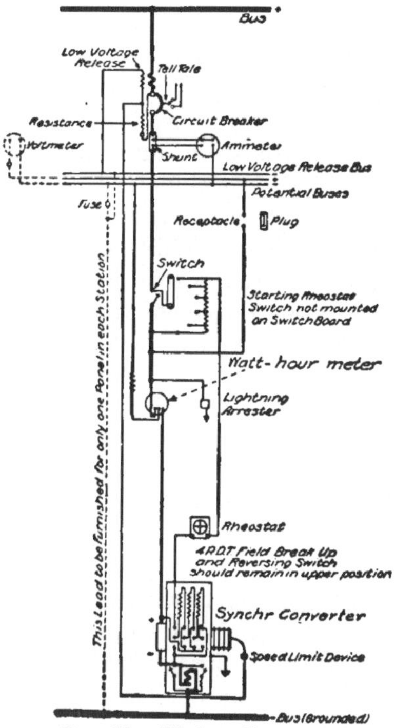

FIG. 13.—Wiring diagram of a single-pole, direct-current inverted converter panel.

4. When the desired voltage is reached close the positive lever switch on the board.

If the converter is to be started on the d.c. side, the switchboard must be provided with a starting rheostat switch. The converter must be synchronized with the line. This must also be done when the machine is started by a separate induction motor. (See Chap. XXI.) To start, proceed as follows:

1. Close the field switch.

2. Close the main switch, allowing the starting rheostat to remain in circuit.

3. When normal speed is reached gradually throw out the starting rheostat.

4. Change the field strength by means of the rheostat until the synchronism indicator shows equal synchronism with the a.c. generator.

5. Close the a.c. oil switch.

Figure 12 shows a compound field converter—hence the equalizer buses. Such machines are used for variable load in traction systems. In case converters with only shunt field windings are to be run in parallel, the equalizer bus is omitted. This type of machine is especially adapted to electric lighting or electrochemical purposes where the d.c. voltage requires special control. This control is taken care of on the a.c. side by a potential regulator (see Chap. XVII), inserted between the low-tension side of the power transformer and the converter.

Figure 13 shows a wiring diagram for a d.c. inverted converter. The speed of this machine, like that of a d.c. motor, depends essentially on the field strength. The speed increases with a decrease in field strength, and *vice versa*. It follows that the series field should be weak, for otherwise we should have a constant change in speed, giving rise to variable frequency in the delivered current. A lagging current weakens the field, thus increasing speed and frequency, and making it possible under certain conditions with inductive load for the machine to run away. Particular attention should therefore be directed toward maintaining sufficient field strength, in order to avoid excessive speed, particularly when the given converter drives a second machine which reconverts the a.c. into a d.c. Speed-limit devices capable of operating an automatic circuit breaker are mounted on the shafts. The starting rheostat is usually mounted separately, or in case the watt-hour meter is located off the board, the starting rheostat is put in its place on the base of the panel.

CHAPTER IV

MERCURY RECTIFIERS

Of late a new system for converting alternating into direct current has been developed. Up to the present this system has been most widely used in electric automobile service, or wherever batteries are charged. Formerly in cases where small batteries were to be charged, and where no low-voltage direct current was available, it was necessary to provide very costly and cumbersome apparatus. The best known devices for this purpose were:

1. A motor-generator set. Of this the disadvantages are high cost, large consumption of floor space, and requirement of higher intelligence for operating. The efficiency at full load for charging batteries is comparatively low, and at light load, very low.

2. Single-phase synchronous converter. This is not as flexible as the motor-generator, particularly as regards voltage, and the higher intelligence requisite for starting and operating adds another disadvantage.

3. Synchronous or mechanically drivenre ctifier. This, though small, requires considerable attention, as the d.c. brushes are apt to spark badly and require constant renewals.

4. Chemical rectifier. This machine has not justified itself in practice on account of the variability and uncertainty of the charge. Its efficiency is low under all conditions.

The mercury rectifier has none of these disadvantages. Its cost is low, it requires little space, and its efficiency of conversion at low or full load is high. It is flexible, safe in operation in that it is impossible to discharge the batteries by a reversal of current, and it has no moving parts. It is on the whole a very simple machine. The only disadvantage lies in the fragility of the glass bulb, but this is compensated for by its low cost. The process of rectifying is based on the fact that it is difficult to excite a cathode in mercury vapor. Since an electrode cannot become a cathode by itself, the current must always have one direction, from the electrode to the vapor. The theory of this phenomenon is treated in different ways by Dr. C. P. Steinmetz and by Peter Cooper Hewitt. The equipments of the apparatus as manufactured by the two companies (General Electric Company and Westinghouse Company) do not differ essentially from each other. Figure 14 shows a general wiring diagram of the mercury rectifier of the General Electric Company. The

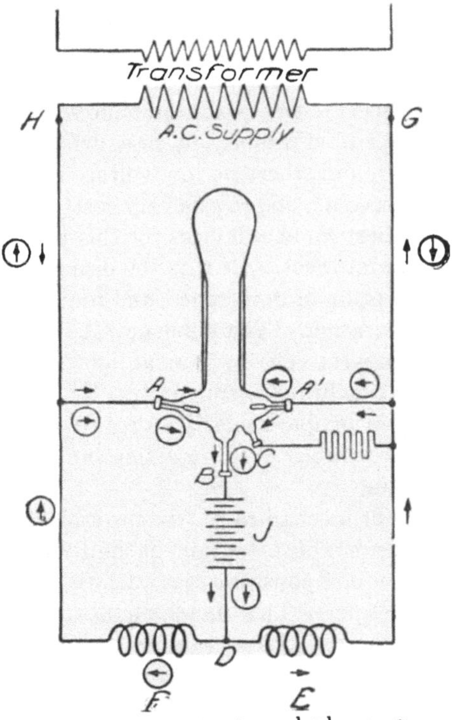

FIG. 14.—Diagram of connection of a General Electric Co. mercury rectifier.

secondary winding of a transformer, which reduces the available alternating e.m.f. to a given value (usually 110 or 220 volts, 60-cycle, single-phase), is connected in parallel with the anodes *A* and *A′* of a glass tube containing mercury vapor in vacuum, and with two reactors, also in parallel. The cathode *B* is connected through the load with the binding post between the two coils.

According to the theory of Dr. Steinmetz, the mercury vapor offers a very high resistance to the passage of electricity; in fact, it may be considered almost a non-conductor. If the vapor is ionized, it becomes a good conductor, but in one direction only. By means of employing a mercury electrode as a cathode, ionized mercury vapor may be liberated. The initial ionization of the mercury vapor is accomplished by a small starting anode, *C*, which is brought into contact with the cathode by a mercury bridge formed by slightly shaking the tube. The breaking of this mercury bridge starts a small initial arc, and the arc thus obtained excites the cathode, giving the necessary ionized vapor,

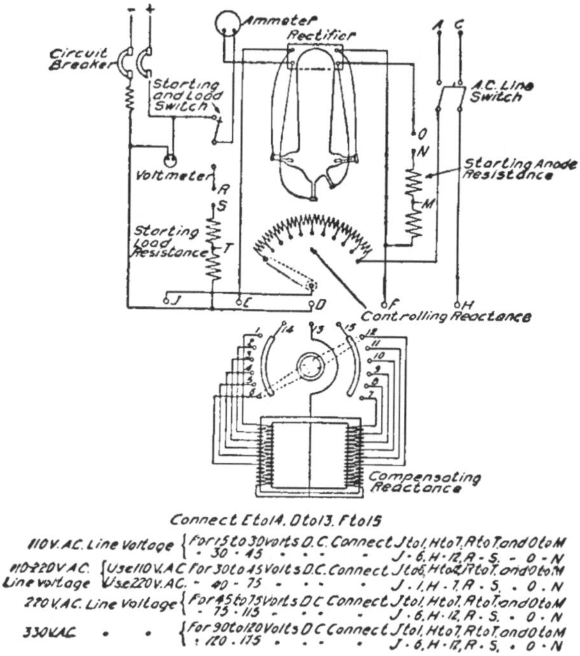

FIG. 15.—Wiring diagram and panel of a mercury rectifier outfit (General Electric Company).

which enables the working anodes immediately to become active and the tube to start. The two anodes, *A* and *A'*, serve as electrodes for the alternating current. The upper halves of the cycles are sent through the ionized vapor alternately by both anodes. Since the displacement between waves sent through both anodes is 180°, the current at the cathode is a pulsating direct current, varying between the values of zero and the maximum. A current of this nature, identical in its characteristics with the alternating current which it replaces, is not serviceable. Although the zero value is but momentary, it is nevertheless sufficient to cause the cathode to lose its excitation. This causes the arc which carries the current between cathode and anode to be extinguished. A device designed to keep the current value constantly above zero is therefore necessary. This is accomplished by the reactors. The coil, E, is charged during the rise of the wave from zero to maximum. The coil discharges according to the laws of induction in the same direction as that of the main current. This has the effect of keeping the value of the current above zero until it meets the rising second wave. The overlapping thus caused maintains the excitation of the cathode and the arc. The same action takes place between the wave of the anode, A', and the coil, F. The resulting current is a pulsating one, but the pulsations are shallower on account of the action of the coils. Figure 15 shows the glass tube with the two anodes, starting anode and cathode, and the wiring diagram of commercial switchboards. On the board are mounted an ammeter, a voltmeter, a double-pole a.c. line switch (for example, to connect secondary winding to transformer), a double-pole d.c. load switch, one double-pole starting switch, one single-pole load switch, necessary fuses and circuit breaker to protect the rectifier from overload. A starting rheostat is mounted on the gas-pipe frame of the panel. The rectifier is started through the rheostat and is then thrown onto the load. A signal lamp connected in parallel with the rheostat is mounted on the board which, when burning, indicates that the load is on the line and that the rheostat should be cut out. The lamp remains dark when only the load is in circuit. The anodes are connected directly with the reactors, and the cathodes with the load. Taps lead from the reactors to a dial switch on the panel, through which the current and e.m.f. can be varied within certain limits. Mercury rectifiers are built for ratings of 10, 20, and 30 amp., for single tubes. They may be operated in multiple by addition of certain auxiliary apparatus and can thus be made to deliver a greater current. They are mostly used with 60-cycle, 110- or 220-volt, single-phase alternating current, and deliver from 16 to 115 volts direct current. At the present time they are coming into use more and more in larger installations for arc lamps, mercury lamps, and magnetite lamps connected in series.

Figure 16 is a general wiring diagram of the Cooper Hewitt mercury rectifier as put forth by the Westinghouse Company. Its action is the same as that discussed above, but the theory of the phenomenon is differently explained by Cooper Hewitt.

The electricity can easily flow into the mercury vapor from a metal or graphite contact connected to a source of power. As soon as the current direction is reversed the solid contact offers a very high resistance to the passage of electricity from the mercury vapor to the contact. This resistance can be overcome by a very

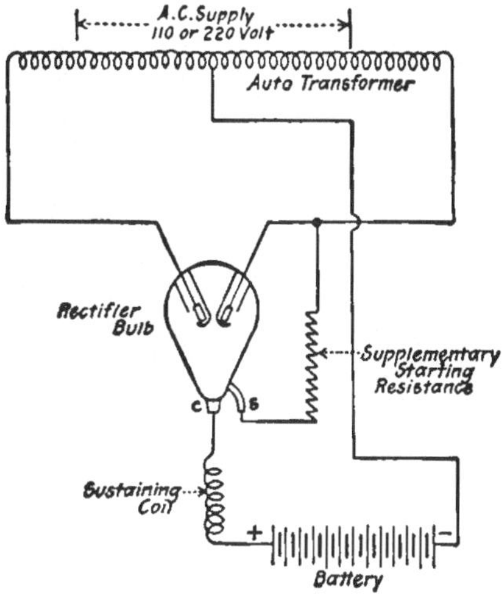

FIG. 16.—Connection diagram for battery charging with a Cooper Hewitt rectifier.

high e.m.f., and as soon as this condition obtains, the normal current can be established from vapor to contact with a low e.m.f., the resistance having practically disappeared. According to Cooper Hewitt's theory, the difficulty encountered in establishing a cathode is due to the great resistance which the latter offers to the passage of electricity at the first instant. As soon as the cathode is established, however, this resistance is minimized. If a single-phase alternating e.m.f. is supplied to the anodes, the action is the same as that described under Fig. 14, where we saw that only one-half of the waves pass through each anode. Since the wave halves are displaced 180° from each other, the resulting current reaches zero value at the end of each wave, which again causes a great resistance at the cathode. The zero value must therefore be bridged over, and this is accomplished by means of the reactors in the connection to the cathode. The applied e.m.f. charges the coil during the rise of the wave and discharges it during the fall of the wave. This causes an elongation of the current waves so that they overlap before reaching the zero value. This overlapping of the rectified current waves reduces the amplitude of the pulsations and produces a comparatively smooth direct current. A momentary metallic contact is brought about

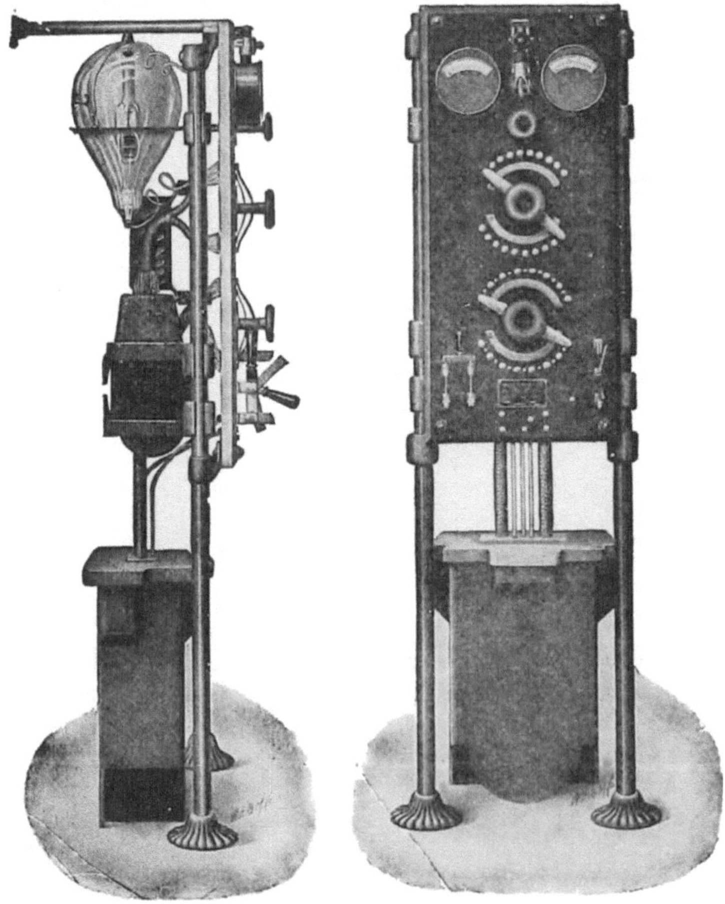

FIG. 17.—Panel outfit of a Cooper Hewitt mercury rectifier.

between cathode and starting anode by tilting the glass tube. When the metallic circuit is opened by bringing the tube back to its original position the current is not interrupted, as the negative electrode resistance is broken down. The apparatus is in no sense a transformer. It does not convert energy from one form into another. Its action is simply that of valves opening and closing gateways and thus allowing electricity of one direction to flow through a given line. Figure 17 shows the Cooper Hewitt mercury rectifier and panel. The auto-transformer is placed on the floor back of the panel. It receives the alternating current. A dial switch on the front of the panel regulates the potential. The "sustaining coils" are fixed on the rear of the panel, as is also a controlling reactor by means of which more precise e.m.f. regulation can be accomplished. The tube is mounted in a ring on the back of the panel. A stem on the ring projects through the panel, and by means of a hand wheel attached to the stem a tilting of the bulb can be secured. These rectifiers are used for from 40- to 120-volt current, connected to 110- or 220-volt, 60-cycle alternating current.

The applications of the mercury rectifier have been steadily growing in number since it was introduced. The earliest and perhaps the most prominent application of the mercury rectifier was the charging of automobile batteries in places where only alternating current was available, but now it is largely used for charging telephone and railway signal batteries, operating d.c. lamps, electroplating, kino-projecting lamps and for various other operations requiring direct current.

CHAPTER V

STORAGE BATTERIES

The scope of this work does not call for any extended treatment of the construction of storage batteries, as this matter is treated fully by any number of authorities. They will be discussed here only in so far as they constitute an essential part of electric power-station switchgear.

Storage batteries are used as emergency reserves to help out a badly engineered d.c. installation, or for taking up peak loads on a system whose maximum load has outgrown the rating of the generating station. But in designing a new installation they are taken into account to assure efficiency, reliability, and economy of investment and operation.

Storage batteries are used for the following purposes:

1. To regulate the station output.
2. To compensate line losses.
3. To act as reserves in case of shut-downs.
4. To act as equalizers in three-wire systems.

The above functions may be called for singly or together, and may be performed at the central station, sub-station, on the line, or in two or more of these places at the same time. Batteries are employed in railway, light, or motor service.

1. Batteries are used for regulating the station output in two cases.

(*a*) In the first case the battery is charged by the machines when the outside load is reduced for some length of time and is discharged on peak loads or at night when the machines are not run. For this reason the generation and transmission of energy up to the point where the battery is installed are made independent of the load variations beyond that point.

(*b*) In the second case the battery is constantly in service, charging and discharging according to the momentary fluctuations of the load. This is a condition which always occurs in railway service. Hence we have the terms "equalizer," "fly-wheel" and "duffer" batteries.

A battery therefore serves to keep the time and rate of energy generation independent of the time and rate of the load fluctuations. In both of these cases the use of batteries has proven highly economical. For maximum efficiency it is essential that all machines should be kept loaded to their full rated power while in service. If storage batteries are not installed, the rating of the machines must be equal to the maximum load, even though such

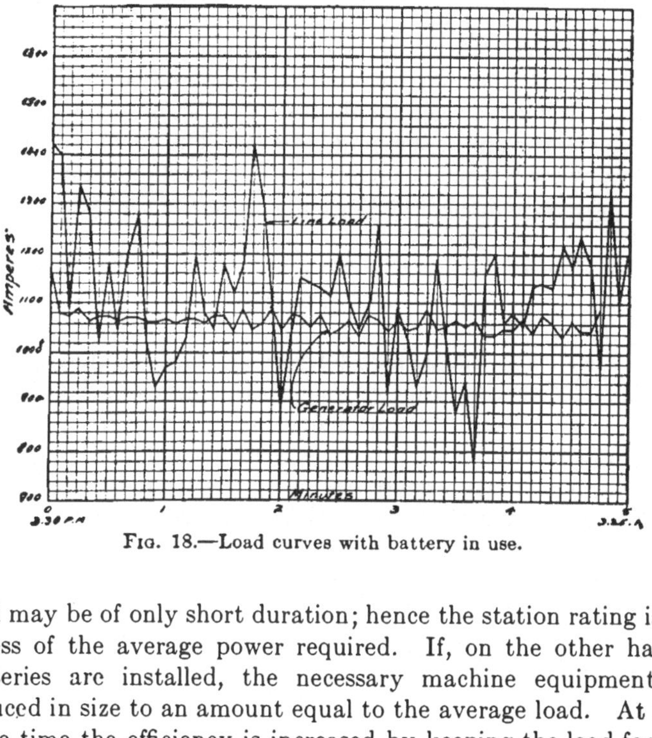

FIG. 18.—Load curves with battery in use.

load may be of only short duration; hence the station rating is in excess of the average power required. If, on the other hand, batteries are installed, the necessary machine equipment is reduced in size to an amount equal to the average load. At the same time the efficiency is increased by keeping the load factor constant, corresponding to the maximum (efficiency). Since, moreover, the battery is best able to deal with the very kind of loads which are imposed upon it, i.e., small loads of long duration (as night railway service) or sudden fluctuations which reduce the efficiency of the machines, a proper division of loading will result in maximum fuel economy. By increasing the load factor of the machines and decreasing their service hours, an additional advantage is gained through decreasing the losses incidental to starting and shutting down.

The diagrams in Fig. 18 show the load fluctuations of a system and the influence of batteries on the generator output. The average load on the line is approximately equal to the generator output, and fluctuations are taken up by the battery. In this case the generator rating is only 1,100 amp., while without batteries it would have to be 1,400 amp.

2. If a battery is joined to the line at a distant point, the line carries only the average load instead of the maximum. With a given line drop this affords saving in copper or a higher allowable voltage for a given size of wire. In some cases, where the average load of the line is small, batteries may be used to replace sub-stations. When installed in sub-stations they afford not only a saving in copper and reduction of line loss, but equalize the load on the converter, which therefore draws a constant current from the central station, so that the high-tension line losses are also reduced.

3. The special advantage of storage batteries is their ability to act as reserves. The following cases are the most important arising in practice:

(*a*) When the central, high-tension transmission line or sub-station is accidentally shut down. In this instance the battery will for a certain length of time maintain the service in the portion of the system affected, allowing the necessary repairs to be made.

(*b*) At momentary, unforeseen, or excess load, such as traffic congestion or increased demand for illumination on dark days, the batteries will take up such loads before it would be possible to put any additional units into service.

(c) They make it possible to disconnect the transmission line and all machines in the central or sub-stations for a considerable length of time during the night so as to allow of inspection and repairs.

(*d*) If batteries are used in systems not generating their own power, they make it possible to buy direct or alternating current from other systems when these are not overloaded, and this at a constant rate. Since the price of energy is figured on the basis of peak load, the gain due to the lower rate of delivery required by batteries frequently offsets the initial cost of the batteries themselves.

If there is any indication of disturbances in the system which might affect the machines, the trouble can be localized by cutting out the machines and throwing in the batteries. The emergency value depends upon the size of the battery relative to the load and the kind and variation of that load.

4. In the three-wire systems, batteries are used as equalizers by connecting the neutral of the system to the middle of the battery and the other buses to the ends.

When connected to an open circuit, a single element has an e.m.f. of 2.08 volts. This tension is called "floating voltage," for if the battery were connected into a circuit of this voltage it would neither be charged nor discharged. When a battery discharges, its voltage drops momentarily, and then remains constant to a point at which it again drops rapidly. In practice it is never discharged beyond a point at which the e.m.f. is 1.8 volts. The initial drop is caused by the internal resistance and partial polarization on the surface of the plates. In charging, the external or impressed voltage must be sufficient to overcome the counter-e.m.f. and internal resistance of the cell. The voltage of the battery on charging rises momentarily, and, as before, remains constant until the battery is completely charged, when it again commences to rise. To charge or discharge a battery at a constant rate the e.m.f. must vary according to the battery's characteristic curve for that particular rate. The rating of the battery is the amount of electricity delivered at a certain rate measured in ampere-hours. The rating of a cell varies according to the time and rate of discharge. A cell which has a rating of 100 per cent., for example, at eight hours discharge will have a rating of only 50 per cent. if discharged in one hour. Batteries are usually rated on an eight-hour discharge basis. For electric automobiles the rate basis is four hours, and for railway sub-stations one hour.

The following table gives the relative ratings, end voltages, and current values of a cell for different rates of discharge:

Hours, discharge	Final voltage	Relative value of current	Relative rating in amp.-hr.
8	1.75	1	8 (100 per cent.)
3	1.70	2	6 (75 per cent.)
1	1.69	4	4 (50 per cent.)
⅓	1.40	8	2⅔ (33⅓ per cent.)

A battery in circuit will require or produce current only when the outside voltage differs from the floating voltage. When this difference becomes sufficient the battery will operate without further assistance, and so take care of peak loads and large rapid fluctuations (a case which often arises with line batteries).

The operation of the battery is somewhat different when strong fluctuations do not exist or are not admissible, and also when precise equalization and quick response are required. In such cases auxiliary apparatus is used to regulate the charging and discharging. Such apparatus is either band-operated or automatic. The former is used when the battery is to be put into commission

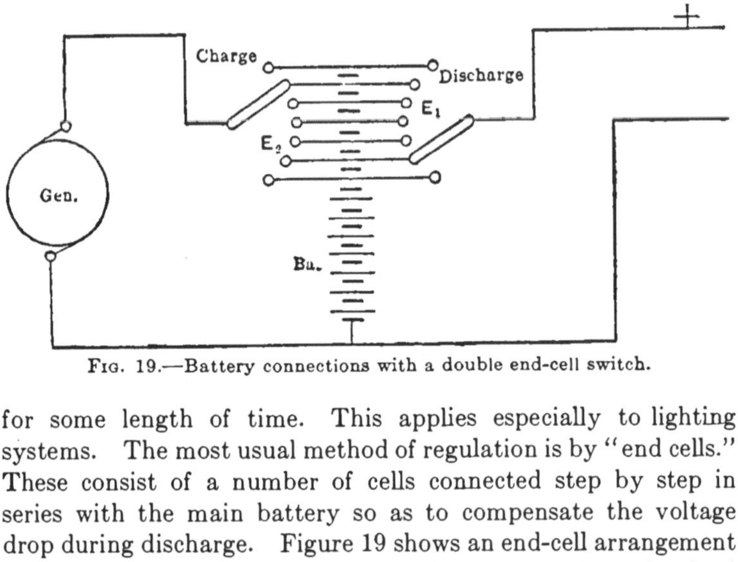

FIG. 19.—Battery connections with a double end-cell switch.

for some length of time. This applies especially to lighting systems. The most usual method of regulation is by "end cells." These consist of a number of cells connected step by step in series with the main battery so as to compensate the voltage drop during discharge. Figure 19 shows an end-cell arrangement with two switches. The required line voltage is maintained through the regulating switch, E_1, while charging is regulated by the generator field rheostat and switch, E_2. With this arrangement the generator charges the battery at a higher voltage and at the same time feeds the line at the normal pressure. The sum of the charging current and line current passes through the end cells, so that these are charged more quickly than the remaining ones; this requires cutting out the cells step by step by means of switch E_2. The battery should be charged only when the outside load is low. The end switches must be carefully designed so that there shall be no short-circuit between adjacent contacts or interruption of the circuit while shifting the contact arm. Sometimes they have an additional arm insulated from the main arm and connected with it through a resistance, so that when the two arms are on adjacent contacts, the short-circuiting current is reduced by the resistance. These end switches are usually in the shape of dial switches or straight-line glide switches. They are mounted on the switchboard and are operated manually or by motor.

Another method of regulating batteries by hand makes use of a carbon pile, whose resistance depends upon the surface pressure on the carbon plates of which it is composed. As this pile is connected in series with the battery, any change in the pressure will change the charging voltage.

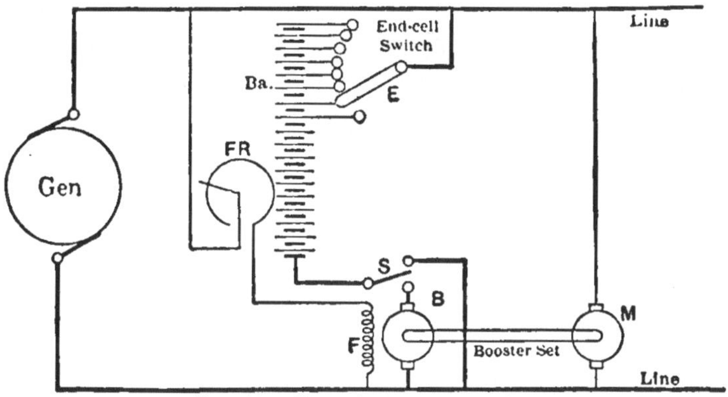

FIG. 20.—Shunt booster connection.

Automatic regulation of storage batteries depends upon the generation of an additional voltage, which causes charging when added to the bus voltage and discharging when subtracted from it. This extra e.m.f. is used to overcome the counter-e.m.f. of the battery. It is produced by a generator called a booster, whose armature is in series with the battery and whose field is automatically regulated through the influence of the load variations.

The simplest case is that of the "shunt booster," whose field-winding is in parallel with the buses. This machine is used only for regulating the charge, and it accomplishes this through a field rheostat. The connection diagram is given in Fig. 20. The booster is driven by a shunt motor and is disconnected while the battery is discharging. For the discharge regulation end cells are used. Shunt boosters with manual regulation are used in plants where the load variations are gradual and uniform, so that the battery may be charged and discharged for longer periods. This is the case for peak loads of long duration, or when the generators are shut down.

In traction systems where the load fluctuates rapidly within wide ranges a booster must be used whose e.m.f. is made to vary automatically with the load, so that it will add or subtract from the battery voltage as required. These requirements are met by the "differential booster." This machine differs from those

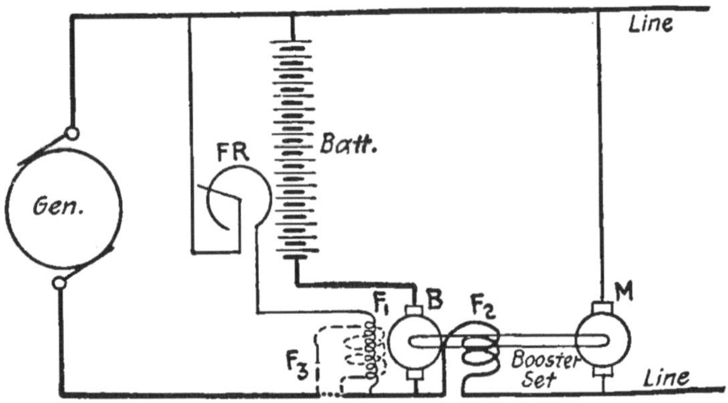

FIG. 21.—Differential booster connection. (Pirani.)

previously described in that it has a series field winding in the working circuit in addition to the shunt field winding. In Fig. 21 F_1 is the shunt and F_2 the series winding. The second winding opposes the first, so that at a certain external load they will balance each other and the resultant e.m.f. will be zero. At this load the generator e.m.f. is equal to that of the battery, so that the latter will neither charge nor discharge. At higher loads the action of F_2 is greater than that of F_1, and the booster e.m.f. is added to that of the battery, causing the battery to be discharged. With smaller load, on the other hand, F_1 predominates, and the booster tends to charge the battery. The booster, therefore, tends to maintain a constant load on the generators by enabling the battery to take up the fluctuations. In order to make this combination as stable as possible, a third field winding is put in series with the generator circuit, so that the increased generator current caused by the increased load produces an additional effect which enables the battery to discharge. The object of this third winding is to secure a more perfect and precise regulation than would otherwise be possible, and to keep the division of load between the batteries and machines at the desired ratio.

Instead of exciting the boosters directly from the line, a separate exciter mounted on the shaft of the motor-booster set, or driven by a separate motor, may be employed. The exciter field consists of two differential windings, one shunt winding supplied at a constant voltage, usually from the station voltage, and one series winding included in the circuit to be regulated. The exciter armature is connected to the booster field through a

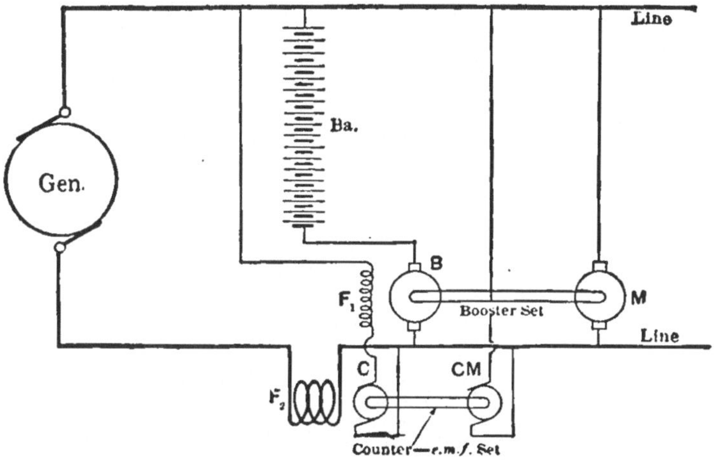

FIG. 22.—Diagram of connections of a booster with counter-e.m.f. generator.

rheostat. The booster excitation will therefore vary with that of the series field coils of the exciter. This type of booster is called "exciter booster." It is essential that the driving motor should be a shunt-wound machine connected across the busbar, for when the battery discharges, the booster becomes a motor to drive the former driving motor as a generator in parallel with the main dynamo. This change could not take place if the driving motor were series-wound.

When the load fluctuations must be kept within certain small limits an arrangement is used which magnifies the influence of the variations on the booster. The auxiliary equipment for this purpose consists of what is termed a "counter-e.m.f. generator." Figure 22 shows the connections of this type of machine with the booster and batteries. The operation is based upon the fact that the counter-e.m.f. generator is excited by a series winding, F_2, built in the main generator circuit, and that the armature winding of this auxiliary machine is in series with the field winding of the booster and across the busbars, but opposing them in voltage. At a certain external load the e.m.f. of the auxiliary generator exactly balances that of the buses. Since the latter is constant, and since the counter-e.m.f. is made to depend on the external load, the booster voltage is inversely proportional to the output of the station. Adjustable resistances are inserted

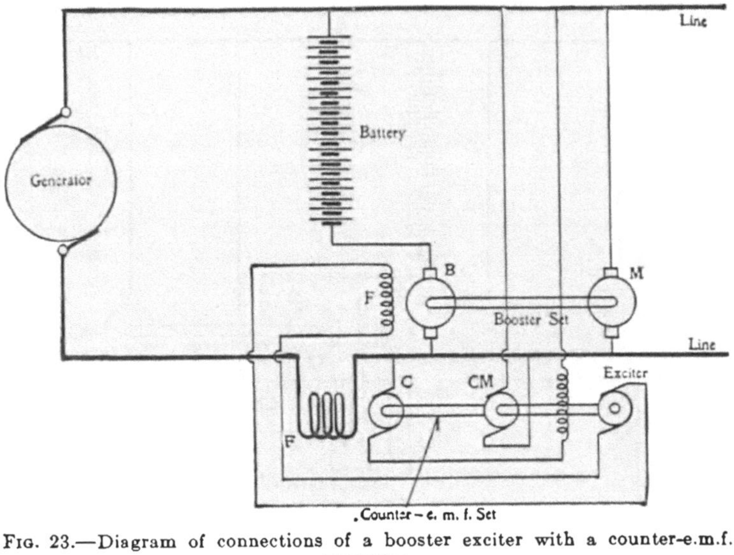

FIG. 23.—Diagram of connections of a booster exciter with a counter-e.m.f. generator.

in the field circuit of the counter-e.m.f. generator, which enable precise regulation of the batteries for different numbers and sizes of generators and different average loads. The auxiliary generator is usually mounted together with the booster or is driven by a separate motor.

In still another case the booster is excited by a separate exciter whose field is in series with the counter-e.m.f. generator. The connections are shown in Fig. 23. This arrangement affords more sensitive regulation, and makes possible the use of a single counter-e.m.f. generator for different sizes of boosters and batteries. This is the method used by the Gould Storage Battery Company.

In cases where the load is composed of a constant load from a lighting system and a variable load from a traction system, a booster called a "constant-current booster" is employed. The field winding of the counter-e.m.f. generator under these conditions is in series with the battery feeder, while the armature winding is so connected to the booster field that any tendency to increase or decrease the current in the battery feeder is instantly opposed by a decreased or increased booster voltage. The booster is inserted between the constant and the variable load, so that the latter is supplied either by current through the booster or by this current augmented by the battery current. The current through the booster feeds either the variable load or the battery, according to the requirement of the feeders supplying the variable load.

FIG. 24.—Carbon-pile regulator. (Entz.)

The small and inexpensive shunt booster may be used for both charging and discharging regulation by inserting a carbon-pile regulator in the shunt winding of the booster.

A carbon-pile regulator is shown in Fig. 24 and the wiring diagram in Fig. 25.

The carbon regulator consists of two or more sets of carbon disks, C, connected sometimes like a Wheatstone bridge where the shunt field winding of the booster takes the place of the galvan-

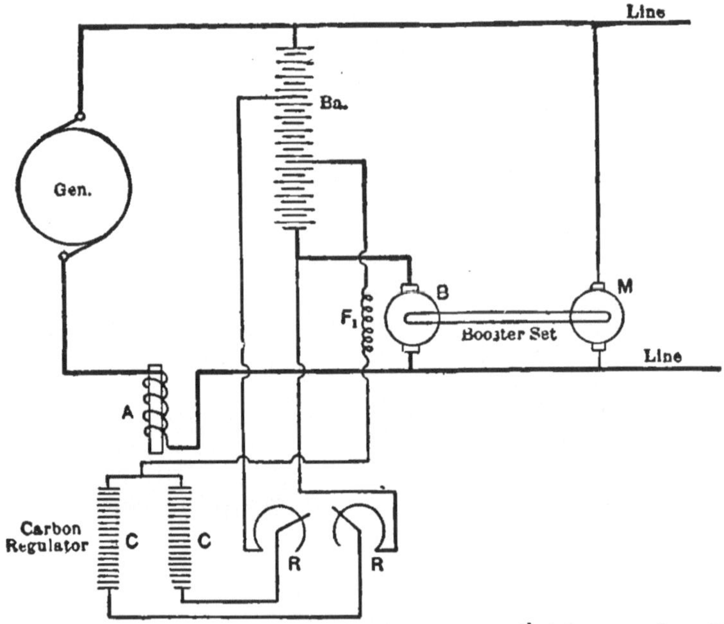

FIG. 25.—Diagram of connection of a booster controlled by a carbon-pile regulator.

ometer. There is a pivoted lever over the top of the piles, arranged so that by changing its position a different pressure is brought to bear on the various piles which causes their resistances to vary. The lever is moved by a solenoid-actuated iron core, whose solenoid is in series with the generator circuit. A spring at the end of the lever opposes the action of the solenoid. At the required average load the pressure on both piles is the same. The ends of the carbon piles are connected to a group of battery cells whose middle point is joined in series with the booster field and to the movable arm on the carbon piles. When the pressure is the same on both piles, their resistances are equal, and there is no current in the shunt winding, F_1. But when the current in the main circuit decreases or increases, one of the piles is compressed more than the other, and there is current through the booster field in one direction or another, which causes charging or discharging of the battery. The piles are set for the required load by adjusting the spring at the end of the pressure arm.

Instead of making the regulator regulate the booster field directly, an exciter may be put between them whose field is influenced by the regulator. The advantage consists in the

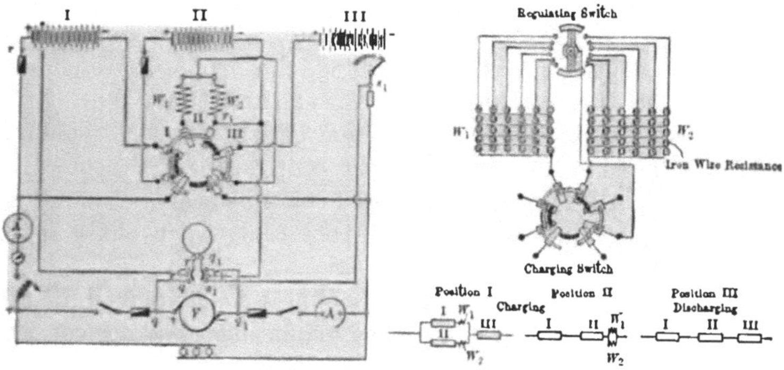

FIG. 26.—Wiring diagram of the Cismeros-Mička three-group method.

decreased energy loss resulting from the use of the smaller-sized machine. This type of regulator is made by the Electric Storage Battery Company. It is mounted directly on the switchboard, with the carbon piles and spring in front and the solenoid in back of the board. The solenoid is in the busbar circuit. With this regulator it is possible to adjust the sensitiveness of the charging side relative to that of the discharging side and *vice versa*. In central and sub-stations, for instance, it is sometimes desirable to utilize the overload capacities of engines, generators, converters, etc., but as soon as the load drops under a certain limit the battery comes into action. With these regulators, also, a zone of non-regulation may be created, extending a certain percentage above and below the average load, whereas for loads above and below this regulation may be as perfect as possible.

In some factories, hotels, hospitals, farmhouses, or other places which produce electrical energy for their own use, there is both a day and a night demand for constant-potential electrical energy to facilitate the operation of motors, elevators, or other apparatus. When no auxiliary charging dynamos or boosters are available, the batteries are charged at the constant potential of the plant by what is termed the Cismeros-Mička three-group method. The battery is divided into three equal groups of cells. The first charge is made while two groups are connected in parallel and the third one in series until the third group has been fully loaded with the aid of a regulating resistance. The charging of the batteries to their full capacity is then completed by disconnecting the third group and changing the connections of the first and second groups from parallel to series.

During the first period of charging, the regulating resistance destroys about 19 per cent. and at the end about 8 per cent. of the charging voltage; during the second period, the corresponding amounts are approximately 20 per cent. and 8 per cent. By this method, the loss of the delivered charging energy amounts to about 14 per cent. as against 39 per cent. when charging in two groups.

Figure 26 shows the diagram of connections employed for the three-group charging system. The regulating resistance consists of iron wire built into glass vacuum bulbs. These resistances regulate automatically in accordance with the temperature rise.

In the last years, storage batteries have found a new field of usefulness, which promises to become of great importance in the near future. This is as equalizer in a.c. stations or in a.c. services.

The general function of a storage battery when used in connection with an alternating-current system is the same as when used with a direct-current system, namely, to relieve the power plant and transmission system of load fluctuations so that the generating machinery and conductors can be utilized to the greatest advantage and maximum economy by subjecting them to a steady load equal to the average load.

Regulating storage batteries are required for such alternating-current situations on long interurban railways with steam operating conditions, that is to say, where few and heavy trains are run at long intervals; or for steel plants with variable loads where gas engines are installed to utilize the furnace gases.

On some alternating-current generating systems with direct-current sub-stations individual batteries have been used to regulate the load fluctuations at each point. In some recent installations, however, it has been found advantageous to concentrate the battery capacity at one point, a single battery being installed to control the combined fluctuations of the load on the entire system. Obviously, this arrangement reduces the first cost because the total battery capacity which is required will be much smaller inasmuch as the maximum fluctuations of the combined load will ordinarily be less than the sum of the maximum individual fluctuations.

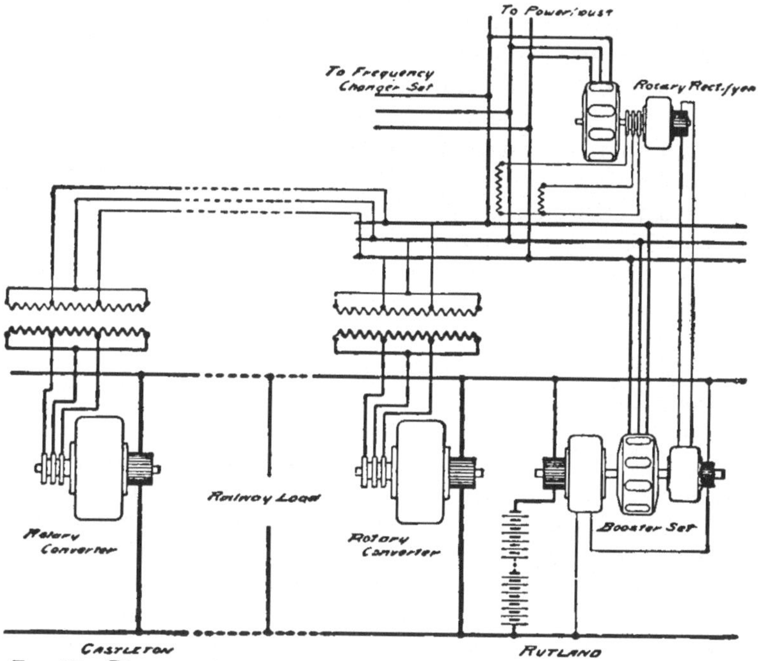

FIG. 27.—Diagram of connection of a storage battery in alternating-current system.

The regulating battery is usually located at a converter station in the power house or in the nearest sub-station, if the whole transmission line passes that sub-station. Several small series transformers are built into the line, and, by means of a synchronous rotating rectifier, the secondary alternating current of the line is converted into direct current. This direct current, which is proportional to the watt component of the alternating line current, excites the field of a counter-e.m.f. generator whose armature winding is in series with the booster field winding and opposes the voltage of the d.c. buses.

Figure 27 shows the diagram of connections. The battery current demanded by the increased load on the line is supplied from the alternating-current transmission line by means of an inverted converter. When the external load decreases, the battery is charged through the converter.

In some cases series transformers are located in the individual generator circuits, their secondaries being connected in parallel to give a combined output proportional to the total generator load. In other cases carbon regulators may be used. The application of the regulator involves the substitution of an alternating-current solenoid for the usual direct-current coil.

In systems where the current is generated and utilized as polyphase alternating current, the battery would be located either at the main generating station or at the center of the load.

The batteries and boosters are joined to the direct-current feeders either of a motor-generator set or of synchronous converters. If a motor-generator set is used, the charge and discharge of the battery may be accomplished by varying the voltage of the direct-current machine in response to the fluctuations of the alternating-current load. The motor-generator set is composed of a synchronous motor and a d.c. generator. (See also Chap. XXVII, Spokane Inland Railway Company.)

On widely varying alternating-current loads a Tirrill or Thury regulator is sometimes used in connection with a booster storage battery. (See Chap. XVII.) The regulator is influenced by the main current of the alternating-current system and, in turn, it regulates the excitation current of a motor-generator set.

By using batteries in a.c. systems the speed of the machines, and hence the station voltage is maintained constant. In a.c. lighting systems this results in efficient voltage regulation. By taking up peak loads in this way the station generating power is increased.

CHAPTER VI

THREE-WIRE SYSTEM

Ever since the introduction of electricity for lighting and power, attempts have been made to obtain two sets of circuits from the ordinary two-wire d.c. generator, so that a lower potential, say 110 volts, could be used for electric lighting, and a higher potential, say 220 volts, for power purposes. The use of higher voltage on the power circuit gives a cheaper installation because less copper is then necessary for the transmission of energy. Such an arrangement of two-voltage circuits can be obtained by means of the well-known three-wire system. The principles underlying this system are as follows: The potential difference between the two outside wires is higher than the normal service potential, being usually twice as great as the latter. The lamps are connected between one of the outside wires and the inner wire, called the neutral, while the motors run connected between the outside wires. The potential between the outer wires and the neutral is equal to the normal service tension. In traction service it is usual to employ the track as the neutral. In this case the saving in copper is comparatively small. The ratio of copper weight for a double track, 5,000 feet long, using 500 volts, to the same track using 1,000 volts, three-wire system, is as 13 is to 11. In lighting systems the saving in copper is much greater, amounting to from 62.5 per cent. to 69 per cent. It admits of the use of both 220-volt and 110-volt apparatus on the circuits. The three-wire system can be built up in different ways:

1. Two generators (Edison three-wire system).
2. One generator with compensator.
3. One generator with balancer set.
4. One synchronous converter.
5. Subdivision of a storage battery.

1. The Edison system is based on two generators connected in series. The two outer terminals are connected to the positive and negative busbars, while the two inner ones are connected to each other and to the neutral bus. The switching diagram is shown in Fig. 28. The fields of both generators are controlled

separately and the voltage of both sides can be adjusted at will, or the generator can be compounded to give a high voltage on one side, where such increase of voltage is desirable to overcome the effects of unbalancing. With this arrangement a large amount of power can be delivered to either side of the system and the extreme degrees of unbalancing can be handled for a short time without disturbance of the lamp voltage. The neutral wire is positive or negative with respect to the true neutral, according

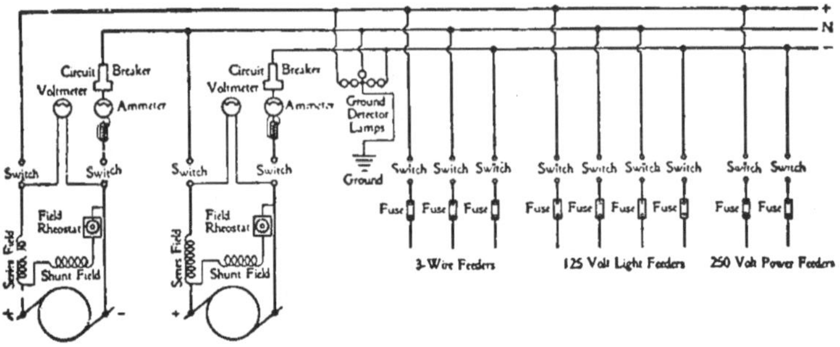

FIG. 28.—Wiring diagram of an Edison three-wire system.

to whether the load is greater on the negative or the positive side. The diagram shows the connections of one three-wire feeder, one 220-volt and two 110-volt feeders. With installations of two or more pairs of machines, two equalizer buses are required, one each for the negative and the positive side of the machines. The disadvantage of this system lies in the higher cost of two generators and their lower efficiency as compared with a single generator of equal capacity and higher e.m.f.

2. *One three-wire generator with compensator.* The generator is compound-wound with two series windings and one shunt field winding. The two series windings are necessary in order to secure with unbalanced loads a compounding approximating that of balanced loads. On the side opposite the d.c. brushes one or two pairs of collector rings are fastened on the shaft; these are joined to two or four points of the armature winding respectively 180° or 90° apart. The collector rings are connected with a compensator, from whose middle point a wire leads to the neutral. (See Fig. 29.) The action of the machine is as follows: With balanced load there will be only the exciting current through the compensators, which are simply auto-transformers. This exciting current is alternating as the relative potential of the taps

to the armature changes from positive to negative. With an unbalanced load, for example, there being a greater load on the positive than on the negative side, the excess current will return by the neutral wire and divide in the auto-transformer, returning to the armature through the collector rings. A circuit breaker and an ammeter are installed on either side of the machine. It is advisable to mount the former near the generators, for in this installation they are closed and tripped electrically from the main switchboard. The equalizer switches can also be mounted near the machine. Such an arrangement affords a saving in copper. The machine is started as follows:

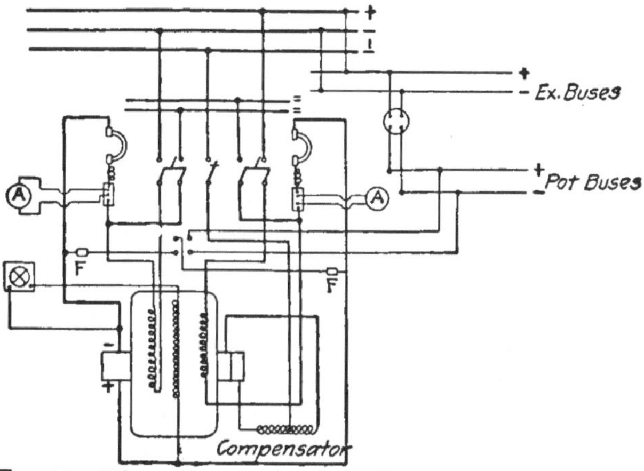

Fig. 29.—Wiring diagram of a three-wire direct-current generator.

(*a*) Close the equalizing switches.
(*b*) Close the main switches on the switchboard.
(*c*) Adjust the voltage and close the circuit breakers by means of the control switches on the switchboard.

3. Generator with balancer set. For a system which requires more energy than can economically be furnished by the Edison system, one generator of normal e.m.f. equal to that between the positive and negative busbars is used in connection with two motors, which are connected to each other in series and to the buses in multiple. The connections are shown diagrammatically in Fig. 30. The neutral wire is connected with the common terminal to both motors. When the system is balanced the set operates as two motors, and as motor-generators when the sys-

tem is unbalanced. The neutral current is unequally divided between the two machines.

If there is an excess load on the positive side of the system, the e.m.f. between the positive and neutral will be less than between neutral and negative. The negative balancer will tend to speed up and will drive the other as a generator. The unbalanced current will divide, part going to the motor balancer to afford the

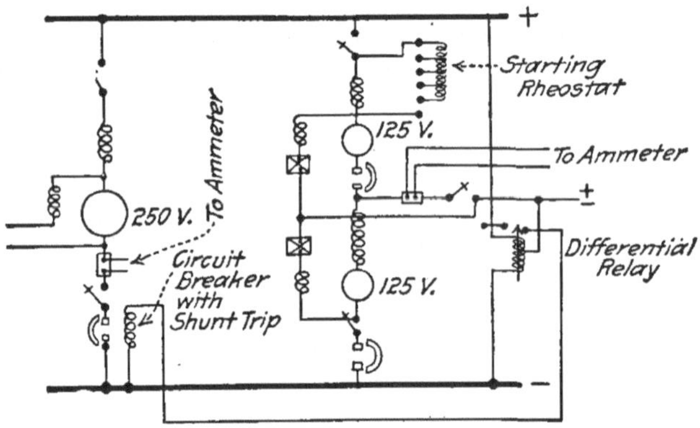

FIG. 30.—Wiring diagram of a three-wire system with balancer sets.

power to send the rest through the generator balancer back to the line. The series winding of the former tends to weaken the field and increase the speed, while that of the latter assists the shunt winding and raises the voltage across the generator. If the excess load be on the negative side, the positive balancer becomes the motor, the negative balancer the generator. It is evident that these machines do not add any power to the system, but serve only to balance the load on the two sides.

Each balancer should have a capacity equal to one-half the maximum unbalanced load that is considered likely to occur. Both machines can be independently adjusted so as to give any desired division of voltage between the sides, and each of the machines can be compounded in such a manner that it will compensate for inequalities of line losses and natural drop when the system becomes unbalanced. The shunt windings of both machines are connected in series, and the middle point is joined to the neutral. The generator and the motors each possess a circuit breaker and the necessary main switches. Fuses can be used to replace the circuit breakers of the motors. Both motors are started by means of a starting rheostat. The method of procedure is as follows: In the first place the generator is started in the usual way and is thrown on the busbars. Then both machines are started together as motors by means of the starting rheostat. When both have reached normal speed, the connection to the neutral is closed, which also throws the shunt windings on to the neutral. Care should be taken not to connect the shunt windings to the neutral when starting. To protect the lamps, etc., against short-circuiting on one side or in case of accidental disconnection of the balancer set, the circuit breaker of the generator is tripped by a differential relay. This relay will trip the circuit breaker in the event of an abnormal rise of potential on either side of the system.

4. Another method of supplying the neutral current is to operate a small synchronous converter as a direct-current motor from the outside conductors, the neutral being connected to the middle point of the compensator operated from the collector rings of the converter. The converter replaces the balancer set.

Where synchronous converters are used to supply a three-wire system, the neutral can be taken either from a common connection of the transformer secondaries or from a compensator connected to the alternating leads.

5. When a storage battery is used, the neutral wire can be joined to the center of the battery. (See Chap. V.) The battery can be then charged as a whole by means of a higher voltage of the main dynamo, or with the help of a booster set each part of the battery can be charged separately to balance the uneven discharge.

CHAPTER VII

FEEDER PANELS

Up to this point we have treated only the methods of generating d.c., and the standard panels controlling the output. Generators, converters, or storage batteries can be used to produce current either independently, in groups composed of like machines, or in mixed groups. This depends upon the number and size of the units and the size and character of the load. We are here concerned with the methods of feeding and the control systems for consumptions of different characters.

Figure 31 shows switchboard wiring diagrams for d.c. feeder panels for railway service. The positive bus mentioned in former chapters, mounted on the back of the board, is identical with the one here shown. In Fig. 4 the positive bus is mounted on the machine, and therefore requires a special cable to connect it with the positive bus on the feeder panel. Three kinds of feeder panels are shown. Figure 31-*A* is a panel controlling one feeder and hence is equipped with one circuit breaker, one ammeter with shunt, one reactive coil as lightning protector, and one lever switch. Figure 31-*C* is a panel controlling two feeders. The equipment is the same as in Fig. 31-*A*, with the exception that there are two lever switches in place of one. Figure 31-*B* is for two feeders with a double equipment of ammeters, reactive coils, and main switches. A modification of Fig. 31-*B* has only one ammeter, which is common to both lines. In all three cases, when the feeders leave the station overhead, the instruments are protected by lightning arresters. Figure 32 shows front views of the three main types of panels and a rear view of Fig. 31-*B*. The same conditions which applied to panel mounting, busbars, and instrument connections for generator panels also apply here. We thus have feeder panels where both positive and negative cables are supplied with energy from the board, corresponding to generator panels with both busbars mounted on the board. Such panels are generally used for voltages of from 125 to 250 for lighting purposes and power distribution. Each panel controls two, three, or four sets of outgoing feeders, and is supplied with

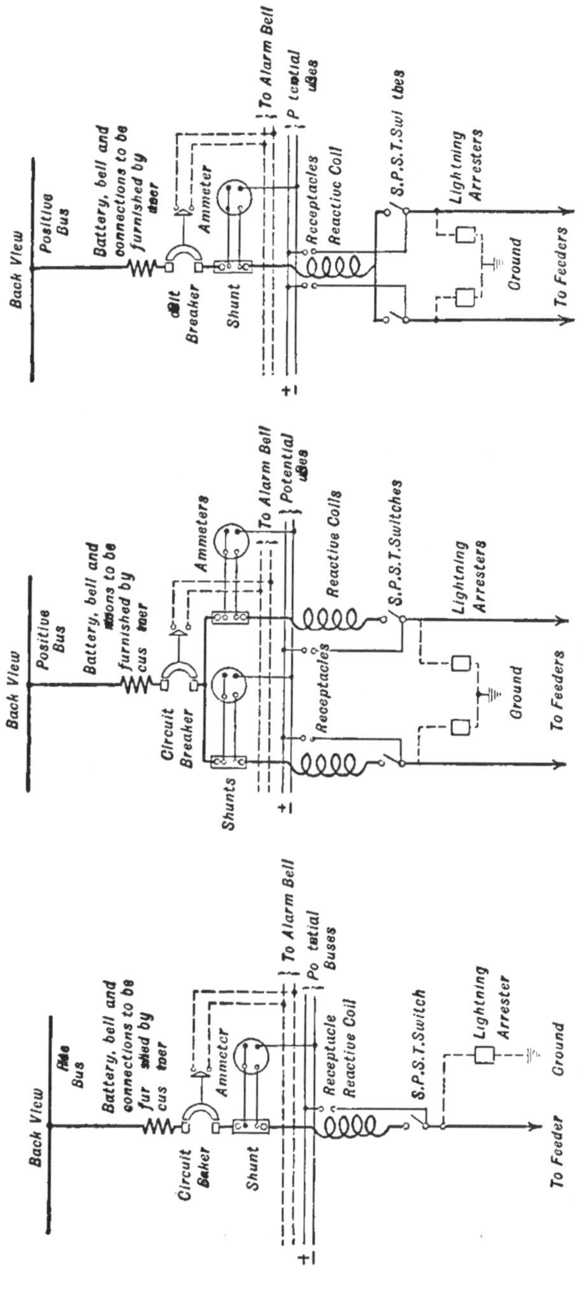

FIG. 31.—Diagrams of connection for direct-current feeder panels.

the requisite number of circuit breakers and double-pole lever switches. Besides the above-mentioned instruments an ammeter may be supplied for each feeder set, and the voltage of all the

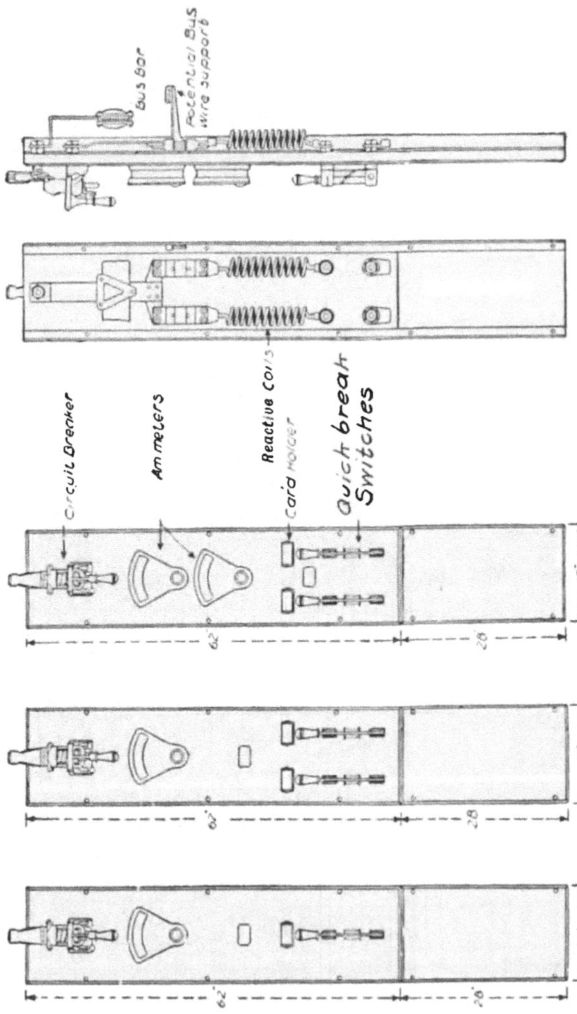

FIG. 32.—Feeder panels.

feeders may be indicated by a single voltmeter. (See Fig. 33.) Fuses can be used for feeders of smaller capacity in place of circuit breakers, to insure against overload. In this case each panel is capable of controlling a larger number of feeders. The

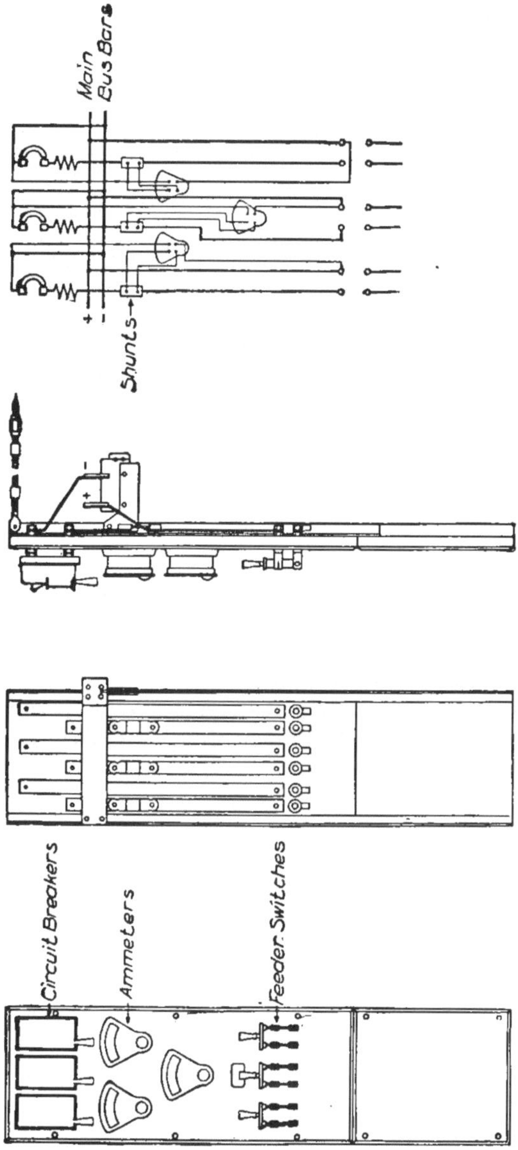

FIG. 33.—Two-wire feeder panels for 125 and 20 volts.

fuses are mounted either on the front of the panel with the lever switches (Fig. 34) or on separate slate bases on the panel rear.

The three-wire feeder panels differ only in the use of three-

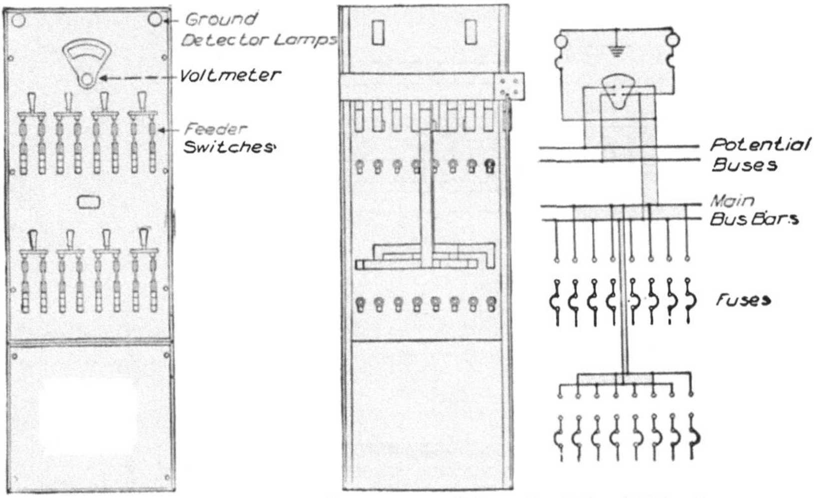

FIG. 34.—Two-wire feeder panels with fuses for 125 and 250 volts.

pole lever switches with or without fuses, or, when an equiptment of circuit breakers is desired, in the use of three single-pole circuit breakers with a common trip attachment which permits them to preform the combined functions of lever switches and circuit breakers.

CHAPTER VIII

DIRECT-CURRENT MOTORS

The amount and distribution of the energy as supplied to the line by the central or sub-stations are controlled from the feeder panels. (See Chap. VII.) The consumption of energy by the individual customers is regulated at the various places of delivery. For lighting installations, lever switches with fuses are employed, which are mounted in groups on cabinet panels. One large double or three-pole lever switch controls the mains, while one or more rows of small double or three-pole switches control the individual lamp circuits. The control for traction systems is much more complicated. This is due to the fact that starting, change of speed, direction, grade, and load, as well as operation of numbers of portable motors, require a very involved switching arrangement. The regulation is accomplished by the controller in the hands of the motorman on the car. It is not the province of this book to give a detailed account of switching arrangements for electric cars or locomotives, as we here treat only of stationary arrangements or such as may be considered stationary in service, for example, portable sub-stations. When electrical energy is required to drive a stationary motor, a special switching arrangement is necessary. This is mounted on a panel near the machine. Figure 35 is a wiring diagram of such a panel, with views of the board. The negative side of the shunt field winding is connected on the line side of the starting rheostat to give a maximum field current on the motor when starting. By tripping the circuit breaker the field circuit is discharged through the armature of the motor. The low-voltage coil of the starting rheostat switch serves the purpose of opening the motor circuit when the source of power is interrupted. The starting rheostat switch arm is not released by the low-voltage coil until after the field is sufficiently dissipated, so that destructive arcing will not occur on the switch. When the motor is shut down a spring throws the arms of the switch back to the starting position. The spring also prevents the switch arm from remaining on an intermediate starting point, which might result in the burning out of the

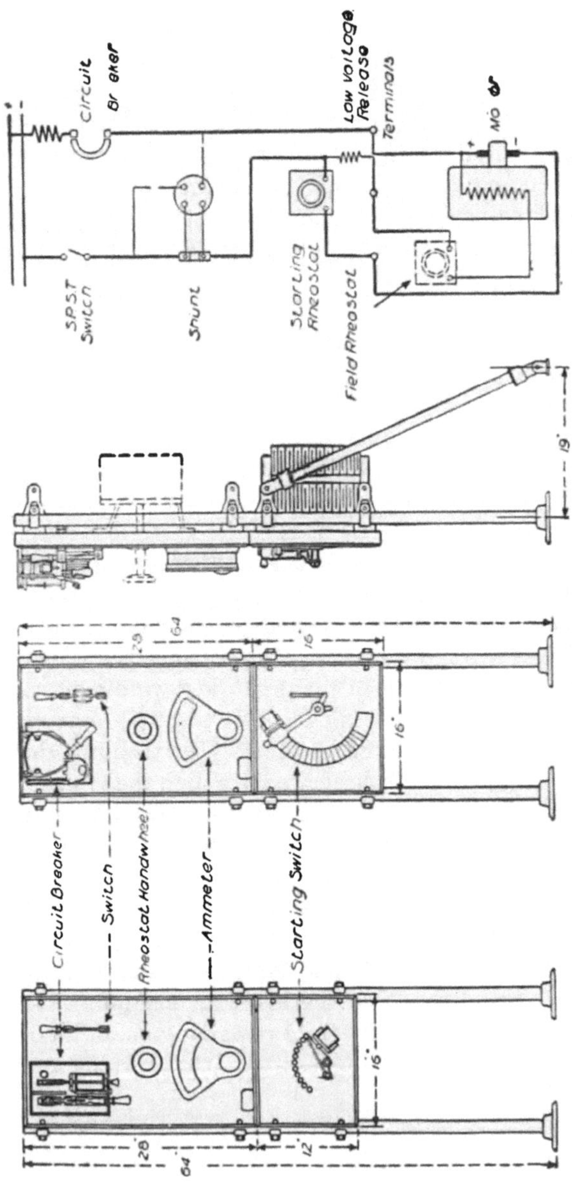

Fig. 35.—Continuous-current motor-starting panels.

starting rheostat. This switching arrangement is used for constant or adjustable-speed motors, from 3 to 15 hp., 125 volts, or for from 3 to 50 hp., 550 volts. The speed of the adjustable-speed motors is regulated by means of a field rheostat shown dotted on the diagram. In place of an ammeter with shunt, a current indicator in series with the circuit can be used whenever minute precision of measurement is not required.

Instead of starting rheostats with sliding contacts as illustrated in Figs. 35 and 36, a multiples-witch system may often be used. The sliding-contact construction calls for a subdivision of the resistance into a large number of steps, in order to reduce to a minimum the sparking which occurs when the contact is moved from one segment to another. Therefore, there is always some sparking with this type of construction. With the multiple-switch system, however, there is no sparking at the switch in starting a motor, whether under load or not, and it is not necessary to divide the rheostat into as many sections. The single switches are pushed into clips before their full carrying capacity is obtained. The first switch is held closed by an electromagnet while all the other switches are held closed by latches. In front of each switch is placed a pivoted metal stop, so arranged as to prevent the closing of any of the switches until the switch next to it on the left has previously been closed. The metal stops therefore constitute the interlocking mechanism. When the supply of current is shut off, the first switch on the left opens and allows the stop on the second switch to drop, so that this switch also opens and releases the catch on the third switch, and so on until all the switches open automatically. This automatic release protects the motor when the current is cut off and then suddenly turned on again, because the sudden rush of current through the motor armature when in circuit might be attended with disastrous results. The automatic release must be so arranged that when the supply of current is interrupted the lever switch will fly back to its "Full-off" position.

Another group of panels comprises those controlling variable-speed motors connected to a three-wire circuit which range in capacity from 2.5 hp., to 20, hp., 125 to 250 volts. The speed of the motors when used with panels shown in Fig. 36 may be increased 400 per cent. above the low speed of the motor. The machine is connected to the 125-volt or the 150-volt circuit by a single-pole, double-throw lever switch. The shunt field-

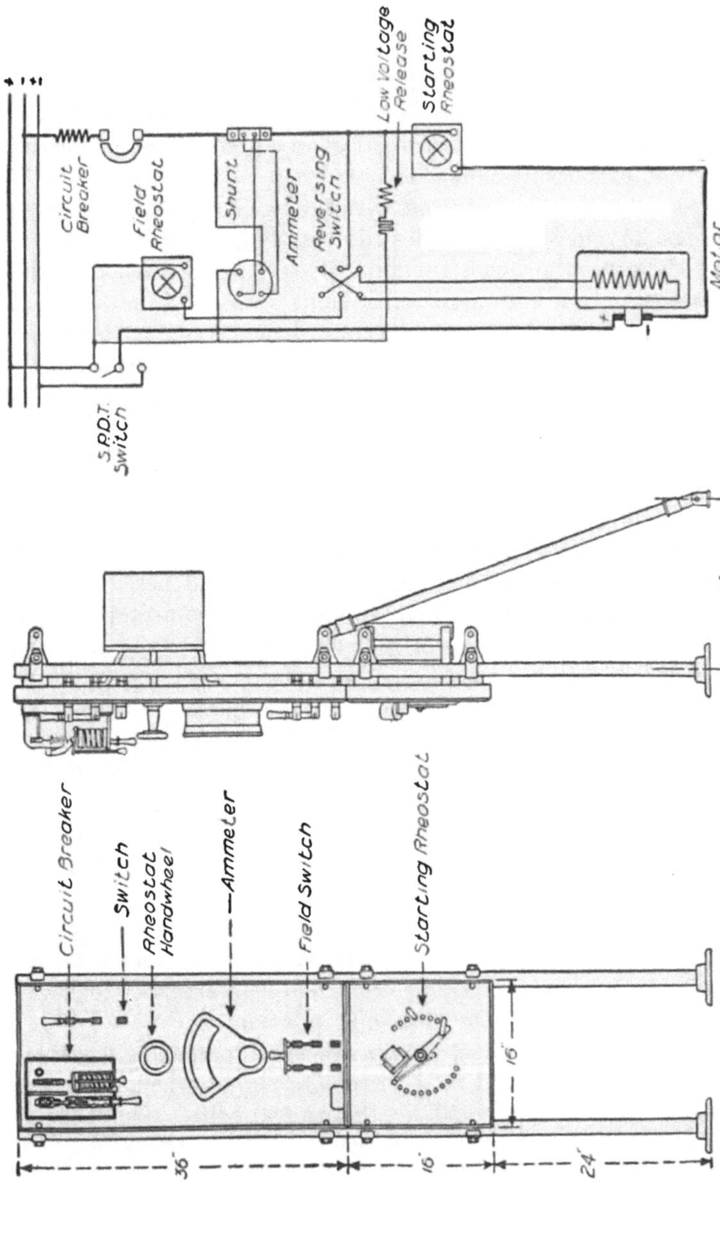

FIG. 36.—Continuous-current motor-starting apels.

winding is always connected to the 250-volt circuit. The field and direction of rotation can be reversed by a double-pole, double-throw switch. However, as long as the motor is in motion, this switch should not be opened. Speed variation is regulated by the field rheostat, which, like the starting rheostat, is mounted on the board. In a modification of this group the ammeter is replaced by a current indicator.

The normal speed of a motor can also be reduced by the use of a rheostat which inserts an external ohmic resistance in the armature circuit. The resistance and the contacts of such a rheostat must possess sufficient capacity to carry the current continuously on any step, provided this current does not exceed the amount required to operate the motor at its rated torque. This means that if the current required to operate the motor at full speed with no rheostatic resistance in circuit is approximately equal to the normal rating of the rheostat and motor, then the motor-speed regulator should have sufficient ohmic resistance to change the speed of the motor from full speed to one-half of full speed when driving a full-torque load.

If the current required to operate the motor under the foregoing conditions is greater than the rated capacity of the rheostat, the rheostat will operate at a higher temperature than that for which it is designed. This condition will cause the motor to run slower than one-half of full speed when all the resistance of the rheostat is inserted, with the consequent likelihood of damage to the motor and rheostat.

If the current required to operate the motor at full speed with no rheostatic resistance in circuit is less than the rated capacity of the rheostat, the motor is likely to operate at a higher speed than one-half of full speed when all the resistance of the rheostat is in circuit. It will be noted, therefore, that the speed at which the motor will revolve depends upon the load which it is driving when an attempt is made to alter its speed by the insertion of external ohmic resistance in the armature circuit.

A shunt-wound motor will operate at very nearly the same speed at variable loads within the capacity of the motor, provided no external ohmic resistance is inserted into the armature circuit. Just as soon as resistance is inserted, the automatic speed-regulating property of the shunt-wound motor is destroyed in proportion to the amount of external resistance used. In order to maintain a standard speed less than the normal speed of the motor, it is necessary to move the lever of the speed regulator by hand each and every time the load on the motor changes.

A peculiarity in speed variation is also the fact that the power required by some kinds of machinery when running slowly is a great deal less in proportion to the reduction of speed than in other classes of machinery. Some machines take almost as much current to run the motor at one-half speed as at full speed when this method of regulation is used.

Probably the most notable examples of these two classes of machinery are the ventilating fan and the hydraulic pump. A motor-driven ventilating fan is controlled by this class of speed regulator in such a way that approximately one-half as much current will be required to drive it at one-half speed as at full speed. A motor-driven hydraulic pump, however, takes practically as much at one-half speed as at full speed with this method of regulation.

Where the load driven by the motor varies greatly, the extensive regulation of its speed by means of resistance inserted in the armature circuit is almost sure to be more or less unsatisfactory except where an attendant is always present as in operating cranes, elevators, printing presses, etc.

Another group of panels includes those controlling small motors up to 5 hp., 125 volts; 10 hp., 250 volts, and 15 hp., 550 volts. These are constant-speed motors and are protected against overload by fuses in place of circuit breakers. The panel may be mounted in any convenient place on account of the simplicity of its parts. Frame supports are shown in Figs. 35 and 36.

As dirt is not a good conductor of electricity, rheostat contacts will arc and burn unless they are kept clean. When a contact once begins to arc, this action will rapidly increase unless the contact is cleaned immediately.

It sometimes happens, especially in using a starting rheostat for the first time or after a long idle period, and when the motor is heavily loaded, that the rheostat will throw off much smoke. This should cause no alarm, however, as it is nothing more than the burning of accumulated outside matter.

The heat generated in the rheostat by the current can be effectively carried away only by supplying thorough ventilation.

To start a motor:

1. Close the main knife switch. This action should not allow any current to pass through the motor.
2. Move the lever of the starting switch quickly and squarely on to the first segment and hold it there for about a second (the same procedure refers also to the first switch of the multiple-switch starting rheostat).
3. Move the lever to the second segment and hold it there for about a second (or the second switch of the multiple switch and so on).
4. Then move the lever to the third segment and hold it there for about a second and so on from one segment to the next until the lever has been moved over all the segments to the short-circuit position, where it should be held firmly by the retaining magnet.

If the motor does not start when the lever of the starting rheostat is on the third segment, open the main knife switch and look for the trouble.

This may be due to any of the following causes:

(*a*) Wrong connections.

(*b*) Excessive load on the motor.

(*c*) An open circuit of some kind.

(*d*) A short-circuit of some kind.

In case of trouble it is always desirable to make sure that the fields of the motor are magnetized.

When stopping a motor open the main-knife switch and let the starting rheostat take care of itself.

It frequently happens that motors are called upon to start a load which requires a starting current much greater than the continuous carrying capacity of the motor even though the motor may operate within its rated capacity after it has been brought up to normal speed. It is also sometimes desirable to take a minute or more to start the motor. A well-constructed motor, however, will withstand temporarily a starting current two or three times as great as its normal rating. When such large starting currents as these are necessary, it simply means that larger starting rheostats than usual should be used.

CHAPTER IX

DIRECT-CURRENT CIRCUIT BREAKERS

We have assumed that the user of this book is familiar with the construction and handling of measuring instruments and the simple forms of lever switches. We therefore treat only of the construction and operation of instruments which embody special features of modern switchboard arrangement. One of the most important pieces of apparatus included under this class is the circuit breaker. This term applies to all devices which automatically interrupt the circuit under special conditions. These conditions are twofold. They depend either on the variation in electrical energy flowing through the circuit breaker or on certain conditions of the machines in circuit, which in turn may produce or be a result of the first. Examples of the first condition are short-circuit, grounding, overload, underload, low voltage, current reversal, and phase reversal with a.c. Running away of an inverted converter is an example of the second condition. In this chapter we will treat only direct-current circuit breakers and low-tension a.c. circuit breakers which differ materially from those generally used for high-tension a.c. A circuit breaker may often displace the combination of switch and fuse to advantage. It insures greater continuity of service and avoids the expense and delays which are involved in the frequent replacement of fuses. The operation of the circuit breaker is absolutely positive, and it can be adjusted to trip at any predetermined point between wide limits of calibration. On the other hand, a fuse is inflexible in this respect; its time lag is appreciable and more or less dependent upon climatic conditions, and it can never be absolutely relied upon to rupture at its rated capacity. The main function of a circuit breaker is to interrupt current rapidly and at the required instant, without injury to itself. When the circuit breaker opens, an arc is formed which keeps up the circuit and is damaging to the apparatus. Devices must therefore be provided to suppress the arc, to divert it from the main contacts, or to blow it out at the instant of formation. The General Electric Company circuit breaker, type C, form K (Fig. 37), is so designed that the arc is diverted to secondary contacts, on top of the breaker, where it is finally broken, thus protecting the main contacts from injury by burning. The auxiliary contacts have carbon tips which are easily renewed. This apparatus is used

Fig. 37.—Type C, form K, carbon-break circuit breaker (General Electric Co.).

for heavy service and special railway work. It is made in two styles, one for circuits up to 250 volts and from 800 to 6,000 amp., and one for circuits up to 650 volts and from 800 to 10,000 amp. The main contacts are bridged by a laminated copper brush, which is pressed against the contacts by a toggle joint.

The automatic operation of a circuit breaker is usually accomplished through the medium of an electromagnet excited by current flowing through a wire or bar-wound coil (or its equivalent), surrounding at least one pole of the magnetic circuit. The coil may be wound for either current or potential, depending upon its method of connection in the circuit.

The current coil is connected in series with the circuit and is designed to carry all the current which flows through the circuit. The power of the electromagnet depends on the amount of current flowing and not on the voltage of the circuit.

The potential coil is connected across the line and has to carry only that fraction of the total current that the voltage of the system can force through its high-resistance winding. Therefore the power of the electromagnet depends upon the potential of the system and does not in any way depend upon the total amount of current flowing in that system.

Overload and underload protection is secured by the use of current coils, while low-voltage release and shunt-trip attachments are actuated by potential coils.

The horseshoe type electromagnet of the overload trip when energized attracts a swinging armature, which armature on its upward motion releases a catch, thus permitting the breaker to open. A flat spring and the weight of the brushes which are pressed against the contacts throw the breaker open.

With the underload trip, when the current flowing through the circuit falls below a predetermined value, the energy of the electromagnet is insufficient to counteract the force of the spring acting on its armature. Hence this armature is released and strikes the tripping lever a hammer blow, thereby opening the circuit breaker.

The principle upon which the tripping of a type C circuit breaker of the Westinghouse Company is based is the lifting of a weight against gravity by the magnetic pull produced by an electric current. The circuit breaker can be adjusted to operate for different current strengths by moving the weight along a graduated scale beam. This adjustment is possible, because for every displacement of the weight a corresponding magnetic pull is required, which is produced by different values of the main current. The laminated contact bridge is similar to that of the General Electric Company breaker described above. When the breaker opens, the current is gradually shifted through the copper shunts to the carbon contacts, and thus no arc is formed until the final break takes place between the carbon contacts at the top. Since the arc is blown out in an upward direction, it is advisable to mount the breakers near the top of the switchboard, in order to protect other instruments against damage.

These breakers are manufactured by the General Electric Company and the Westinghouse Company for d.c. as well as for a.c. use. They are built single, double, or triple-pole, separated from each other by marble barriers, and are tripped independently by automatic tripping coils. By means of special devices the tripping coils can be interlocked in such a way that the

FIG. 38.—Type M, magnetic blow-out circuit breaker (General Electric Co.).

circuit breakers are closed and opened together, or that the closing is independent and the opening simultaneous. For these breakers the service voltage should not exceed 750.

Figure 38 shows a General Electric Company circuit breaker, type M, where the arc is blown out at the instant of formation, by means of a magnetic field. This type is recommended for use with generator or feeder panels connected to circuits where violent overloads are frequent. The secondary contacts and the coil of the blow-out magnet are in parallel with the main contacts. Owing to the comparatively high resistance of the secondary contacts there is virtually no current through them until the main contacts open. Then the whole current is shifted to the magnet coils, and the strong magnetic field extinguishes the arc as soon as it is formed on the secondary contact. The secondary contacts and coil are inclosed in a fiber box over the main contacts. The laminated copper contact bridge is pressed against the main contacts by a toggle joint. The tripping is accomplished by means of a horseshoe magnet which encircles one of the main studs on the rear of the panel. Its action is similar to that of the breaker first described. When the breaker is released by the armature of the magnet, the bridge is thrown open by the action of a spiral spring and its own weight. This type is adapted to

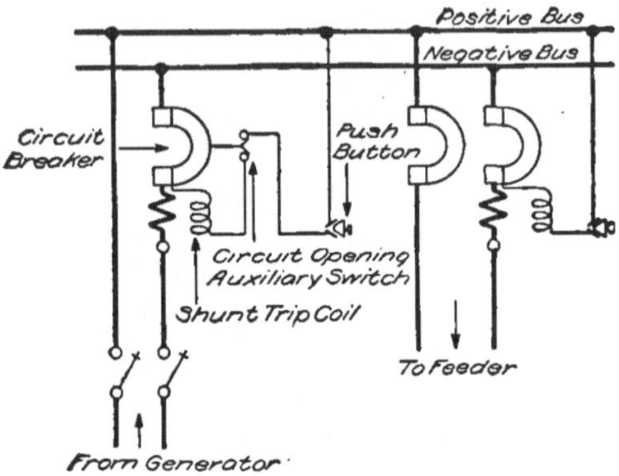

FIG. 39.—Connection of shunt-trip coil with and without circuit-opening auxiliary switch.

650 volts and from 3,000 to 10,000 amp. Several other forms of breakers of the types mentioned are on the market, corresponding in their construction to the various requirements of current, voltage, and character of service. In most of these forms the tripping coil is in series with the line which is to be protected. All of the breakers are constructed so as to interrupt overload. Through additional devices they may be tripped also at low load or low tension at over-voltage and on reverse current. They can also be operated by push button from any given place, or they may be tripped together. Figures 12, 13 and 40 show the switching arrangements for the low-voltage release of the breaker connected with the speed limiter of a converter. Its object is to trip the circuit breaker when the line voltage drops to 50 per cent. or less of the normal pressure. It also performs the function of a shunt trip when used in conjunction with a push button, auxiliary switch, or speed-limiting device. Figure 39 shows the diagram

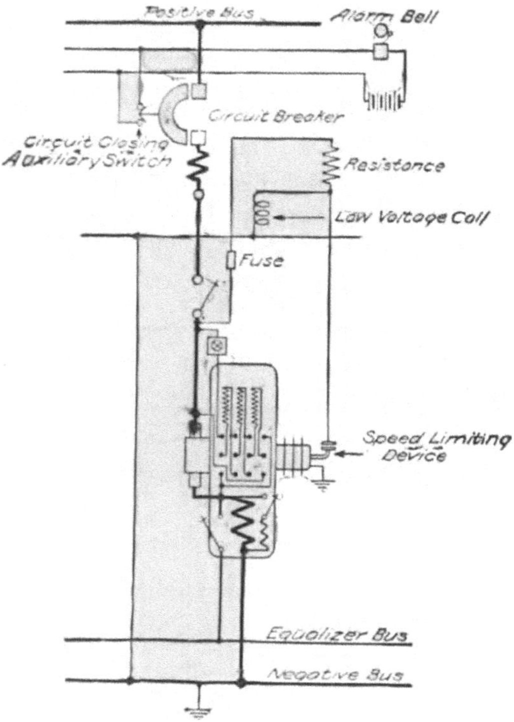

FIG. 40.—Connection of circuit breaker with low-voltage coil with circuit-closing auxiliary switch.

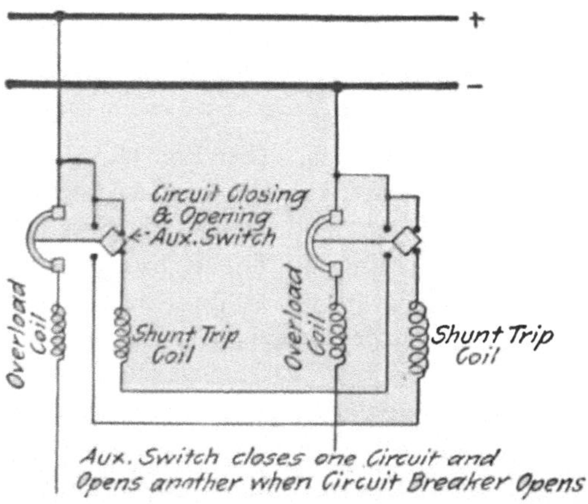

FIG. 41.—Connection for interlocking two circuit breakers by means of shunt trip and auxiliary switches.

of a shunt trip with and without circuit-opening auxiliary switch. It can be operated by means of a push button from any desired point. At the moment of opening of the circuit breaker, the auxiliary switch opens the shunt circuit. The circuit-opening auxiliary switch may also be employed to trip another circuit breaker having a low-voltage attachment, permitting another circuit breaker to remain closed only when the circuit breaker equipped with the auxiliary switch is open. Closing auxiliary

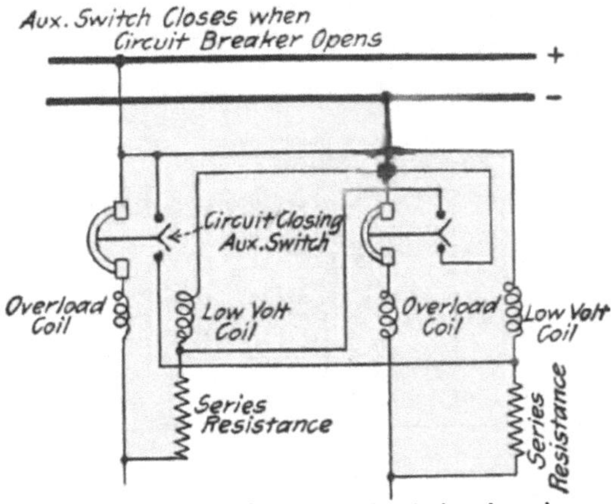

Fig. 42.—Connection for interlocking two circuit breakers by means of low-voltage releases and auxiliary switches.

switches are also constructed, built on the same line as opening auxiliary switches. They are used for closing tell-tales or signal lamp circuits at the instant of main current interruption. (See Figs. 1, 5, 12, 13 and 40.) Circuit closing and opening auxiliary switches are employed when two or more breakers are interlocked, these being used for simultaneous operation or to fix a predetermined sequence of operations. (See Fig. 41.) At the moment of tripping of the circuit breaker, the auxiliary switch interrupts the circuit of its own shunt coil, but closes that of the second circuit breaker, thus causing the circuit breaker to be tripped. Figure 42 is a similar diagram of two interlocked circuit breakers with the use of circuit-closing auxiliary switches and low-voltage coils.

The line is protected against current reversal by a reverse-current relay in the breaker. These relays are especially necessary when storage batteries supply the line in conjunction with a motor-generator set or synchronous converter. They prevent the batteries from delivering energy back into the motor-generator set or converter in case of disturbance on the high-tension side. This type of relay consists of a horseshoe magnet encircling one of the contacts of the circuit breaker. A movable armature connected to the busbars is inserted between the poles of the

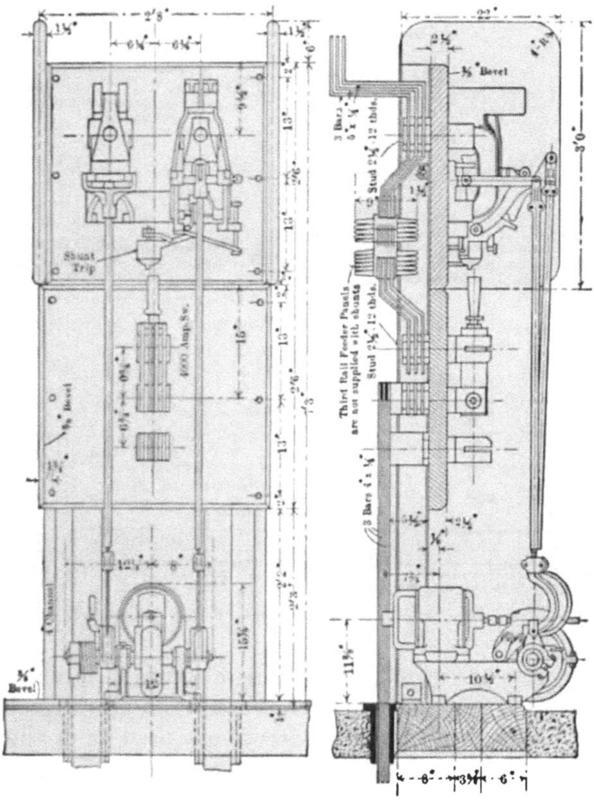

FIG. 43.—Motor-operated circuit breakers.

magnet. With normal current direction the armature will move in one direction. A stop is provided to prevent movement beyond a certain point. By reversal of current the armature revolves in the opposite direction, which closes the shunt-winding circuit, thus tripping the breaker. There are many cases where, for convenience of operation or economy of space, remote-control circuit breakers are very desirable. These may be installed at considerable distances from the operating switchboard, as best suits the particular conditions in any given case. Substantial reduction in the length and cost of the main cables is usually effected. Several breakers may be operated on a single control circuit or one breaker may be operated from a number of

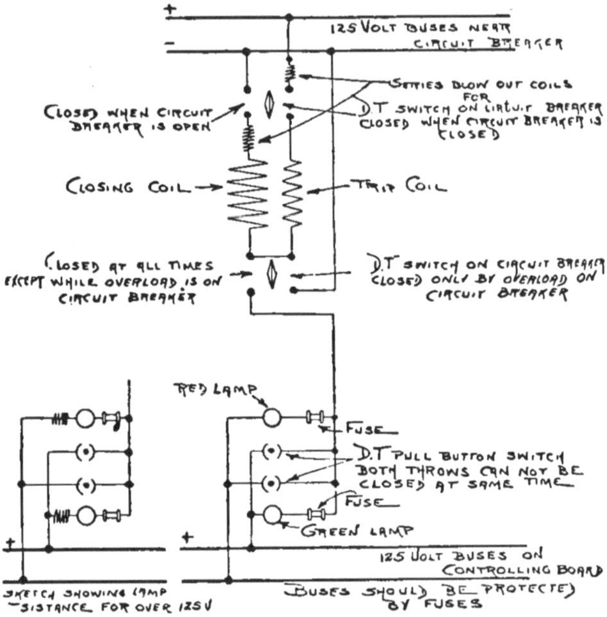

FIG. 44.—Wiring diagram of solenoid-operated circuit breaker with trip coil.

control points as desired. Two general types of mechanism are employed, namely, solenoid-operated and motor-operated. The solenoid mechanism is more extensively used than the motor mechanism, which comes in combination with large breakers. The main line to be opened by the circuit breaker may be either direct-current or alternating-current and of any voltage up to 650. The controlling circuit for operating the solenoid or motor mechanism must in every case be direct-current With each solenoid or motor-operated circuit breaker or combination of two or more breakers, one or more single-pole, double-throw, twin-pull button-control switches, with the necessary indicating lamps, red and green lenses, are required. Figure 43 shows the arrangements for two circuit breakers for 4,000 amp. operated by an electric motor. These breakers are connected in series with each other and with a single-pole, double-throw switch. The opening and closing are simultaneous. The double-throw switch is operated by the same motor, and is shut down only after closing of the circuit breaker, so that in case the feeder is closed on a short-circuit the breaker can immediately open. The arrangement described is that used by the New York Central Railroad Company in its circuit-breaker sub-stations. Figure 44 shows the connection of an 800 or 1,200-amp. breaker with an overload device operating the trip coil of the solenoid mechanism.

Breakers for compound-wound generators should always be connected on the side opposite to the series field. If this is not feasible, as with three-wire generators having two series fields, they should be connected between the armature and the equalizer connections. If this plan is not followed, the armature cannot be properly protected and the breakers are liable to trip out on account of the equalizing currents, at the time when it is most important for them to stay closed.

For the protection of shunt-wound two-wire generators on two-wire or three-wire lighting systems, single-pole overload, single-pole reverse-current or a combination of both breakers are suitable. Compound-wound two-wire generators on two-wire or three-wire systems should always be equipped with overload protective devices. Railway generators are almost invariably compound-wound and have the negative side grounded. The main switchboard panels should be of positive polarity, with single-pole overload or single-pole overload and reverse-current circuit breakers, the series field being on the negative side. A single-pole overload circuit breaker should be connected in the negative lead between the brush and the equalizer tap, in order to protect the generator against internal troubles, unless the machine is motor-driven, in which case reliance can be placed on the motor protective devices.

Three-wire generators are usually compound-wound and should always be provided with overload protective devices.

With motor-generator sets one of the following protective devices should be provided:

1. When the motor-generator set has a speed-limit device, the generator panel should contain a circuit breaker with either a shunt-trip or a low-voltage attachment.

2. When the motor-generator set does not have a speed-limit

device and the motor is protected by fuses, a reverse-current circuit breaker should be used on the generator panel.

3. When the motor-generator set does not have a speed-limit device and the motor is protected by an automatic oil switch, the latter should be equipped with an auxiliary switch while the generator panel should have a circuit breaker with either shunt-trip or low-voltage attachment, so connected that when the motor switch opens, the auxiliary switch on the same will trip the generator breaker.

When motor-generator sets are reversible, low-voltage coils should be used to protect the d.c. generators when operating as motors.

For synchronous converters the same d.c. protective devices are required as for d.c. generators, but no negative circuit breaker is necessary for grounded railway service as the a.c. protective devices take care of the internal troubles in the machine. Speed-limit devices are usually furnished with synchronous converters and require the use of shunt-trip coils or low-voltage-release coils on the breakers.

For d.c. motors single-pole or double-pole overload circuit breakers should be used for protection. Low-voltage coils should be provided, either on the breakers or starting rheostat switches, to cut off the motor if the source of power fails and for use with speed-limit devices when furnished.

Railway feeder boosters are ordinarily series-wound and motor-driven, speed-limit devices being provided for the sets. Single-pole overload circuit breakers are required, and, when no speed-limit device is provided, they should be mechanically or electrically interlocked with the motor breakers, so that in case the motor breaker trips out the boosters cannot operate as series motors without load. Breakers good for the bus voltage plus the booster voltage should be used.

Protective devices for batteries should be designed for the equipment as a whole, including boosters and regulators, and for the operating conditions of the particular battery.

Railway feeder panels for grounded negative systems are connected in the positive side and should contain a single-pole overload circuit breaker for each circuit.

Two-wire feeder circuits on two-wire systems should be equipped with single-pole overload circuit breakers. On small-capacity circuits double-pole breakers can be used to save space.

Two-wire feeder circuits on three-wire systems can be protected by the same devices as are used on two-wire systems, unless connected to the "outsides" of a grounded neutral system. In this case two single-pole overload circuit breakers mechanically or electrically interlocked (or a double-pole, double-coil breaker) should be used.

Three-wire feeders should be protected by two single-pole overload circuit breakers mechanically or electrically interlocked (or a double-pole, double-coil breaker), regardless of whether the neutral is grounded or ungrounded.

No carbon circuit breaker is suitable for operation in cement or flour mills, plaster or furniture factories or any similar industry where acid fumes or excessive dust and dirt are present. For operation in such location the installation of oil circuit breakers is strongly recommended. (See Chap. XV.) Where carbon circuit breakers are installed in places subject to an accumulation of dust and dirt, special care should be taken to keep the contacts clean and bright. Circuit breakers which in the course of operation are rarely opened should be regularly opened and cleaned by the attendant to insure good contact.

CHAPTER X

DIRECT-CURRENT STATIONS

We will use this term to include only those direct-current plants where natural or derived mechanical energy is converted into electrical energy. Converter stations do not come under this heading, inasmuch as they interconvert the different forms of electrical energy only. Direct-current stations are constructed to furnish energy for traction, for stationary motors, for lighting systems, and for chemical or metallurgical purposes. Plants for power and lighting are often built in one, while traction plants may also be used for different classes of work. Direct-current stations are not used for high-tension transmission systems in this country. European exceptions to this rule are the Thury d.c. transmission systems, for example, St. Maurice-Lausanne operating at 27,000 volts, and Mountier-Lyon at 57,600 volts.

Traction.—Direct-current central stations for from 550 to 650 service voltage are profitable only when the traction system is confined to a small area, and when it is possible to locate the power house at or near the load center. The reason for this is found in the fact that for larger systems the cost of copper for feeders materially increases the first cost, thus making the investment unprofitable. We therefore see that 600-volt d.c. stations are limited to small street railway systems or to isolated traction systems for industrial purposes. When a system has outgrown the area for which the plant was originally designed, independent plants may be added to supply the different sections. The choice between adding independent stations and changing the method of supply to another system depends upon conditions.

When the service voltage is doubled (say to 1,200 volts), the economic limit of operation is correspondingly increased.

After the characteristics of the proposed line are known, such as length, direction, curvature, and grade, the next duty of the engineer consists in determining the average load in different sections of the line, from known data as to size, occupation, and shifting of the population of the adjacent territory. From these points he is enabled to fix upon the load center of the average load. Since the load center determines the minimum weight of copper necessary for feeders, the advantage of locating the power house at this point is evident. Value of real estate, proximity of water and coal supply, and methods of feeder installation must naturally be taken into consideration. The feeder system is calculated and distributed after the maximum load of different parts of the line or the load variation per day and per season have been determined. All of these calculations and estimates serve to fix the number and size of the power units in the central station. The location of the power house is often predetermined by the proximity of water power to the system. Other sources of power must often be added to that afforded by water power in order to meet the requirements of the service. The size of the building is completely determined by the number and size of the generators, by the choice of motive power, and by the estimated expansion necessary in future time. The portion of the building with which we are concerned is that part which is reserved for the installation of generators, auxiliary electric machines, switchboards, and cables. In practice, the electrical engineer must work hand in hand with the mechanical and civil engineers and architect. In d.c. central stations the switchboard, which is generally a direct-control board, is located in a place whence the operator can easily overlook all the machines. It is therefore placed in a gallery at one end or along one of the main walls of the machine room.

High-tension Traction.—The recent tendency has been to increase the service voltage. This movement is advocated by Frank J. Sprague (*Street Railway Journal*, Dec. 23, 1905), Mr. Hobart (*Electrical Review*, London, vol. 46), and Dr. Louis Bell ("Power Distribution for Electric Railways"). The following paragraphs sum up the arguments in favor of the high-tension current brought forth by the above-mentioned engineers:

One of the most important factors in the investment and cost of operation for an electric traction system is the value of copper in the feeders and trolley wires and the drop in voltage due to their resistance. With a given service voltage, as for instance 550, a certain drop in the line and a corresponding minimum cross section of feeders is permissible. Any extension of the service area or load will cause an increase of drop in voltage, which must be compensated for by allowing a larger cross section of feeders, or by the use of additional feeders. This increase in copper weight can be avoided by increasing the service voltage, thus decreasing the current.

The weight of copper varies inversely as the square of the voltage. That is, if we double the voltage from 500 to 1,000, a line four times as long can be supplied, using the same amount of copper and allowing the same per cent. drop. The actual results are slightly better even than these figures indicate, since the track return gets relatively better and better as the voltage rises and the current diminishes. A larger percentage can therefore be allowed for the copper line drop. This matter is of even greater importance for a.c. central stations and d.c. sub-stations. (See Chap. XXVI.) For voltage between 750 and 1,000 the construction of generators of less than 1,000 kw. rating is entirely feasible. If the cost of one large machine should be too great, this machine may be replaced by two smaller ones of lower voltage connected in series, or by two synchronous converters. The increased voltage may also be obtained by making use of a booster, as will be explained later. The motors used for this system are also easy to construct. A 650-volt machine is guaranteed for 750 volts. The motors used for the Berlin Elevated System and for the Interurban Railway of Zweisimmen-Montreaux are built for 800–850 volts, but are actually wound for 1,000 volts. By a series-multiple connection, motors for 500–600 volts can be used in 1,000–1,200-volt circuits. There are a number of railway systems in the United States which now employ or contemplate employing a direct e.m.f. of 1,200 volts.

Lighting Systems.—The economic considerations involved in this class of plants are similar to those discussed under traction systems. Small system d.c. stations are built at the load centers. They are often called upon to supply energy for motors, for which purpose separate machines are sometimes used. If the number of units for various purposes is increased above a certain point, the efficiency of the station will become less than would have been the case had the small generators been replaced by a few large a.c. machines used in conjunction with converters. The most common case where d.c. is used for lighting is that of isolated plants, for office buildings, theaters, hotels, etc. Such plants are of the most economical form, because they are located near the receiving apparatus, and the cost of wiring is therefore reduced to a minimum. Besides supplying energy for the lighting system, the plant also supplies energy for elevators, fans and other light machinery. The predominating system for light and power distribution is the three-wire system. Where higher voltage for certain purposes is not obtainable with this system, separate units must be used, as mentioned above.

By a system of feeders and mains, direct current is usually distributed at 220 volts from a station located at the center of power demand. When the limit of economical distribution from a single station is reached, additional stations are built and all the stations are connected by tie lines. Characteristic of these stations are:

1. Good voltage regulation, because the tying together of the mains results in the exchange of energy between the feeders and thereby equalizes the voltage.

2. High economy resulting from the parallel operation of all stations and the consequent reduction of line losses to a minimum. As the reactive drop and magnetic screening are absent with direct current, a large conductor can thus be used and a greater density of load can thereby be cared for.

3. Parallel operation of all stations to guard the system in case of the breakdown of one or several stations. By limiting the resistance of the mains and the feeders, the effect of any breakdown at any point of the system is confined to its immediate vicinity. Thus no local breakdown involves or even seriously affects the entire system.

4. Use of a storage battery reserve, which greatly increases the service reliability and maintains service even during a complete shut-down of all generating stations.

For certain classes of service, such as high-speed elevators, direct current is more advantageous.

It frequently happens in medium-sized European towns that after the municipal gas works have reached the limit of their capacity the future wants of the population must be provided for either by extending the works and the pipe lines or by constructing an electric power system. In most cases the latter alternative is given the preference as electrical energy is better adapted than its competitor to supply the light, heat, and motive power required under the highly variable conditions of public service.

In such towns direct-current systems with a storage battery are most frequently adopted. In choosing the prime mover for the dynamos in cases where no water power or cheap coal is available, oil engines of diesel type have been favored very frequently in recent years because of their small space requirements, easy superintendence, high thermal efficiency (37 per cent.), absence of smoke and ash and immediate readiness for service. In fact, the diesel installations often make it possible for small and medium-sized direct-current power houses to maintain themselves despite the competition of much larger generating systems.

CHAPTER XI

TYPICAL ELECTRIC POWER STATIONS

Memphis Street Railway Company (Tennessee).—The power house is an old building with an annex of modern construction. In this annex is installed a 2,000-kw. G. E. generator, driven by an Allis-Chalmers vertical two-cylinder, cross-compound engine of 3,000 hp. Figure 45 shows a load curve of the system for January 27, 1908. Figure 46 is a plan of the founda-

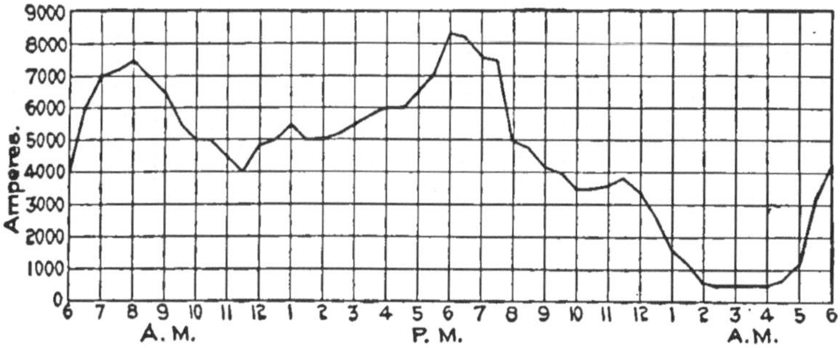

FIG. 45.—Load curve of the Memphis Street Railway Company's power house.

tions of the three engines on whose shafts the generators are mounted. The right wall is that of the old power house, which communicates with the annex by only one door. The plan shows the cable distribution. The positive cables lead from the foundations of the machine, where they are inclosed in the ducts, under the engine-room floor, and to the front wall, whence they pass between the windows up to the switchboard gallery. The plan indicates the cable disposition for possible fourth and fifth generators. The negative and equalizer buses are conducted in tile tubes through the foundations and are connected by means of cables with their proper switches, which are in turn connected with the terminals of the machine by cables running through the foundations. Figures 46 and 47 also show the manner of conducting the feeders from the main building into the underground

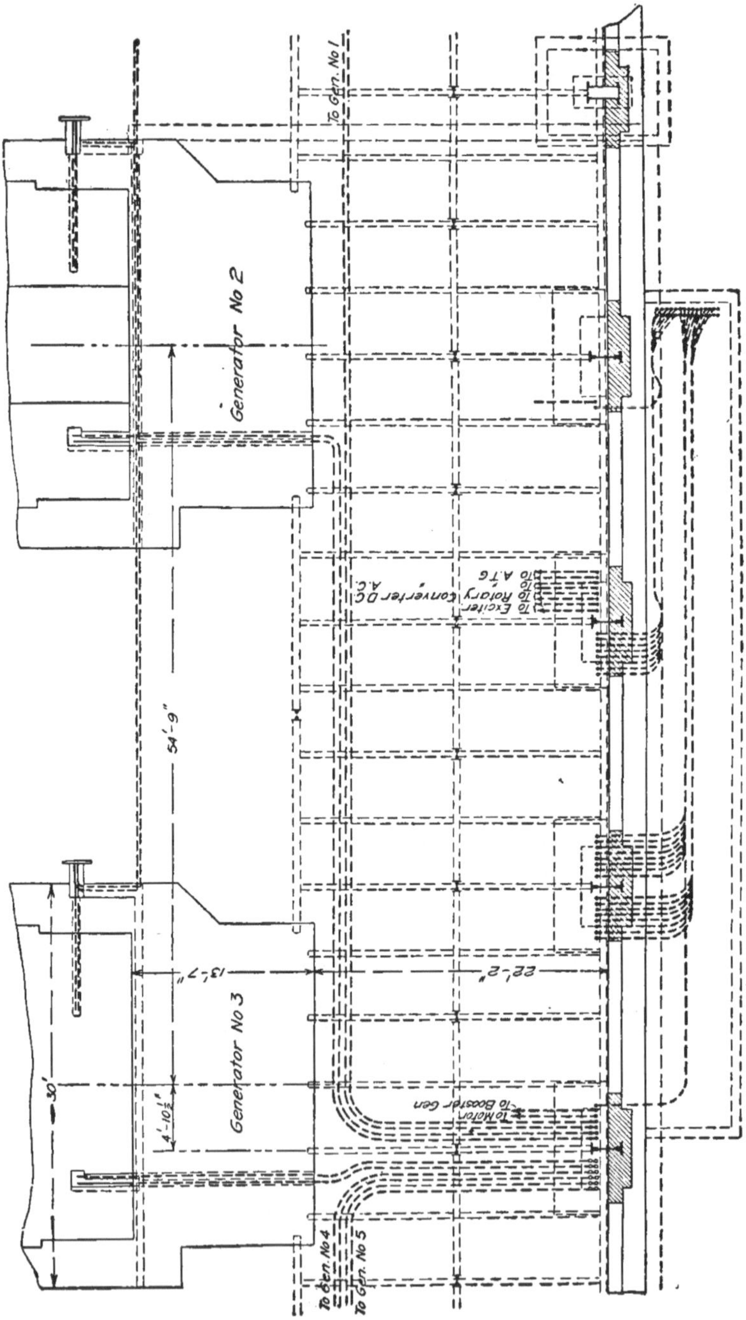

FIG. 46.—Plan of engine room, showing layout of generator cables (Memphis Street Railway Company).

passage which runs along the main wall. The feeders are supported on racks in three rows, one on the main wall and two rows on either side of a pipe support in the center of the passage.

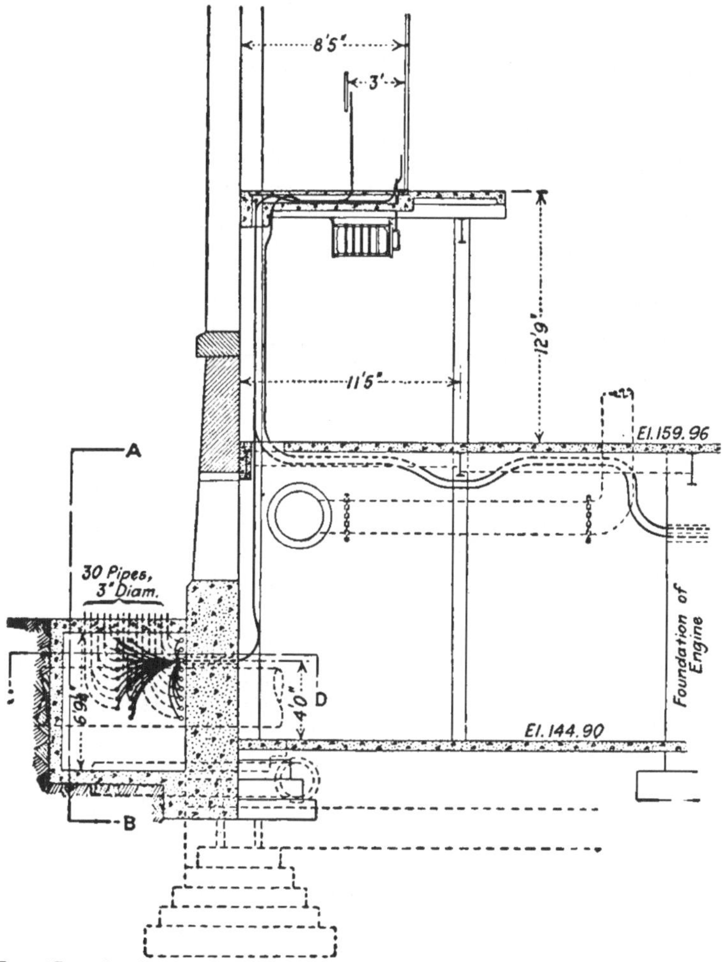

FIG. 47a.—Section through switchboard gallery (Memphis Street Railway Company).

At the end of the passage the feeders pass through the ceiling to the overhead transmission line. A number of ducts are built into the end of the passageway, to be used in case it is desired to extend the feeders underground instead of overhead. Figures 48 and 49 show front, rear, and side views of the switchboard

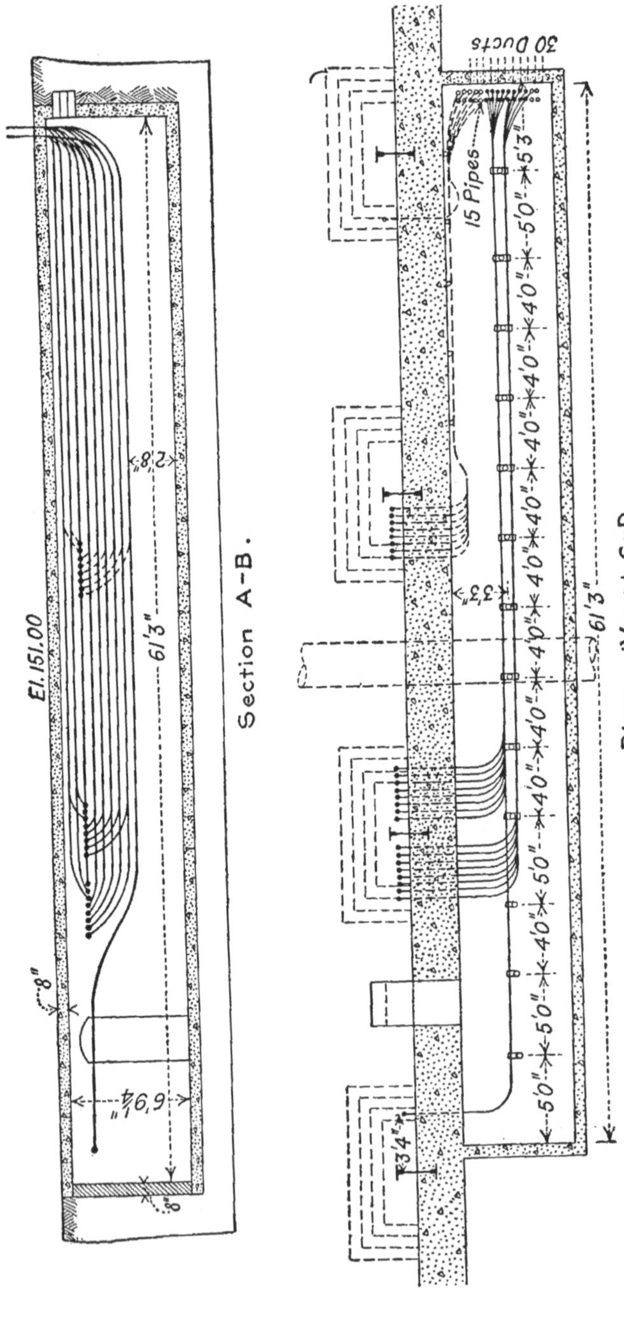

FIG. 47b.—Layout of cables in [illegible] (Memphis Street Railway).

and the wiring diagram of the station. The positive bus of the generators is mounted on the board, and the machines are equalized on the negative side. A general output ammeter, a recording wattmeter, and a watt-hour meter are connected with the main bus between the generator and feeder connections. A motor booster set is connected to the buses, whose functions are as follows:

In traction systems the service area sometimes expands to such an extent that there is a considerable voltage loss at the extreme points. This is also the case in systems where at certain points

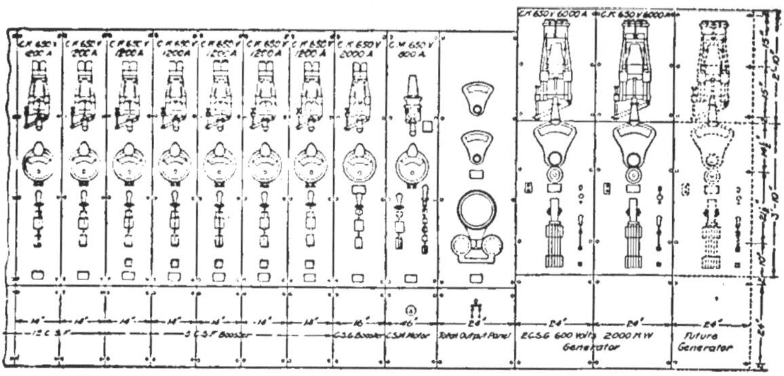

FIG. 48a.—Switchboard, front view (Memphis Street Railway).

the load rises to an excessive degree. In order to overcome this difficulty, the cross section of the feeder conductors to these points must be increased, so that the voltage drop may be kept under a certain maximum. Several such feeders of large cross section and increased weight are sufficient to add materially to the necessary investment. One of the methods of diminishing the cost due to this source is to increase the initial pressure of the feeder in question to such a point that the resultant voltage will at all times be above the required minimum. The voltage of the generator itself can be raised only to a limited value by means of over-compounding the machine. If this value is exceeded, the higher voltage at other parts of the system, as, for instance, near the station, becomes a serious disadvantage. Over-compounding the generator also causes increase in cost. In cases where the probable growth of the system can be estimated a.c. plants with converters are often recommendable. However, we have another means at our disposal for raising the voltage of the system

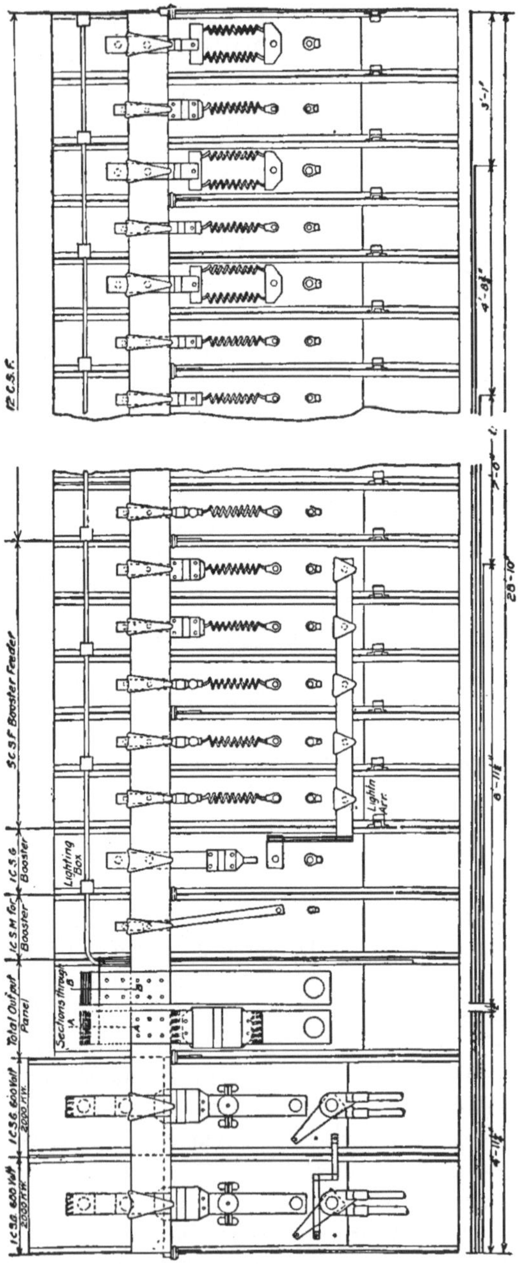

FIG. 48b.—Switchboard, back view (Memphis Street Railway).

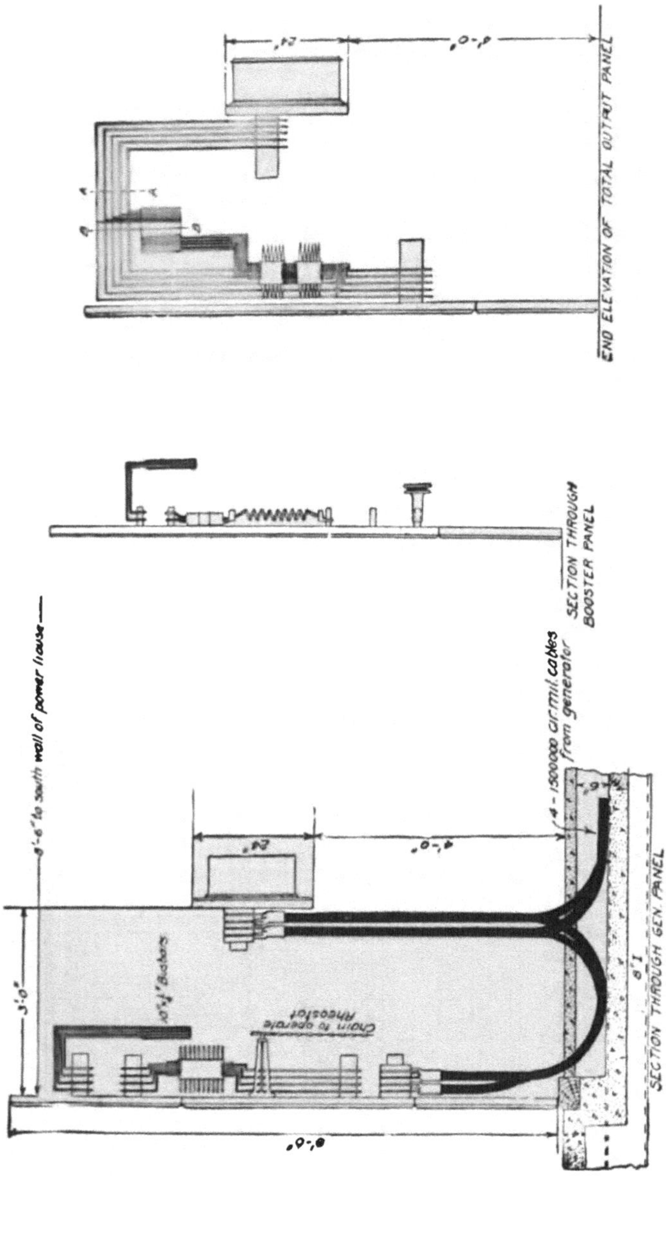

c d

FIGS. 48c and 48d.—Switchboard, ew (Memphis Street Railway).

at certain points or in the whole line, and this is the application of the booster. (See Chap. V, case 2.)

Figure 50 shows a booster in series with the main circuit. The

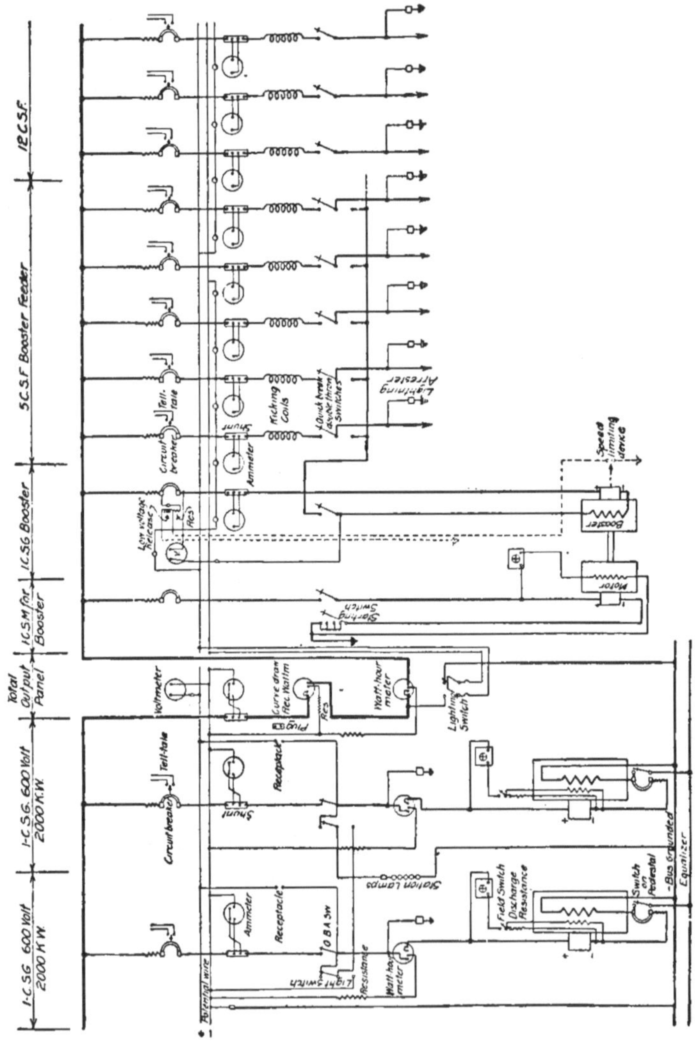

FIG. 49.—Switchboard wiring (Memphis Street Railway).

case taken is that of a generator feeding only one main at an e.m.f. of 550 volts. The booster voltage is assumed at 200, and the current at 500 amp., so that the combined voltage amounts to 750. Since the initial voltage is 750, a loss of 300 may be allowed on the line. The reduction in cross section of the feeder thus made possible is considerable when compared with the cross section necessary with an initial voltage of 550 and an allowable drop of 150 volts. This method of allowing the booster to feed directly into the line is economical only when used about three hours per day of service at full load. It is therefore well suited to help over the time of unusually heavy traffic. For, although the initial cost of the booster, which is generally driven by a motor or other mechanical power, is small when compared with the saving in copper, its operation for a longer period than three hours at full load is nevertheless very expensive and uneconomical

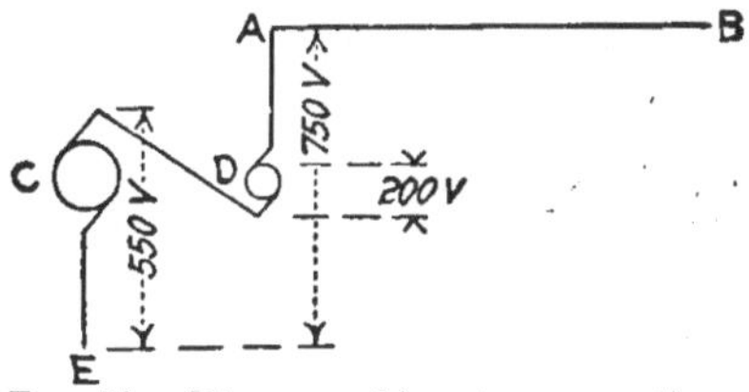

FIG. 50.—Diagram of booster connections.

on account of the energy lost in the line. This machine is used to its best advantage for the independent operation of the special feeders required, not including the entire system. This is the case assumed in Figs. 48 and 49. The booster, which is driven by a d.c. motor, is connected through double-throw switches with five feeders, thus making it possible to feed these cables with either the normal voltage of 600 from the main bus or the higher voltage from the booster bus. The side and the rear views of the switchboard show the busbar connections and mountings. The four cables of 1,500,000 cir. mils each for each generator are run up the wall to the gallery floor, where they are laid in brick partitioned channels. (Fig. 51.) They are then connected to the watt-hour meters mounted on separate slate slabs in back of the board. From the watt-hour meters they lead back under the floor to the main switch on the generator panel. The three spaces between the main wall, watt-hour meter panels, main switchboard, and gallery edge are kept to a minimum in order to keep down the width of the gallery, but are nevertheless wide enough to admit a safe access to all parts. The switchboard stands on a wooden block 1 inch above the floor. The channels for the generator and feeder cables are covered with concrete

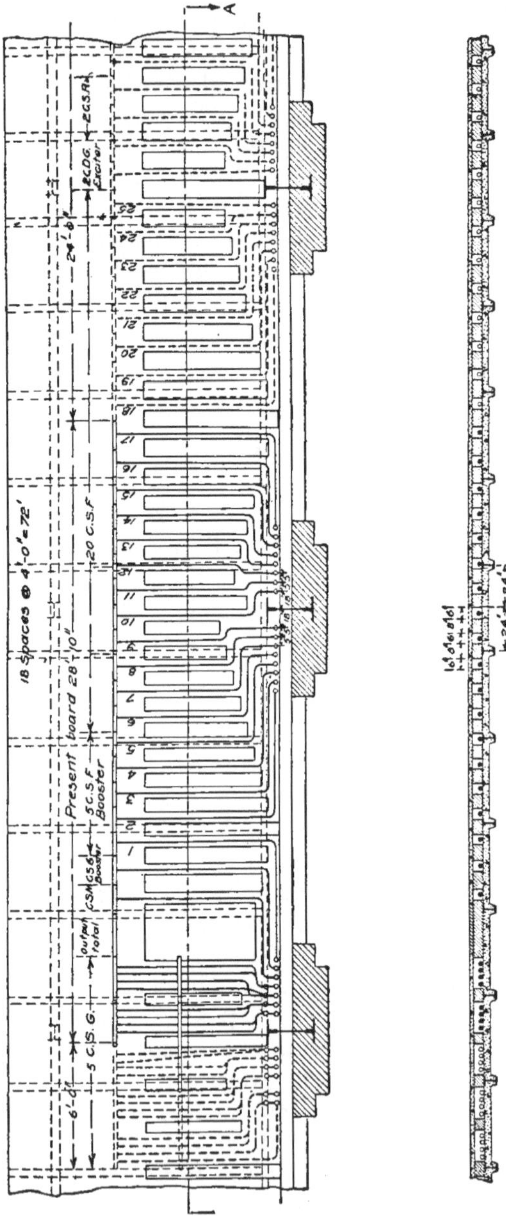

Fig. 51.—Switchboard gallery floor construction.

slabs. The positive bus is composed of copper bars 10 inches by 0.25 inch, the number of which decreases with the increase in distance from the generator panels, since the amount of feeding decreases as we recede from the panels. Note the special construction of the gallery floor, which is lowered in back of the board on account of the placing of the cables. The construction is reinforced concrete resting on 8-inch I-beams, 4 feet center to center, running perpendicular to the main wall. These I-

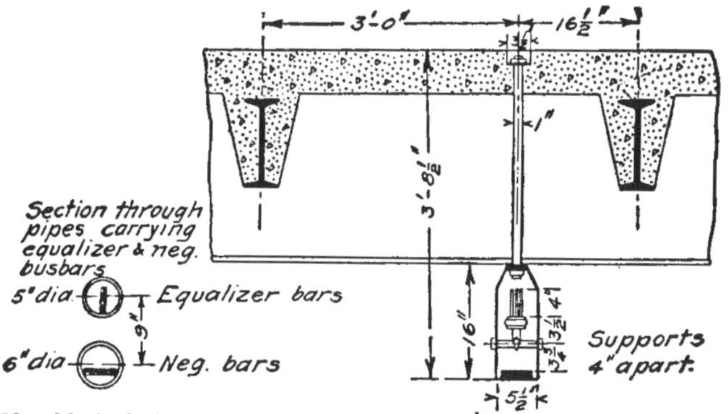

FIG. 52.—Method of supporting negative and equalizer busbars between engine foundations.

beams are fastened on one side to the crane columns by means of channels, and at the other ends to longitudinal I-beams supported by independent columns. The space between the channels and the masonry affords a passage for the cables leading to the gallery floor. The feeder and positive generator cables are lead-covered and are therefore fastened directly to the wall. All the feeders are 1,000,000 cir. mils for each panel. The negative and equalizer cables are double-braided. All the outgoing feeders lead to the above-mentioned underground passageway through ducts built into the concrete foundation of the wall. The passageway communicates with the basement of the engine room by a door toward the end of the passage. The field rheostats are mounted under the floor of the gallery and are easily operated from their respective boards. The negative bus is made up of 5-inch by 0.25-inch uninsulated copper bars. The equalizer bus is of similar dimensions, but is wrapped with insulation. The method of mounting these two buses between the foundations of the units is shown in Fig. 52.

Indianapolis and Louisville Traction Company.[1]—The first 1,200-volt interurban railway in the United States began operation on November 6, 1907, between Seymour and Sellersburg, Ind., 41 miles apart. The central station is located near the center of the line. The trolley wire of No. 0000 copper wire is fed by feeder cables running 17 miles in each direction from the station. They are made up as follows:

10 miles of 500,000 cir. mils
20 miles of 300,000 cir. mils
4 miles of 211,000 cir. mils

The electrical energy is generated by two units. Each unit consists of two generators driven by an engine. The generators for each unit comprise two 600-volt General Electric multipolar (type M P) units with armatures, mounted commutator to

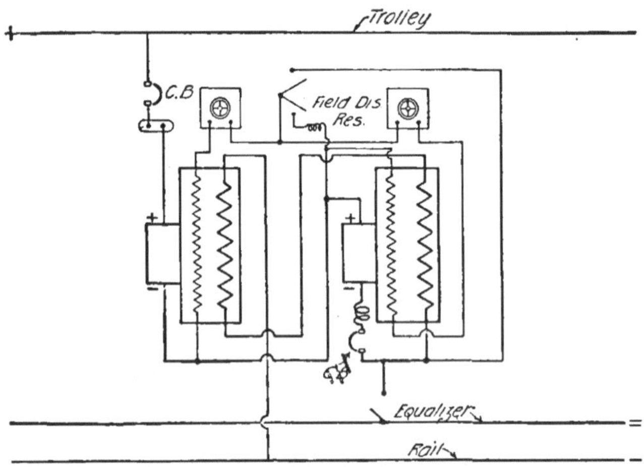

FIG. 53.—Indianapolis and Louisville 1,200-volt railway: method of connection of two 600-volt generators.

commutator on the engine shaft. Figure 53 indicates the switching arrangements for both generators of one unit. Their armature coils and series field coils are connected in series as though for one generator. The shunt field for each machine is connected across a circuit having 600 volts difference of potential. Each of these 300-kw. machines is compound-wound, having a flat output curve. The switchboard for controlling the output of the two 1,200-volt generating units comprises two machine and two

[1] Abstracted from *Electric Railway Review*, Nov. 30, 1907.

feeder panels, with two 1,500-volt voltmeters on swinging brackets and a watt-hour meter on each machine panel. For use in the nearby repair shop, and for lighting the power house, 600 volts is obtained by connection between the two machines of one unit. The absence of sub-stations such as are necessary with a.c. generating stations and of high-tension feeders is an advantage of this system.

Another road (the Pittsburgh, Harmony, Butler and New Castle Railway) uses a different scheme for supplying the line with 1,200-volt direct current. It generates at the main power house alternating current at 13,200 volts, 60 cycles, and feeds two sub-stations at this potential. A sub-station in the power house takes care of the central portion of the system. Synchronous motor-generator sets in the sub-station transform the high-potential alternating current to the 1,200-volt direct current which is the working potential of the trolley.

Lighting Switchboard.—Figures 54 and 55 show a combined system where an independent generator delivers a 600-volt

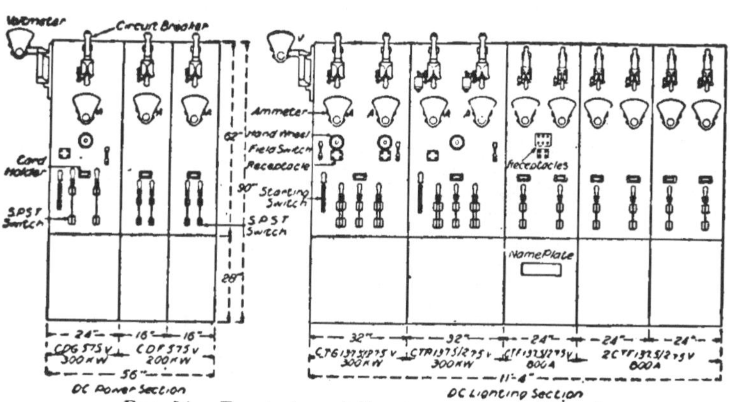

FIG. 54.—Front view of direct-current switchboard.

direct current, and an independent three-wire system is used for lighting purposes. The d.c. generator of 300 kw. rating and 575 volts furnishes energy for two sets of feeders. Both busbars are mounted on the board, which therefore requires two single-pole lever switches on each board. The generator is a part of a motor-generator set and is to be arranged for d.c. starting. A four-throw starting switch should be mounted on the generator panel. The three-wire system is fed by two low-tension generators connected in series (Chap. VI, case 1), and by a synchronous converter (Chap. VI, case 4). Both generators are coupled to an a.c. motor (induction or synchronous) and are started on the d.c. side. If they are driven by an engine, the starting switch

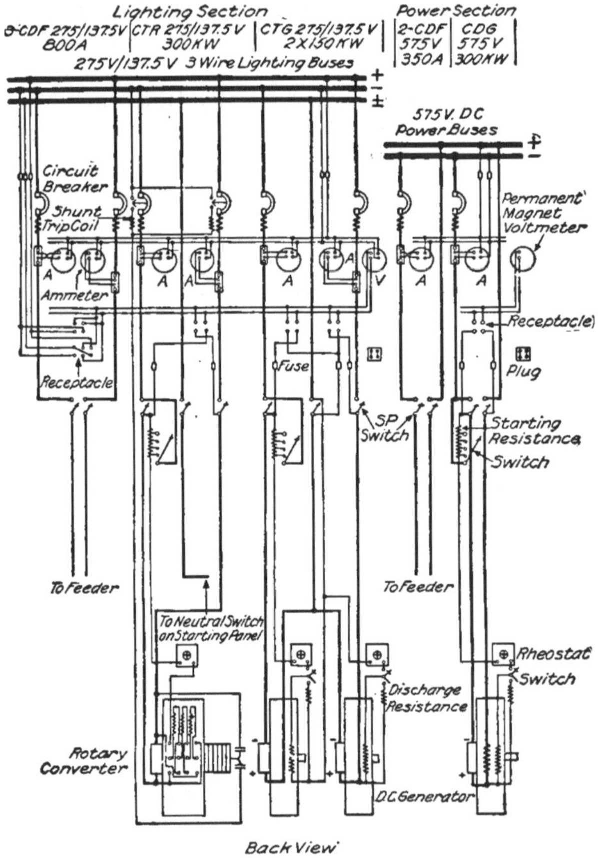

FIG. 55.—Wiring diagram of board shown in Fig. 54.

can be omitted. The equipment for both generators is mounted on one panel. The converter, as well as the generators, is shunt-wound since it is thus better suited for lighting purposes. Circuit breakers with shunt coil and ammeters are inserted on both the positive and the negative side of the converter. The diagram indicates that the converter is started on the d.c. side. Note that the shunt field windings of both generators are in multiple

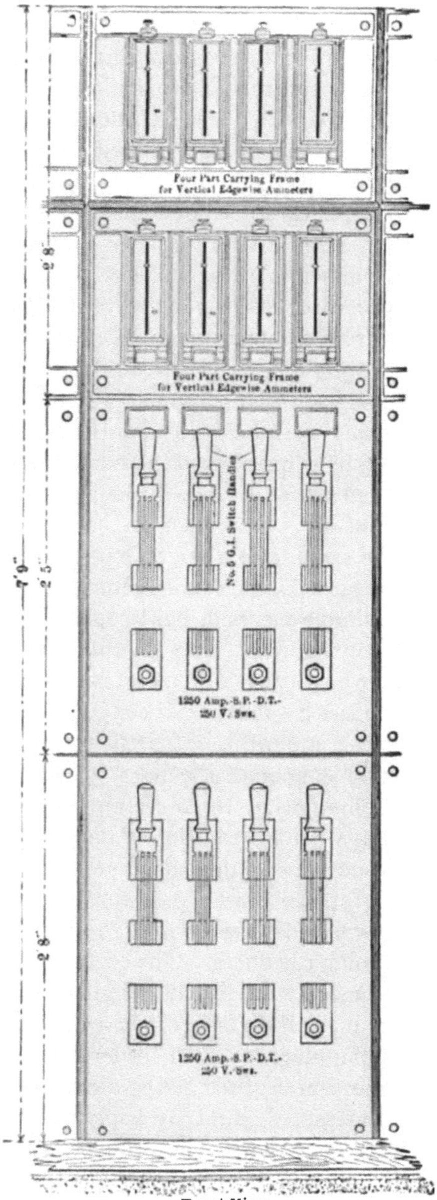

Front View

FIG. 56*a*.—Front view of 1,250-amp. feeder panel (Commonwealth Edison Company).

with the side which the respective generator supplies. The feeders can be connected to the positive and negative bus, or to either of these and the neutral. Both mains of a feeder set are supplied with a circuit breaker and an ammeter. The voltage between any two buses can be ascertained by using an eight-point receptacle and a voltmeter.

The design of the panels for a low-tension feeder system for lighting purposes in large cities, like New York and Chicago, obviously requires particular care. One of the main objects in such a design is to control safely and conveniently a maximum number of feeders from a minimum number of panels. Figure 56 shows front, rear, and side views of the feeder panels used by the Commonwealth Edison Company in its sub-station for feeding the three-wire lighting system. The board controls four sets of feeders, which can be connected to the main or to either of the auxiliary buses by means of double-throw switches. All buses are mounted horizontally. This insures the best separation of potential and, at the same time, the greatest accessibility to the rear of the switchboard. The connections between studs and buses are made by simple pieces of bent copper, which facilitate the interchange of connections. The horizontal mounting of the buses has the further advantage of making it easy to increase the cross section of the buses by addition of copper bars. The poorer ventilation due to this mounting is partly compensated for by a liberal allowance of copper in the bars. Such allowance has the advantage of eliminating to a certain degree the unsafe handling of live buses when an addition of bars is required. The middle studs of the positive double-throw switches are connected to the fuses on the opposite rear wall through bent pieces of copper. The positive feeders lead from the lower terminals of the fuses to the outgoing tile ducts. The cost of the copper bends is more than compensated for by the elimination of the cable racks and the gain in walking space between panels and wall. Two ammeters are supplied for each feeder set. The bar connections between the center studs of the switches and the fuses can be used as an ammeter shunt by applying a compensating coil for temperature correction. The connections for the center studs of the negative switches run down below the floor line and up to the negative fuses on the back wall mentioned above. The floor is made of 1-inch removable slate slabs protecting the negative connections. The neutrals of the feeders terminate

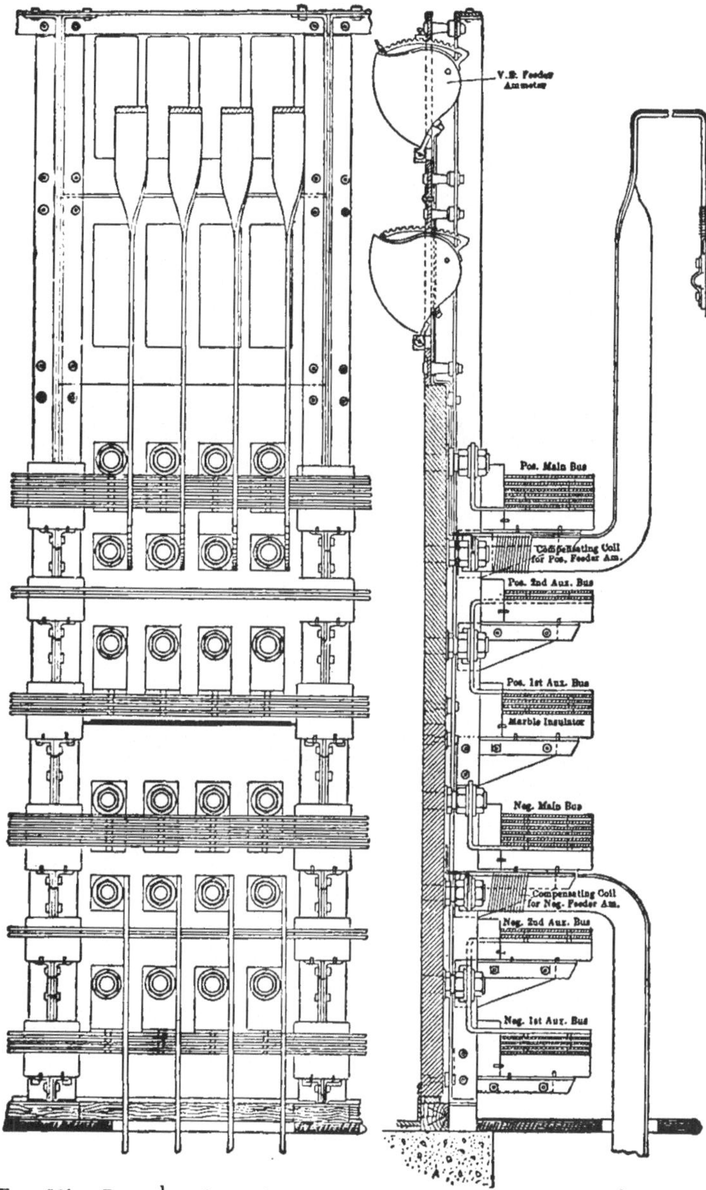

FIG. 56*b*.—Rear elevation and cross section of 1,250-amp. feeder panel (Common-Edison Company, Chicago).

at the neutral bus located in the basement. Figure 57 *a–b* shows front and rear elevations of battery, balancer, and booster

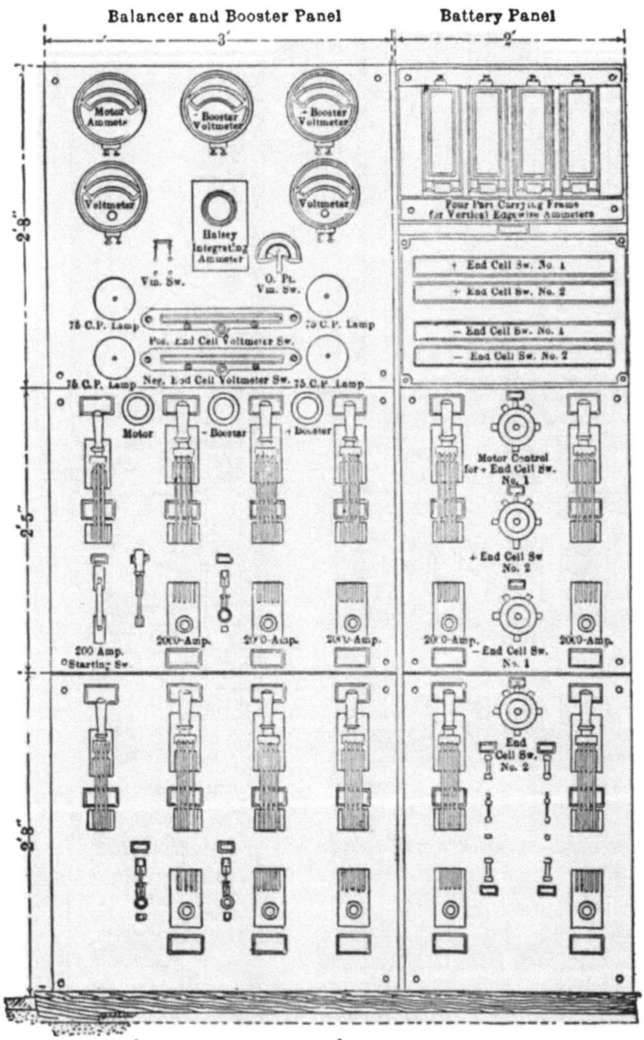

FIG. 57*a*.—Front elevation of battery, balancer, and booster panels (Commonwealth Edison Company).

panels, using the same method of busbar mounting. It will be seen that the studs, nuts, and bolts are as accessible in these as in the feeder panels.

Current-carrying contacts on switchboards should receive careful periodic attention. Considerable heat loss may be due

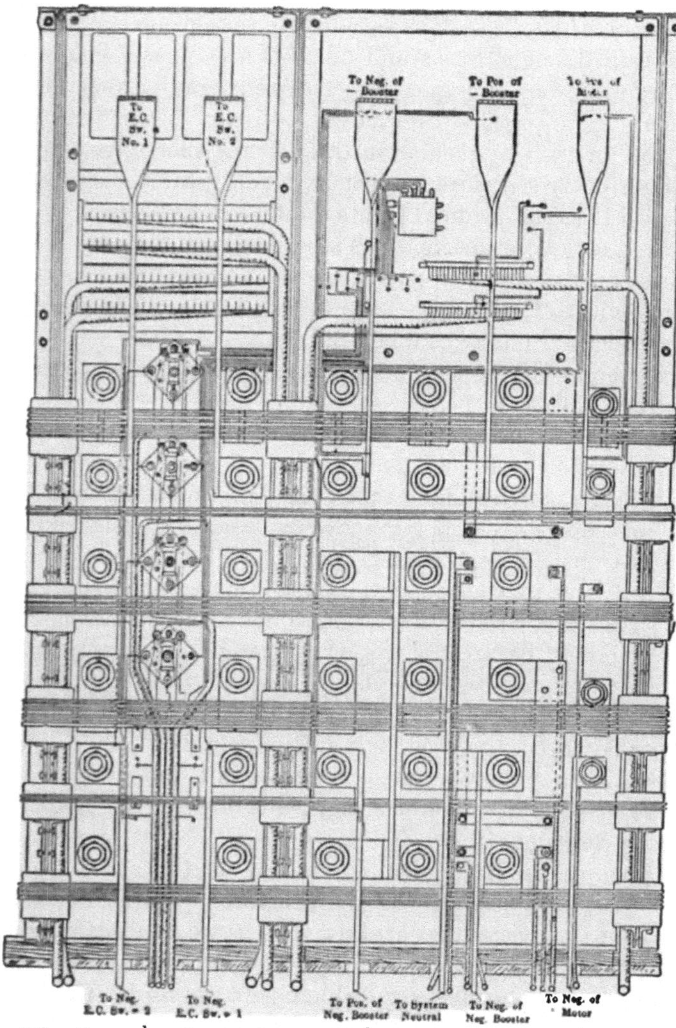

FIG. 57b.—Rear elevation of battery, balancer and booster panels (Commonwealth Edison Company).

to contacts not perfectly planed or to loose or dirty contacts. This applies to switch and circuit-breaker contacts, as well as to permanent connections back of the board.

CHAPTER XII

LOW-TENSION ALTERNATING CURRENT

Alternating currents cover a much more varied field of application than does direct current, since they are employed in practice in different forms and quantities of phase, frequency, e.m.f., and power, and since direct current may be regarded as no more than rectified alternating current. Their treatment from the moment of generation through the successive steps to consumption is therefore subject to much greater variation in regard to the four cardinal points established in Chap. I than is treatment of direct current. The electrical equipment of a.c. plants depends mostly upon the output voltage and kw. rating of the units. Alternating-current systems may be classified with respect to voltage as follows:

1. Systems of the same e.m.f. as those employing direct current, 240 to 600 volts (110 volts).
2. Systems whose e.m.f. ranges from 600 to 33,000 volts, which are those most commonly used.
3. Systems of extra-high e.m.f., ranging from 33,000 to 100,000 volts and upward, used for the most recently designed long-distance transmission systems.

We will first treat classification No. 1' since it differs entirely from the other two. Classes 2 and 3 have many points in common and will be treated together, emphasizing only those differences due to extra-high voltages.

Generator (240–600 Volts).—Alternating-current generators of these potentials are used for small industrial plants or mining purposes. They supply three-phase current and are rated at from 20 to 400 kw. Figure 58 shows a wiring diagram and front and side elevations of a switchboard for a 240-volt generator of this type. A separate exciter mounted on the shaft delivers direct current for the field of the generator. With large units the exciter is driven independently, and, as we shall see later on, constitutes a very important part of the central station. A separate rheostat regulates the shunt field winding. The value of the current supplied to the field circuit of the alternator is regulated by a second rheostat and is indicated by a field ammeter. A double-pole field-discharge switch closes the field circuit. This is provided with carbon break and discharge switches for connection to the discharge resistance. With the low tension used, the connection of the generator cables with the buses is easily accomplished through a three-pole lever switch,

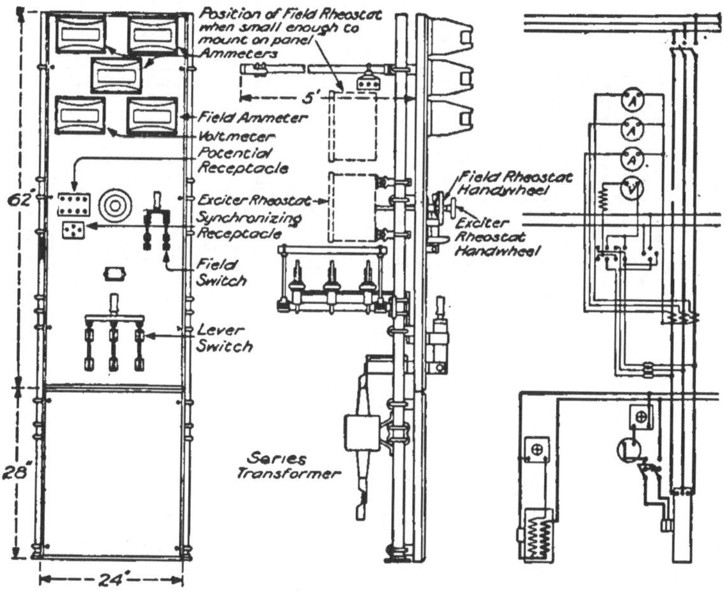

FIG. 58.—240-volt, three-phase generator panel.

as it is recommended that no automatic protection be provided for a.c. generators or their field circuits. With unbalanced lighting load, the value of the current of each phase is indicated by separate ammeters, which are connected directly to the line for currents below 300 amp.; above this point series transformers are used. A voltmeter, receptacles, and corresponding plugs are used to indicate the voltage across any two phases. By this means the voltage between any phase and the neutral of the machine can be indicated in case the neutral of the generator is used for lighting purposes at 125 volts. Instead of three ammeters one ammeter with a three-way switch can also be used. Before an alternator can be thrown into parallel with other machines in service it must be synchronized with them. Synchronism indicators and lamps in connection with receptacles and plugs serve to indicate the difference in phase of the machine in starting. The side elevation shows how the three busbars are mounted. They are protected from foreign objects which might cause short-circuiting, such as tools, by means of a screen. If the buses are very heavy because of larger output of the generators, the supporting arm is stayed by a gas pipe running up from the floor. The board is mounted and fastened to the rear wall with angles, tees, or gas pipes. These supports also carry the series

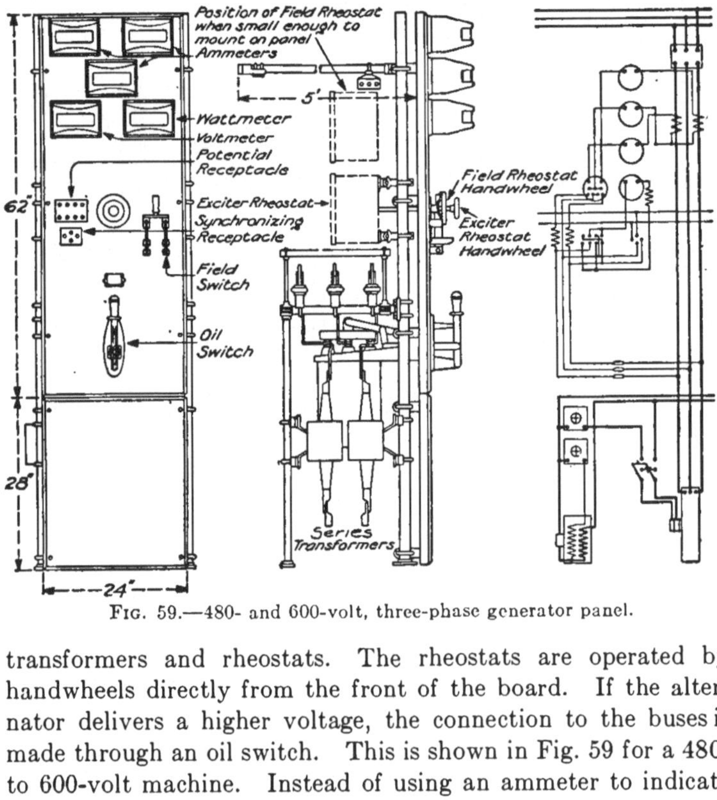

FIG. 59.—480- and 600-volt, three-phase generator panel.

transformers and rheostats. The rheostats are operated by handwheels directly from the front of the board. If the alternator delivers a higher voltage, the connection to the buses is made through an oil switch. This is shown in Fig. 59 for a 480- to 600-volt machine. Instead of using an ammeter to indicate the necessary amount of field regulation, the voltmeter or synchroscope may be employed to perform this function. For balanced loads and current under 50 amp. one ammeter in the middle leg suffices. For larger current values the ammeter must be connected to two series transformers in the outside legs. For unbalanced load the arrangement is the same as that described for Fig. 58. A polyphase wattmeter indicates the total output of the machine, and by comparison of the ammeter and wattmeter indications the power factor of the circuit can be determined. The oil switch interrupts the current under oil in order to eliminate the danger due to the arc between the contacts of the switch at the moment of interruption. It is mounted on the rear of the board and is operated by a handle from the front. (See Chap. XV.) All voltage-carrying instruments are protected with fuses against overload. One synchroscope may be used for several generators, in which case it is mounted on a swinging bracket on the generator side of the board.

A. C. Feeders.—The energy is drawn from the three busbars through an oil switch or circuit breaker. Both of these pieces

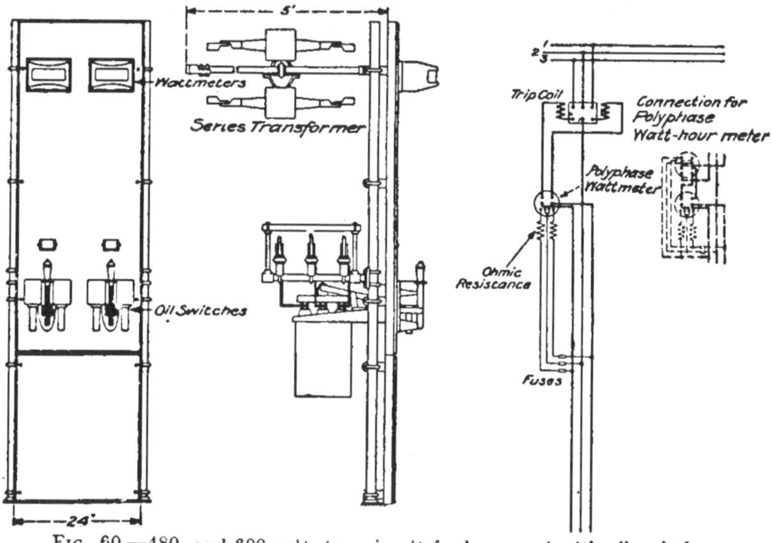

FIG. 60.—480- and 600-volt, two-circuit feeder panel with oil switches.

of apparatus are equipped with tripping coils which operate to interrupt the circuit at overload. Figures 60 and 61 show the wiring diagram and switchboard for a.c. feeders. As we wish to ascertain the amount of power consumption of each feeder system a polyphase wattmeter must be connected into the line. A special polyphase watt-hour meter with series transformer may be used to measure the energy output. The feeders are generally run from the buses and oil switches or circuit breakers to the nearest wall, whence they lead to the place of consumption in the factory. The series transformers are therefore mounted on pipes near the top of the board. This differs from the method of mounting for generator panels, where we saw that the apparatus was mounted near the base of the board. This was due to the fact that the three feeders from the generator reached the board below.

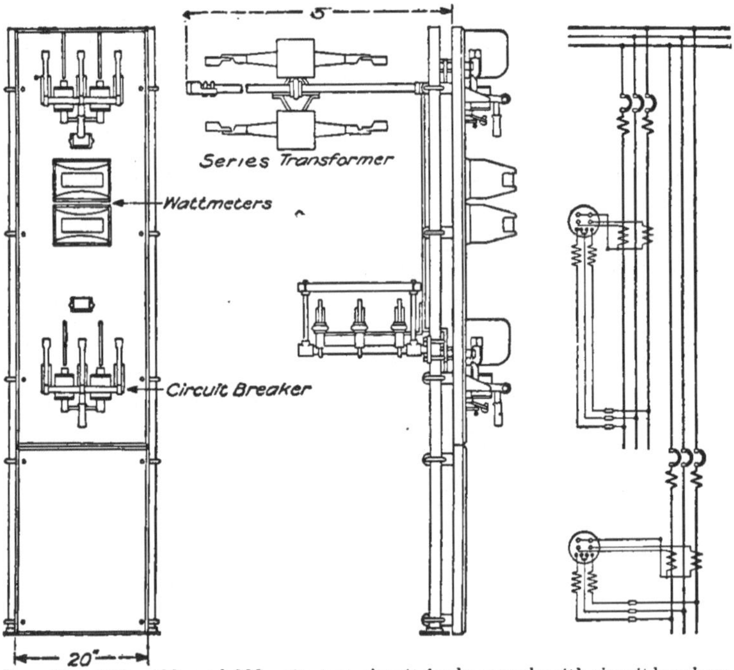

FIG. 61.—240-, 480- and 600-volt, two-circuit feeder panels with circuit breakers.

Alternating low-tension switchboards are identical with d.c., boards in respect to material of boards, method of setting up copper connections, spacing between busbars, and mounting of same, as illustrated in the accompanying figures. Up to this point we have had to dėal with voltages requiring no special safety devices, with the possible exception of circuits requiring interruption in oil on account of induction phenomena.

CHAPTER XIII

HIGH-TENSION SWITCHING ARRANGEMENTS AND METHODS OF CONNECTION

In order to make a correct choice of the various switching arrangements for a plant, an investigation of the following points is essential:

1. Cost.
2. Reliability and continuity of the service.
3. Greatest protection of life and property.
4. Available space.
5. Voltage and capacity of plant.

The order of investigation of the above-mentioned points will depend upon the case in hand. It may be a question whether more consideration is due to the stockholders or to the public, or whether the system is one for extended electric traction service or only street railway service, etc.

1. In regard to cost it must be kept in mind that improvements, safety appliances, continuity of service, and future extension will involve not only the first cost, but additional expense for maintenance, repair, and operation. After taking all these matters into consideration, our final estimate must show that a reasonable return on the investment may be expected.

2. The reliability and continuity of the service are in fact the main objects of the plant. If these conditions of stability do not exist, and if frequent prolonged interruptions of service occur, the consumers will suffer and the public become prejudiced. Such conditions will give material advantage to competitors and may eventually lead to forfeiture of franchises or other concessions. These important considerations together with those discussed above afford the engineer a clue as to what methods to adopt in order to insure the desired conditions of reliability, continuity, and adaptability of service. The amount of useful information thus obtainable depends upon the conditions which show to what extent our object may be accomplished by the best machines and material in the market, by installation of a greater or less number of auxiliary apparatus with their connections, or by establishing

a reserve of units and parts. It should be noted that the greater the adaptability, the greater is the amount of apparatus necessary, and hence, the greater the liability for disturbances.

3. Since legal requirements as to protection of life and property must be complied with, devices guarding against lightning and fire, special arrangement of apparatus and the use of special insulating material become important factors in the plant.

4. Here we have two cases to consider. In the first case the building site need not be bought. As an example, an old d.c. station may be converted into a modern a.c. plant on the same site, or an addition to the old station may be built on land already owned. In the second case the site must be bought, and we here again have two distinctions to make. The choice of building sites is determined either by the location of a given source of water power with respect to the central station or by the location of load centers with respect to sub-stations, or the choice is limited only to certain minor considerations, thus allowing greater freedom of selection. (See a.c. plants, Chap. XXIV.)

5. The power rating of a plant depends, on the one hand, upon the object of the plant, which includes quantity and character of the output, and on the other hand, upon the available mechanical energy. The voltage of the generators to be selected increases with the size of the machines, for instance, 2,300 volts up to a power rating of 2,000 kw., and 6,600 volts for higher ratings. Machines of 5,000 kw. rating may have a voltage of 11,000, 10-cycle, or 13,200, 25-cycle. Ane. m.f. of 22,000 volts is admissible with certain powers, speeds, and frequencies. For long-distance transmission lines where the copper weight of the line adds materially to the cost of the system, the voltages of the generators are stepped up, giving e.m.fs. of 11,000, 15,000, 22,000, 33,000, 44,000, 66,000, 88,000, 100,000, 110,000, and 140,000 volts. The most usual frequencies employed are 15, 25, and 60 cycles, the former two for traction and motor-power distribution, and the latter for lighting. Units supplying energy to motors should be so chosen that their maximum overload capacity is of such value that in case one of the units is momentarily thrown out of service the output of the station and the voltage are not materially affected. It is therefore desirable to have one reserve unit, to be used only in case of emergency or at overload, thus making it possible to repair disabled machines. The determination of the capacity of the individual units is a

very important matter. For although the cost per kilowatt varies inversely, the capital invested for the reserve varies directly as the power rating of the machine. If the number of the machines is large, the ratio of reserve cost to total cost is small and is partly compensated for by the greater adaptability of the service and the decrease in line loss which the reserve accomplishes. The relation of the area included between the load curve and the time axes to the area under the rated output curve of the machine is called the load factor of the plant. A high degree of efficiency, if obtained for a long period of time, requires the use of larger power units, which again decreases the overload limit of the machines. If the load curve is flat, a smaller reserve unit may be used to take up the overload. If water power or steam turbines are used, all machines, including the reserve, may be run at all times on low load, since in this case no material losses in water or steam power need be considered. This method of operation differs from that employed with steam engines, which run economically only at full load and require the use of the reserve only from time to time. (See Chap. XXIV.)

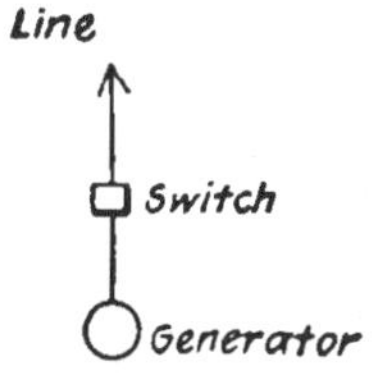

FIG. 62.—Simple single-unit system.

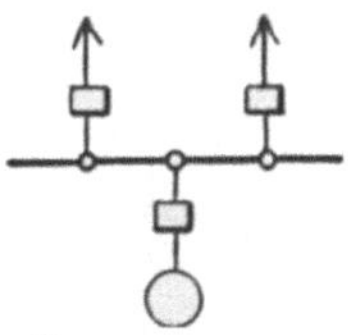

FIG. 63.—One bus system.

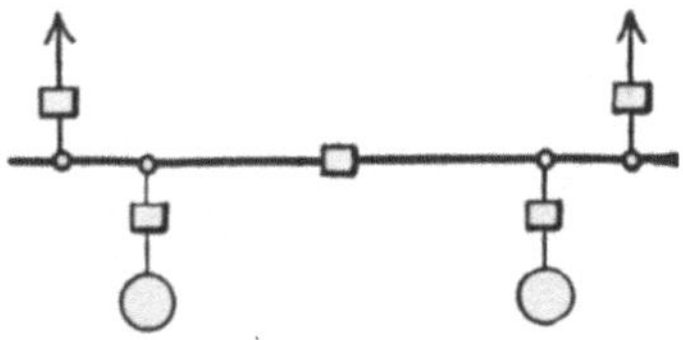

FIG. 64.—One bus with sectionalizing switch.

In settling on the design of a power plant, it is necessary to consider carefully the relative importance of flexibility and simplicity, and to determine the relative value of apparatus which would be needed only in emergencies and the saving which would result from omitting it. Consideration and careful examination of the nature of load, capacity of station both for initial and ultimate installation, and size of generator and transformer units available, lead to expressing the general problem for all plants as the obtaining of the maximum amount of flexibility and safety with the minimum cost of apparatus and building to house the apparatus.

In the diagrams herewith the cable connections between machines and pieces of apparatus are represented by single lines.

They therefore show the arrangements without regard to phases of the machines. The simplest case is that in which the generator feeds directly into the line (see Fig. 62), which is connected to the machine through an oil switch. If a machine is to supply several

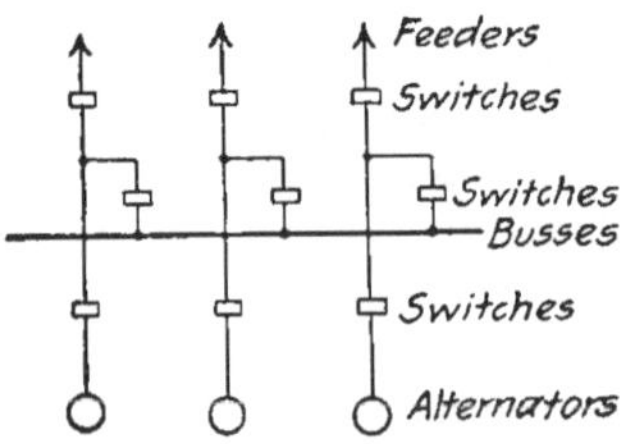

FIG. 65.—Relay system.

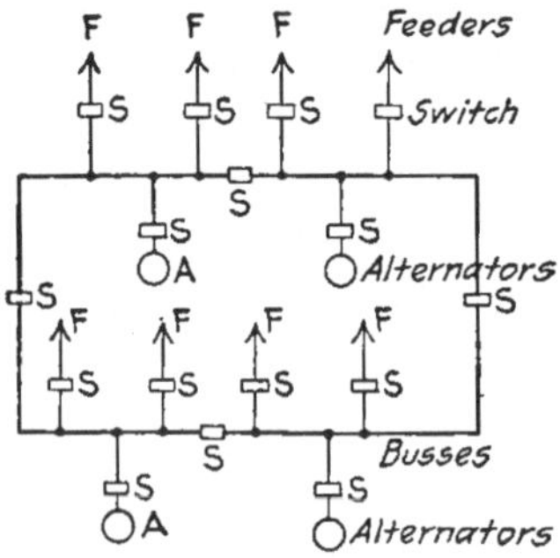

FIG. 66.—Ring system.

feeders, the energy is first led to busbars, whence the necessary amount for individual feeders is drawn off through oil switches. (See Fig. 63.)

If two or more generators are used to supply the feeders, the

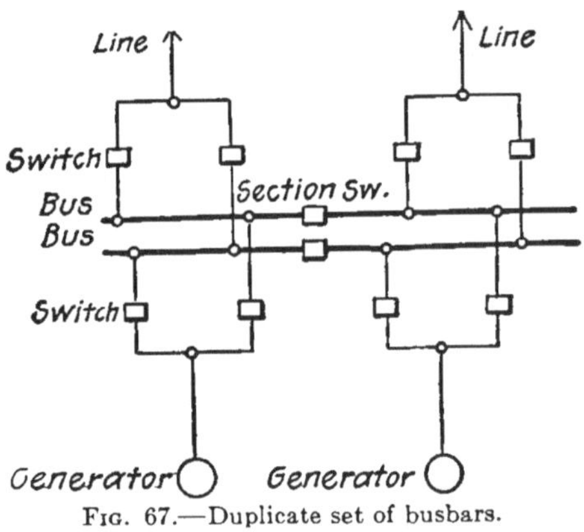

FIG. 67.—Duplicate set of busbars.

switching arrangement becomes more flexible by inserting a bus-sectionalizing switch in the busbars between generator connections. This switch enables the attendant to feed any set of feeders from any machine. By its use a disturbance on one side of it is localized and cannot affect the generators or connections on the other side. (See Fig. 64.) In another system each generator feeds one line directly, as shown in Fig. 62, but the feeders are connected by an auxiliary bus, so that, in case of need, one feeder can be supplied from the other generators. This is the relay system. (See Fig. 65.) Figure 66 shows a ring system where sectionalizing switches are inserted in the buses between each group of feeders and their corresponding generator. The advantage of this system is that any disturbance can be confined to a small part of the station, but a greater number of generators and feeders are required to maintain the service. The best and most

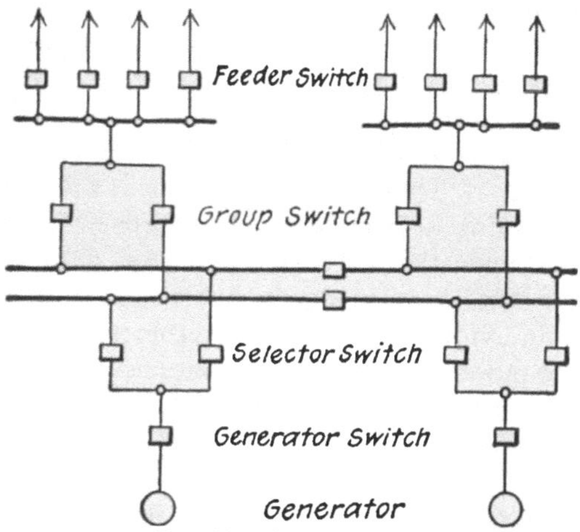

FIG. 68.—Group system.

common switching arrangement is that where a double set of busbars is employed, one called main or operating bus, and the other auxiliary bus. (See Fig. 67.) This method is more expensive because it requires two oil switches for each feeder and generator, and an additional weight in copper for busbars and connections. On the other hand, it affords an easy means of feeding any group of feeders from any group of generators. Such an arrangement has the further advantage that repairs of any part of the busbars can be made without danger and without interrupting the main service, since such part is easily disconnected from the live parts. This system is used when a considerable number of feeders are in service all day, or when it is desired to keep separate two different kinds of load.

For street railway systems in large cities where the central and sub-stations are connected through a large number of feeders, a system called the group system is employed. (Fig. 68.) The essential point of this system is the grouping of the feeders, which are connected to auxiliary buses. These buses are supplied from the main buses, which are in turn fed by the generators. Systems of this kind are used in plants of the New York Street Railway Company at 96th Street, and at Kingsbridge, in the 74th Street station of the Manhattan Railroad and in the power house of the Long Island Railroad, to be described later. The advantages of such a system are:[1] (a) "It affords an additional means of opening a feeder switch that fails to open its circuit when operated for that purpose." This applied mainly to oil switches in their early days when two switches in series were required to insure their proper operation. This consideration is no longer of importance, on account of the advance made in the construction of these switches. "They act as an assurance against a shutdown in a more important service." (b) "It affords means of reducing aggregate load upon the power house, in case of necessity, more rapidly and is otherwise less objectionable than the usual method of cutting off individual feeders. It will sometimes happen in the operation of a power plant that it becomes necessary suddenly to shut down one of the generating units. If the load carried at the time be such that the shutting down of the generator implies reduction of the external load, this can be accomplished most conveniently by operating one or two group switches." An objection to this point is that a group of feeders can also be simultaneously disconnected in other cheaper ways. The oil switches for a group of feeders can be electrically or mechanically interlocked in such a way that by the operation of a control switch all the feeder switches of a group can be opened simultaneously, while the several circuits still retain their individual control. (c) "Where duplicate main busbars are used it facilitates transfer of load from one set to the other, in case it becomes necessary suddenly in operation to make such transfer. If the feeders were connected to two main buses, it would become necessary to supply two oil switches apiece, or two-, three-, or four-pole double-throw switches for low tension. In the present case there suffice one switch per feeder and two oil switches per feeder group, which cuts down the number of switches necessary to be operated for transfer." (d) "The grouping of the external feeder circuits in group units, being a simple fixed relation to the

[1] L. B. Stillwell, "The Use of Group Switches in Large Power Plants," March 25, 1904, Proceedings, A. I. E. E.

generator units, establishes a symmetry and proportion most useful to the operator, particularly in time of emergency. The portion of load per group of feeders is known from the number and size of the feeder groups in relation to the number and size of the generators. With partial load it is easy to determine if a generator may be cut out of service, thus throwing its load on to other machines, or if it is better to disconnect a whole group of feeders, thus throwing the load of the sub-stations onto the other feeders." The above applies to city service, where several independent feeders lead to each sub-station.

Arguments against the group switch are: (a) "It introduces additional apparatus, and therefore in itself increases the risk of interruption due to failure in switch insulation, etc., and of disturbances which may often be more harmful than those which the switches are supposed to prevent." (b) "It implies an increase in cost of the plant. In the case of the Manhattan plant this increase is about 10 per cent. of the cost of the switchgear and measuring apparatus and about 0.4 per cent. of the cost of the plant. So that in a large plant the cost is negligible in comparison with the gain."

Generally no definite rules can be formulated as to when and where such a system is of advantage. A choice of systems depends upon the conditions of the case in hand.

Figure 69 shows connections for alternator, transformer bank, switch, and line. It represents the case where a number of plants feed into the same system.

A layout for several feeders in connection with one transformer bank and one generator is given in Fig. 70. An oil switch is inserted on the high-tension side of the transformer. In Fig. 71 two sets of busbars are used, one for the low-tension and one for the high-tension side of the transformers. The first set is fed by the generators, and the second is connected to the feeders. Both sides of the transformer are protected by oil switches. In Fig. 72 we have two or more transformer banks and a large number of generators and feeders, and desire to make their operation independent. This is accomplished by insertion of sectionalizing switches in both high- and low-tension buses between the transformer connections. We are thus enabled to divide all units into groups with corresponding transformers and feeders. The arrangement corresponds more or less to that of Fig. 64.

One of the most modern switching arrangements is represented

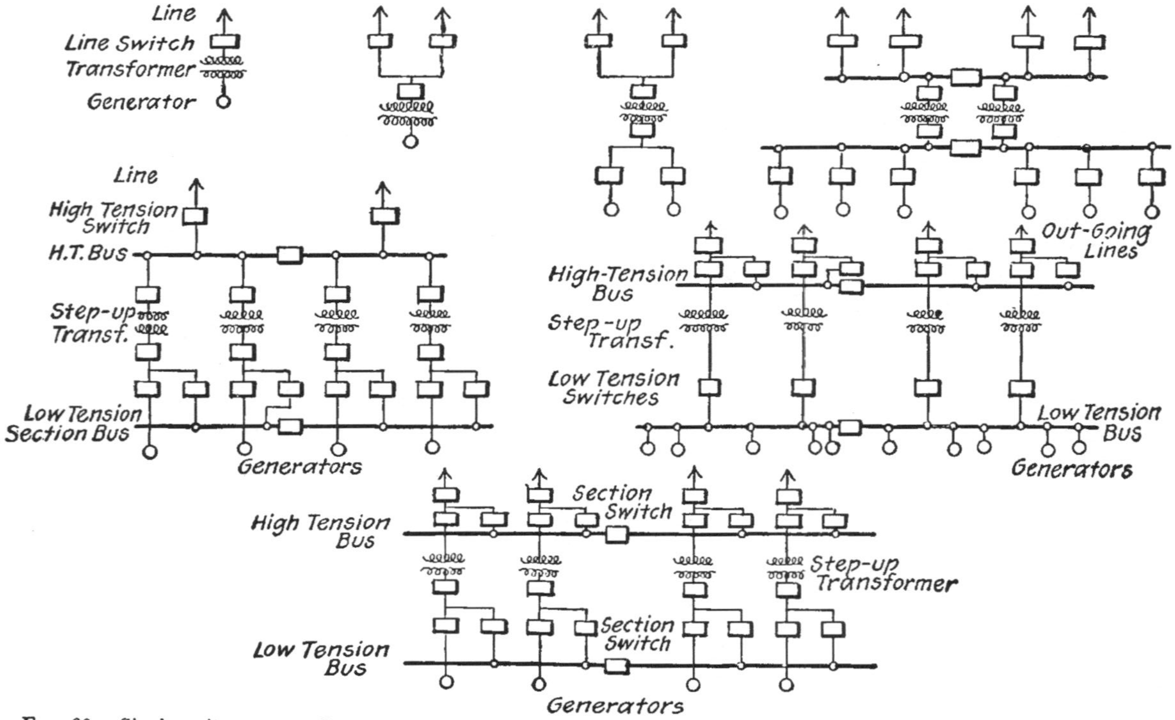

FIG. 69.—Single-unit system. FIG. 70.—Single generator-transformer unit multiple feeders. FIG. 71. Single transformer with multiple generators and feeders. FIG. 72.—Units like the one in Fig. 70, connected together with sectionalizing switches. FIG. 73.—Generator-transformer unit system. FIG. 74.—Transformer-line unit system. FIG. 75.—Generator-transformer-line unit system.

in Fig. 73. Each transformer bank is directly connected with its generator and with the low-tension busbar. By running in parallel on the low-tension side only, any generator can be run with any transformer. The whole station can be run in parallel or two parts run separately. This represents a case where the generator power is sufficiently large to allow of a generator-transformer-unit combination with the duplicate outgoing line. If, however, the generators are of smaller power, and we wish to form a transformer-line unit, the arrangement of Fig. 74 is used. The feeder is directly connected with its transformer and also has a connection to the high-tension busbar. The connection of any feeder with any transformer is thus made possible.

Figure 75 is a combination of Figs. 73 and 74. The unit is made up of generator, transformer, and line. Each unit can be operated independently as shown in Fig. 68, or any desired combination may be employed.

The systems described above are only typical arrangements, and may be modified by using varied combinations of different systems or by addition of auxiliary buses or other apparatus. As already stated, the arrangement is subject to no fixed rules, but to the judgment of the engineer.

CHAPTER XIV

CIRCUIT-INTERRUPTING DEVICES

In this chapter we shall discuss the principles involved in the application of instruments and apparatus of standard make.

One of the most important parts of a high-tension system is the circuit-interrupting device. Its functions are the same as those of a d.c. circuit breaker, i.e., to open or close the circuit at certain critical moments. Such instants may be foreseen or they may occur suddenly, necessitating instantaneous operation of the apparatus. In the first case the breakers are operated manually, and in the second case automatically. The automatic type can also be operated manually. With medium voltages where, on account of insulation, the size of the apparatus does not become excessive, such apparatus may be mounted on the board or in its vicinity, on supporting frames or in cells. With high and extra-high tension more space must be allowed for the apparatus on account of the high insulation. They are therefore located away from the board. In the former case they are operated mechanically from the board, and in the latter instance their operation is either manual, electric, or pneumatic. Four groups of circuit-breaking devices may be differentiated according to their make, operation, voltage, and the load to which they are connected. These are:

1. Disconnecting switches.
2. Plug switches.
3. Fuses.
4. Circuit breakers and oil switches.

The general conditions for which these breakers must be designed are as follows:

1. The rated current of the device ought to be carried with negligible drop and no heating. This depends upon the material of which the switch is constructed, upon its dimensions, and upon the construction and pressure between the surfaces of the contact pieces.

2. Insulate all live parts for maximum potential, both in an electrically and mechanically permanent manner. The best cri-

terion for the judgment of the efficiency of a circuit breaker is the extent of its ability to maintain perfect insulation in spite of the formation of arcs. The higher the voltage the more difficult does it become to obtain perfect insulation in service. Every breaker is therefore rated for a maximum voltage.

3. Mechanical means have to be provided for opening the circuit, which may be either automatic or non-automatic, as stated above. Fuses are always automatic, plug switches non-automatic, and disconnecting switches, oil switches and circuit breakers may be either automatic or non-automatic.

4. The formation of arcs has to be prevented or rendered harmless. This requirement may be met in various ways. The contacts may be widely separated, the interruption may be made to take place in oil, the arc may be blown out magnetically or through air expulsion, or may be quenched in an inclosure. Choice of the above-mentioned methods depends upon the voltage, space, conditions of service, etc., as we shall see under the discussions of the various types of breakers.

The d.c. circuit breakers and the a.c. breakers up to 600 volts were treated in Chap. IX.

Disconnecting Switches.—This type of breaker is seldom used by itself for main current interruption, being usually used in conjunction with an oil switch. If the breaker is to be used independently to break the line current, it has one great disadvantage. In order to break the arc the contact studs must be placed far apart, thus consuming a great deal of space for installation and operation. It can be used only when the voltage drops to a low value, for otherwise the arc would produce a high-potential oscillation in circuits of high inductance and capacity. These potential oscillations are quite liable to destroy the insulation at some part of the circuit, especially in the transformers. When used in conjunction with an oil switch it serves to disconnect the terminals of the latter from the live parts of the line, the oil switch having previously been opened. It is used to connect lightning arresters or potential transformers to the main line. Figure 76 shows a 12,000-volt, 300-amp. disconnecting switch mounted on a slate base. One of the terminals is in front and the other on the back side of the base, or the positions of the terminals may be varied according to circumstances. The hinge studs are incased in porcelain or glass bushings, with the surface distance between metal parts and

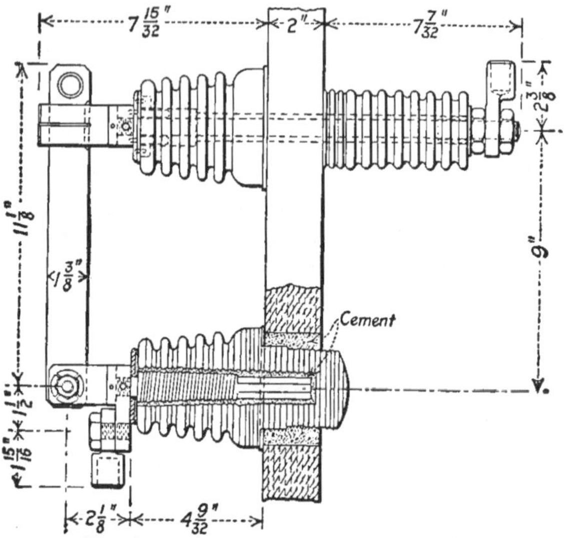

FIG. 76.—12,000-volt disconnecting switch.

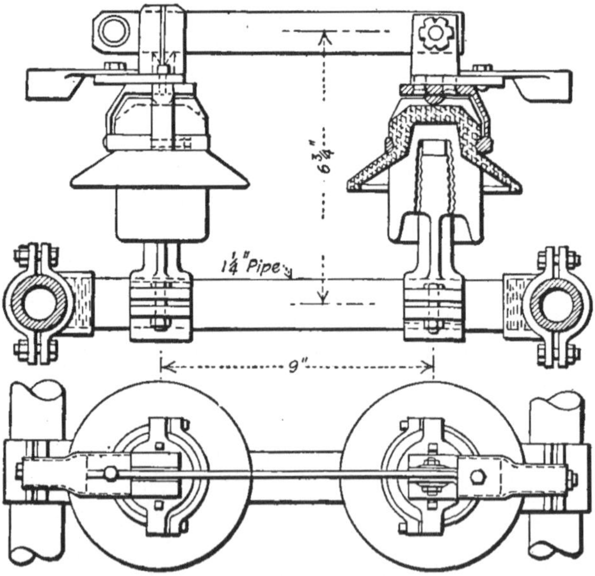

FIG. 77.—16,000-volt disconnecting switch on line insulators.

base corresponding to the voltage, usually being put at 1 inch per 1,000 volts. The air space between metal parts, and between these and ground, and the dimensions of the metal parts are dependent upon the voltage and capacity of the line. The air-line distance equals half the surface distance. Another method of mounting disconnecting switches is shown in Fig. 77. The porcelain insulators are fixed on a pipe support. A metal cap is fastened on the top of the insulators, on which are mounted the clips of the disconnecting switch. The terminals can be

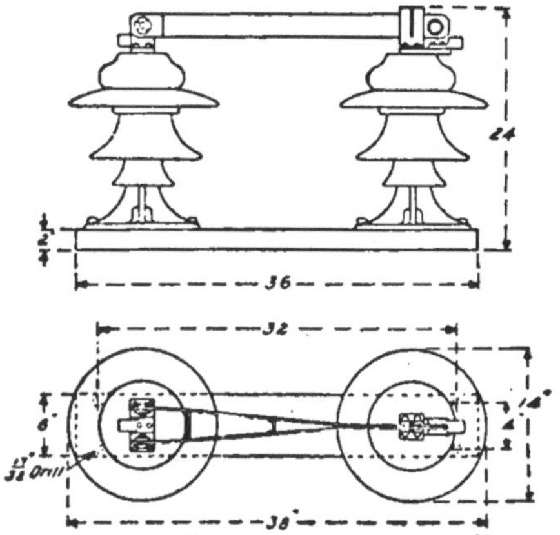

FIG. 78.—60,000-volt disconnecting switch on line insulators.

located in the front only. The illustration shows a 16,000-volt, 300- to 800-amp. switch. The support may be slate instead of piping, as shown in Fig. 78, for a 40,000 to 60,000-volt switch.

All these switches are mounted out of reach, so that operators cannot accidentally come into contact with live parts. It is advisable to separate the switches of different phase by slate, marble, or asbestos barriers in order to protect them against arcing or a short-circuit across the phases. The barriers become superfluous when there is sufficient room for a safe separation of the switches. All disconnecting switches are operated with a hooked pole. The attendant must therefore have sufficient room in which to use the pole, which is from 2 to 3 feet for low tension, and 10 feet for voltages up to 60,000. The cable connection should be such that when the switch is open the handle will be dead.

Another type of disconnecting device is the bus-sectionalizing switch. The busbar is broken for 6 inches, and both ends are in the form of clips. A copper link with an eye connects the two parts. A horn type of open-air disconnecting switch has been developed which is in common use on the Pacific high-voltage systems and responds satisfactorily to the condition of operation. The horns of the switch are of galvanized steel and are circular and hollow in section. The circular form reduces wind stress to the minimum; moreover, it presents no edge on which the arc can hang. The tubular construction helps the heat of the arc to be carried away both from within and from without, providing a double radiating surface. The current-breaking mechanism consists of a hinged blade operated in a vertical plane. The actuation of the mechanism can be both automatic

FIG. 79.—150,000-volt horn-type open-air switch. (Bowie Switch Company.)

and non-automatic. The arc built at the moment of current rupture is transferred from the blade end to the diverging horns, where it is driven rapidly upward both by the heat of the gases and by the magnetic effect of the steel horn.

Figure 79 shows an electrically operated 150,000-volt horn-type open-air switch of the Bowie Switch Co., San Francisco. The operating insulator which gives motion to the blade rotates about a vertical axis.

For severe service double-break switches are employed, each pole being equipped with a double set of horns, thus making the break in series. Each operating insulator operates simultaneously both blades of the same pole. The operating insulators are joined by a connecting rod which causes simultaneous operation.

The automatic horn-type switch is actuated by a trip coil built in directly in series with the line circuit.

For operating the horn-type switch from a distance there are in use mechanisms with manual, electric, pneumatic, or hydraulic control.

Instead of a hinged blade a high-tension fuse may be used to bridge the gap between the horns, thus building a high-tension fused switch.

Plug Switches.—A plug switch consists of one fixed and one movable part. The fixed part carries the contacts between which

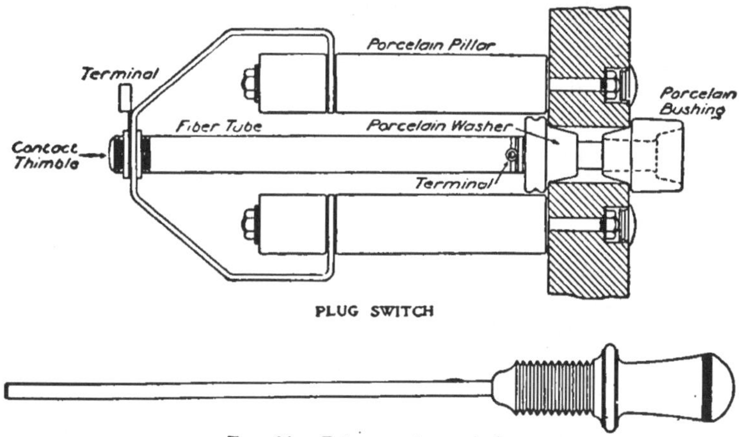

Fig. 80.—Primary plug switch.

connection is to be made. The movable part is a metal plug with insulated handle which connects the contacts when inserted in a porcelain bushing. The plug switch with plug is shown in Fig. 80. The switch is from 5,000 to 10,000 volts, 10 amp. One of the contact terminals is mounted on a fiber tube on the end near the switchboard, and the other on the farther end of the tube. The tube forms a receptacle for the plug. The bushing on the front of the panel prevents accidental touching of live parts of the switch, and at the same time serves as insulator for the fixed part and the board, as do also the porcelain pillars on the back of the board.

Another type is shown in Fig. 81. Two copper buses insulated from each other are mounted on the fixed part back of the board. Two terminal contacts are provided on the farther side. As long as the plug is inserted in the receptacle, it separates the two cable contacts, but connects them at the same time with their respective buses near the board. When the plug is removed, the contacts are connected together. The buses are permanently connected with the series transformer of the ammeter. The contact

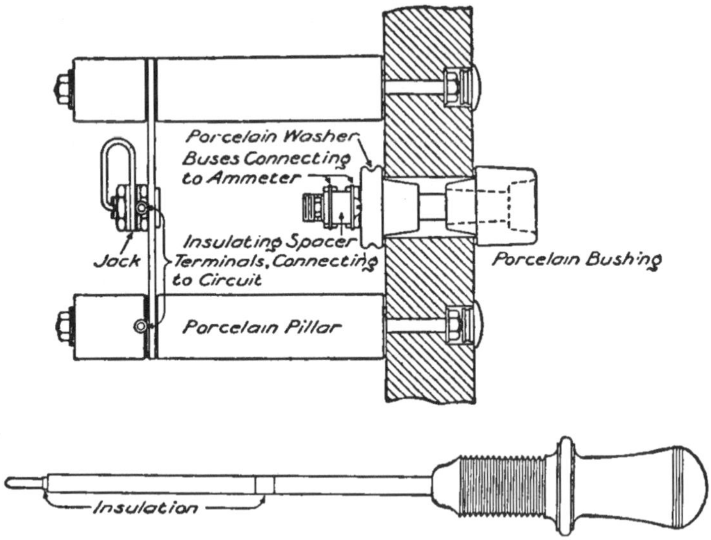

FIG. 81.—Ammeter jack.

terminals are on the side of the main circuit, and, as mentioned above, are connected when the plug is out. Plug switches are used for series alternating systems of arc or incandescent lighting in conjunction with constant-current transformers. (See Chap. XVIII.) They close the circuit on the primary side of the transformer, which generally carries from 1,150 to 2,300 volts, and which is protected by fuses against overload. The secondary side of the transformer is generally measured by the number of lamps connected to it in series, this number varying between 15 and 100 for different transformers. By means of the plug switches the circuit can be opened, short-circuited, or transferred to other lines. Any arc forming in the fiber tube (Fig. 80) is quickly quenched. On the secondary side of transformers they are used only up to 10 amp., since with larger current the arc is stronger and harder to extinguish. They are therefore often replaced by oil switches.

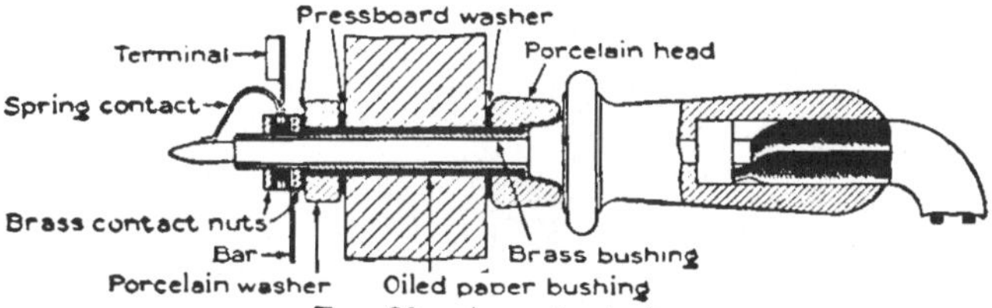

FIG. 82.—Ammeter jack.

Figure 82 shows another method of connecting the ammeter to the circuit. Two cables in the handle lead to the series transformer of the instrument. The plug makes contact between the cables and the metallic springs connected to the circuit. When the plug is taken out, one of the springs snaps back against the other contact, thus restoring the circuit.

The use of this type of switch where many bare parts are connected to high voltages requires very careful handling and special fireproof mounting.

As we have seen, plug switches are preferably used to connect the series transformer of one ammeter with different lighting circuits. As long as the plug is not inserted in the receptacle, the primary winding of the series transformer is open. To avoid this the General Electric Company has designed a switch so constructed that the plug connects the primary wind-

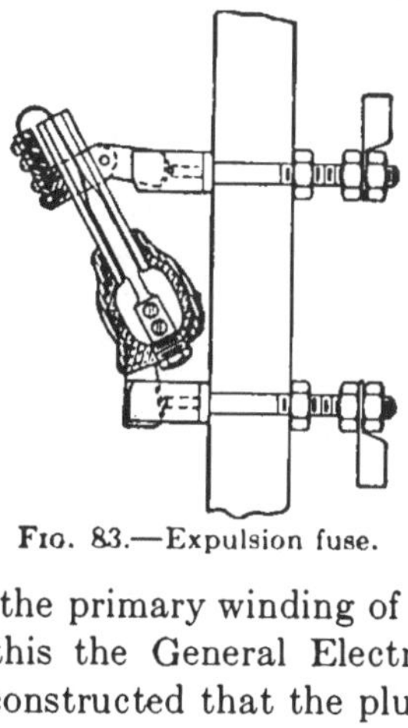
FIG. 83.—Expulsion fuse.

FIG. 84.—Expulsion fuse for shunt transformer.

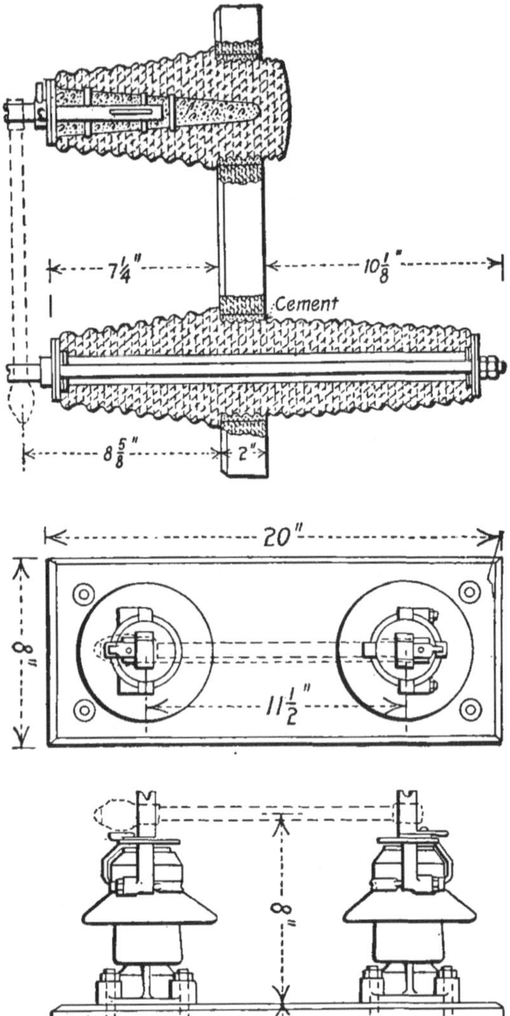

FIG. 85.—Expulsion fuses for shunt transformer as disconnecting switches.

ing with the various circuits, or short-circuits the winding, according to the depth to which it is inserted in the receptacle.

Fuses.—We have noted in treating of direct current that all voltage-carrying instruments and apparatus must be protected against overload or short-circuit. The simplest and cheapest of such devices is the fuse. For direct-current use fuses are constructed in open link or in closed types. Their purposes for alternating current are similar to those for direct current, but their construction differs on account of the high voltages employed. They act automatically and must be replaced by hand before the circuit can be restored. They can be used in series with disconnecting switches, rendering the breaking of the switch at high voltage less troublesome. Fuses are used for very high voltage and amperage. Their construction for high voltage is as follows (Fig. 83): The body of the holder consists of an insulated metallic chamber into the upper end of which is screwed a fiber tube. That part of the fuse in the chamber is of smaller cross section than the remainder, to insure rupturing at that point. The expansion of the gases formed by the arc in the chamber expels the fused metal and effectually opens the circuit. The illustration shows the cross section and method of slate or marble mounting of a General Electric Company expulsion fuse for 2,300 volts. The lower stud of the holder should be connected to the source of power, and the upper one to the load. All high-tension fuses are fastened on insulated bushings. Figure 84 shows the cross section of a fuse used for shunt transformers for high voltages. They are often fastened to clips similar to disconnecting switches and are used to insure safe access to the shunt transformers. (See Fig. 85.) Another type of a high-potential fused circuit breaker consists of a long hardwood pole on which a movable arm consisting of a reinforced fuse tube is mounted. At the bottom of the fuse tube is a brass expulsion chamber which is connected to the lower terminal of the breaker by a flexible copper shunt.

When the load on the line exceeds the capacity of the fuse, the latter blows and the arm of the breaker swings by its own weight away from the upper-line terminals, thus giving a positive indication that the fuse has blown. It is recommended that the following minimum distances be used between centers for successive fuse holders:

Volts	Distance Between Centers in Inches
2,300	12
6,600	18
11,000	42
22,000	48
33,000	54
44,000	60
66,000	72

As noted above, fuses can be used only for automatic current interruption with short-circuit or overload in service which permits of their easy replacement. That is, they are useful only when the time required for replacement causes no inconvenience due to interruption of service. Fuses, however, cannot be made to open all lines of a circuit at once, nor can they be arranged for underload or other variation; further, they introduce some fire risk and are expensive to replace.

CHAPTER XV

OIL SWITCHES

Development in oil-switch construction has resulted in the production of a number of invaluable pieces of apparatus without which it would be difficult to handle high-tension currents safely and economically. The utilization of oil has made it possible to interrupt circuits of high and extra-high tension and power with ease and safety, under most severe conditions of short-circuit or overload, and this without damage to the device itself. The latter point is of great importance, since the breaker must be capable of passing current as soon as the outside disturbance is over. A comparison is made by E. M. Hewlett of the relative merits of oil-break and air-break switches:

1. "Abnormal rise in pressure.—Owing to the fact that in oil switches the circuit is opened at the zero point of the wave, the rise in pressure found in the air-break switch is not experienced. This point is of considerable importance in high-pressure long-distance lines and in cables carrying considerable energy." The interruption in oil is therefore no quick-break action. The effect of the oil as shown by the oscillograph is to make the arc last through several wave lengths, being broken at the zero value. The efficiency of oil as an arc-suppressing medium is due principally to its cooling effect upon the arc and to its exclusion of air or oxygen, the former corresponding to pouring water on a flame, the latter to removing the fuel.

2. "Power.—Experience has proved that oil switches may be designed to break circuits of practically unlimited power."

3. "Length of arc.—Owing to the smothering action of the oil on the arc, the length of arc in oil is only a fraction of its length in air."

4. "Insulation.—The insulating qualities of the oil decrease the distance required to prevent leakage and arcing."

5. "Size of switch.—Owing to the fact that the arc length is materially decreased and the value of the oil as an insulation reduces the creeping surface, an oil switch can be made very much more compact than an air switch."

6. "Remote control.—The design of the oil switch lends itself readily to operation by control from a distance."

7. "Arc confined.—The fact that the arc is ruptured under the oil within the switch has two advantages; first: switches can be placed close together without danger of short-circuit; second: in case of emergency, confusion is avoided, as there is no visible arc to disconcert the attendant."

8. "Station arrangement.—The flexibility of the oil switch places no limitations on the station arrangement, permitting the circuits and buses to be arranged in the most advantageous manner."

9. "Isolation of phases.—The possibility of complete isolation of the phases in a reasonable space is easily secured by the use of the oil switch."

Ideally, the current of the circuit should not reëstablish itself after passing the first zero value following the separation of the contacts. This means that in 60-cycle service not more than $\frac{1}{120}$ (0.0083) of a second should elapse before the contacts have separated sufficiently to make the reëstablishment of current impossible. With such a switch the minimum energy will be dissipated between the contacts. The heat of the arc formed at the moment of separation of the contacts disintegrates some of the oil surrounding the contacts, forming a gas bubble. The size of the arc and of the resultant gas bubble depends upon the current flowing at the instant of separation and the duration of the arc. The ability of the gas bubble to rise far enough so that the arc will be quenched during the zero point of the wave is controlled by specific gravity of the gas bubble in relation to the head and volume of oil and the viscosity of the oil. Oil having specific gravity, sufficient head and low viscosity will have enough pressure to force the bubble up away from the contacts, irrespective of their position. The relation of the horizontal surface of the oil to the cross-sectional area of the contact and terminal arrangements will also determine the ease with which the gas bubble will clear itself of the contacts. To insure a plentiful supply of cool oil at the point of arcing, the contacts are sometimes arranged so that a jet of oil is thrown against the portion of contact on which the arcing occurs. This jet had an additional function of wiping the arc off the contact piece.

A well-designed oil switch should have a liberal supply of oil provided above and around the point where the disturbance occurs in order to insure the exclusion of air from the arc and to displace the gas bubbles as rapidly as possible. In order to keep the oil from being ejected from the tanks owing to the great disturbance of a short-circuit, it is necessary to close the tanks almost completely. To minimize the gas pressure, a gas chamber of considerable size should be provided above the oil level and arranged with an outlet that will offer free passage to the gas but will baffle the ejection of oil. The gas above the oil, being easily compressed, will act also as a cushion and will tend to absorb the shock of the break. The qualities essential in an oil good for use with oil switches, besides those mentioned above, are high-voltage breakdown, (i.e., good insulation), high flashing and burning points to reduce fire risk, small evaporation, and no sedimentation.

The interruption with one phase can be accomplished in a single point, double or manifold. Since in the latter case the breaks are in series, the construction for high voltage can be quite compact and the action of the switch is positive, direct, and rapid. High insulation of phases and poles of the same phase must be maintained, the phases for high tensions being broken in separate tanks. The insulation must be of a permanent character, one, for instance, that will not deteriorate because of the action of the oil or ozone which is nearly always present where high-tension currents are handled and where static discharges are in evidence.

Oil switches may be classified, according to operation, as automatic and non-automatic. In order that the former may interrupt the circuit at short-circuit or overload, they must be self-actuated and must be able to break a current of several times the generator rating. The non-automatic breakers are not used for such large currents, yet they must be able to interrupt the current when the generator is short-circuited. As to operation we may divide oil switches into three classes:

1. Manually operated.
2. Electrically operated.
3. Pneumatically operated.

1. Manually operated switches of small power rating are mounted on the back of the switchboard and are operated by means of linkages from the front of the panel. It may sometimes be more convenient or advantageous to mount them on brackets or framework away from the board on the wall or in cells, which

may be the case with switches of larger power rating. Under the circumstances they are operated from the board through rods with bell cranks or wire rope.

A handwheel and rotating shaft, with bevel gears or universal joints for changing directions of motion, are used to considerable extent by foreign manufacturers. The connecting rods are commonly made of ¾-inch gas pipe; wooden rods are sometimes used because of their light weight.

An oil switch should close readily with the ordinary effect exerted by the average man using but one hand. It is impracticable to operate a switch satisfactorily at a horizontal distance much in excess of 35 feet, or a vertical distance of 20 feet, with standard commercial apparatus. This distance may prove too great for satisfactory operation if more than two bell cranks must be used or if the direction of motion must be changed, as in clearing columns or other obstructions. A long heavy mechanism will considerably increase the inertia of the moving element and consequently slow down the action of the oil switch. Therefore an accelerating device such as a spring combined with a dash-pot to absorb the shock of opening should be applied to regain the inherent speed of the switch. When the mechanism operates vertically, as when controlling an oil switch on a floor above or below the switchboard, the mechanism may be counterweighted to take the weight of the vertical rod off the operating handle; but, as the inertia of the mechanism is thereby increased, the operation with counterweight is limited.

The advantages gained by the use of remote mechanical control over those of the direct mounting on the back of the switchboard may be summarized as follows:

(a) All high voltages are removed from the panels, thus permitting ready inspection of the instrument and control wiring, eliminating danger of injury to attendants from contact with live parts and permitting the location of the board to much better advantage as regards the remainder of the installation because less space and less protection are required.

(b) Panels are not subjected to the mechanical strains due to automatic operation or to the dead weight of the apparatus.

(c) In case of marble panels their appearance is not marred by stains from creeping oil.

(d) Violent explosions, attendant upon the opening of heavy currents or the failure of an oil switch, will not injure the panels and if the switches are enclosed in fireproof cells a fire in any cell can be localized.

(e) The panels may be much narrower, the reduced cost thus gained offsetting to a considerable extent the additional cost of the remote-control feature. Moreover, the decrease in the total length of the board may result in a very material saving in cost.

(f) A more compact arrangement of the apparatus is of great assistance to the operator. Since the attendant has a safe and isolated place of operation, he is not so easily disconcerted in case of accident when quick action is imperative.

(g) Much shorter main connections are made possible and high voltages kept away from certain floors, or certain rooms, by locating the remote-control structure properly. A saving in copper may thus be attained by placing it near the source of power and the busbars.

(h) Where a wall is used for supporting the apparatus, the cost of the complete outfit may be materially reduced. Moreover, accessible arrangements of apparatus with ample spacing may be obtained easily.

(i) Where a steel or masonry structure is used, access may be had to either side of the structure, and an arrangement of this kind will satisfactorily accommodate the maximum amount of apparatus ordinarily used for either single- or double-throw arrangements.

2. Electrical operation is substituted for manual when the distance from the board to switch and increased size of switch render the above arrangement clumsy or inconvenient. The advantages of such an operation are mostly the same as for the mechanical distant control, only of still greater degree. The switch can easily be located so that it cannot be disturbed through any machine parts like steam pipes, etc. For higher power and voltage the switches are so large that they can be managed more easily electrically then manually, and if operated manually they would occupy too much space.

In large stations it is important that the control and operation of all generators, exciters, transformers, and feeders be concentrated as much as possible, so that one switchboard attendant may control the entire system. This will result in more advantageons management and decrease of operating expenses. In such stations the power units are so large that the distance between machine centers is considerable. Hence, if cables were to be run from these machines to mechanically operated oil switches, a very undesirable position for the switchboard might be necessitated, if cable cost and ease of installation are taken as criterions. The great danger incurred in handling high-tension apparatus compels us to mount them in such a way that no high-tension wires lead to the switchboard.

Two methods of electrical oil-switch operation are in use. One is by means of solenoids and the other by means of motors. Motor-operated oil switches are built by the General Electric Company, while the solenoid type is made by the General Electric, Westinghouse and other manufacturing companies.

3. *P*neumatic oil switches are sometimes employed in place of motor-operated switches. Their manner of operation is much more complicated and not so safe as that of the other types since they require special machinery to produce air pressure. The operation of the valves of the pressure cylinder on the oil switch, moreover, is not reliable. The following synopsis gives a survey of the action, operation, and manipulation of the various types of breakers in use. For relays see Chap. XVI. In the diagram of automatic control there are included overload and underload switches for direct current.

Methods of Operation

(All of the following: Without electric trip or with electric trip)

- Manual
  - On panel
  - On wall
  - Remote
    - On flat surface
    - In cell
    - On pipe framework
- Electrical
  - D. c. motor
    - Std. sw. op. mechanisms
    - H 3. forms
  - Solenoid
    - D.c. standard mechanisms
    - A.c. special switches
- Mechanical
  - Pneumatic
  - Float
  - Press reg.
    - Air
    - Liquid

Automatic Control

- Without series transformers
  - Series trip
    - Direct relay
    - D.c. trip
    - With or without time — Constant / Inverse — Overload
  - Auxiliary trip
    - Push button
    - Low voltage
    - With or without shunt transformers or series resistance

- With series transformer
  - Depending on load in secondaries — Without relay
    - Direct trip a.c.
      - On panel
      - On shaft
  - Depending on time element — With relay
    - Without time element — Instantaneous
    - With time element — Constant or inverse
  - Short-circuit, Overload — A.c. trip, D.c. trip
  - Reverse current — D.c. trip
  - Reverse phase — A.c. trip, D.c. trip
  - Differential, Low voltage, Underload — D.c. trip
  - Overload — A.c. trip, D.c. trip
  - Reverse current — D.c. trip

Attachments:
Auxiliary switches (circuit-opening)
Indicating switches (circuit-closing)
Interlocks { Electrical, Mechanical

For voltages up to 2,500 series trips are used to open automatic oil switches, one for a double-pole (single-phase) and two for three- or four-pole switches (three-phase or quarter-phase). (See Fig. 86.) The trips are in series with the side of the switch which is to be protected. Mechanisms for automatic oil switches

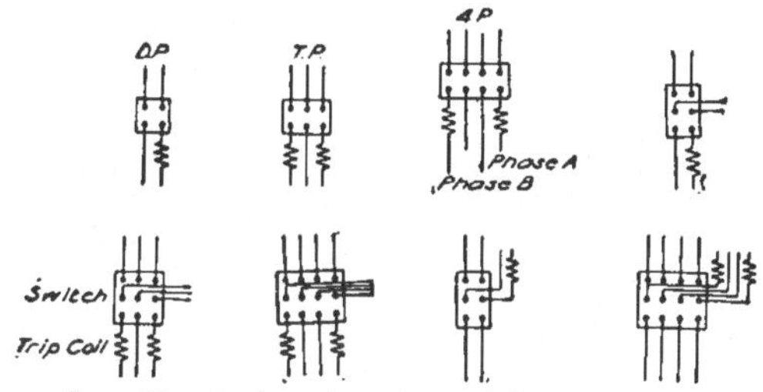

FIG. 86.—Series trip coils for K3 oil switches.

are of two kinds, those in which the mechanism is stationary during tripping of the oil switch, in which case the oil switch trips free from the mechanism, and those in which mechanism returns to the open position at the time of the opening of the circuit. The latch and trip coils may be mounted at the oil switch for both kinds, but it is most usual with the latter kind to mount them at the operating handle. The arrangement of mounting the trip coils and latch at the control handle with the control mechanism and tripping free from the handle is the one most commonly used. The trip coils or relays, if used, are usually actuated, at higher voltages, by current transformers, the same which operate the ammeter. This arrangement permits all secondary wiring from the current transformers to be made at the switchboard. Moreover, as the auxiliary lever indicates the open or closed position of the oil switch the operator does not require a signal device at the switchboard for this purpose. Where the distance from the board to the oil switch is considerable

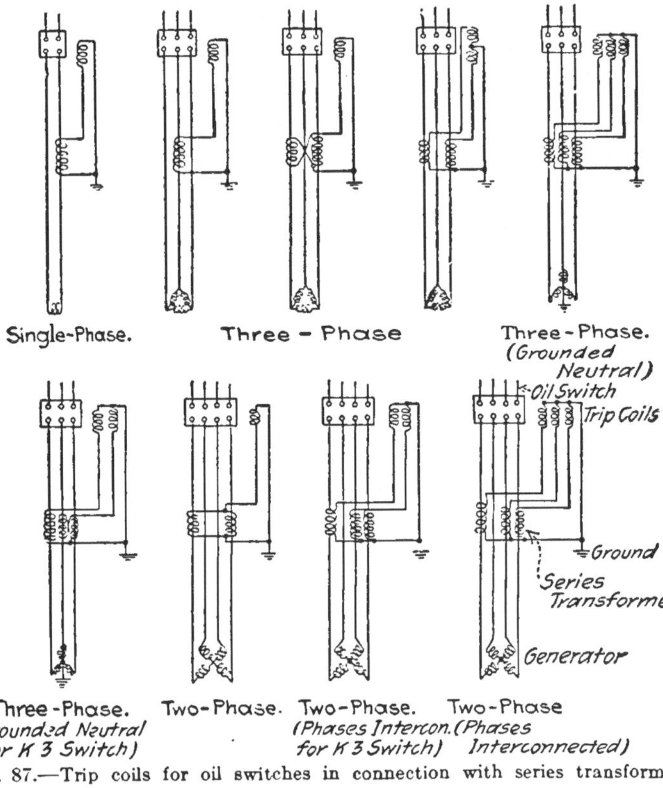

FIG. 87.—Trip coils for oil switches in connection with series transformers.

it is advisable to place the trip coils, latch, and relays, if used, at the oil switch so that the latter will trip free from the heavy mechanism. In this event an indicating device at the switchboard, such as lamps or mechanical signal, is desirable even though the instruments may indicate, to a certain extent, the condition of the circuit. Figure 87 shows the various connections for different phases. The number of tripping coils used depends upon the size of the switch to be opened. The number of series transformers depends upon whether or not the load is balanced and the neutral of the machine by star connection is grounded, and also upon the number of phases. Transformers actuating tripping coils must not be connected to instruments with series and shunt windings, such as wattmeters or power-factor indicators.

ULTIMATE BREAKING CAPACITY OF OIL SWITCHES, IN KW.
General Electric Company
For Three-phase

	Rating	Line voltage	Non-automatic or automatic with time-limit relay with ½ second delay		Automatic (instantaneous)		Remarks Number of poles[1] and throws. Operation
			On panel, wall, or framework	In cells	On panel, wall, or framework	In cells	
	Amp.	Volts	Kw.	Kw.	Kw.	Kw.	
P	100	4,500	3,000	......	......	......	Built: Double-pole, single-throw.
	100	4,500	4,000	......	......	......	Triple-pole, single-throw.
	100	4,500	6,000	......	......	......	Four-pole, single-throw.
	200	7,500–15,000	1,500	......	......	......	S. P. S. T.
	200	7,500–15,000	3,000	......	......	......	D. P. S. T.
	200	7,500–15,000	4,000	......	......	......	T. P. S. T.
	200	7,500–15,000	6,000	......	......	......	4 P. S. T. Operated manually. Mounted on poles or wall.
P_4	200	7,500	3,000	......	......	......	Built: 1, 2, 3 poles, single-throw. Operated manually. Mounted on wall or manholes.
K_7	50–200	600	3,500	......	1,400	......	Built: 2 or 3 poles, single-throw. Operated manually. Tripping coil in series with main circuit.
	300	600	4,200	......	1,700	......	
	50	2,500	3,500	......	1,400	......	
K_8	100–200	2,500	5,300	......	2,100	......	Built: 2, 3, or 4 poles, single- or double-throw. Operated manually or electrically. Tripping coil in series with main circuit or in connection with series transformers
(K_6)	100–200	4,500	4,700	......	1,900	......	
	300–500	2,500	5,900	......	2,400	......	
	300–500	4,500	5,300	......	2,100	......	
K_{20}	60–200	600	3,500	......	1,400	......	Built: 2, 3 poles, single-throw. Operated manually. Tripping coil in series with main circuit. Mounted on wall.
	300	600	4,200	......	1,700	......	
	60–200	2,500	3,500	......	1,400	......	
	300	2,500	4,200	......	1,700	......	

[1] The number of poles and throws built in one oil vessel.

Ultimate Breaking Capacity of Oil Switches, in Kw. (*Continued*)
General Electric Company
For Three-phase

	Rating	Line voltage	Non-automatic or automatic with time-limit relay with ½ second delay		Automatic (instantaneous)		Remarks Number of poles[1] and throws. Operation
			On panel, wall, or frame-work	In cells	On panel, wall, or frame-work	In cells	
	Amp	Volts	Kw.	Kw.	Kw.	Kw.	
K_2	300–500	2,500	5,900	6,800	2,400	2,800	Built: 1, 2, 3, or 4 poles, single-throw. Operated manually or electrically. Tripping coil in connection with series transformers.
	300–500	4,500	5,300	6,000	2,100	2,400	
	300–500	7,500	4,300	4,900	1,700	1,900	
	300–500	15,000	2,100	2,400	900	1,000	
	800–1,000	2,500	7,700	8,800	3,100	3,500	
K_4	300–800	2,500	12,000	13,800	4,800	5,500	Built, operated, and tripped as the K switch.
	300–800	4,500	11,000	12,600	4,400	5,000	
	300–800	7,500	9,500	11,000	3,800	4,400	
	300–800	15,000	6,500	7,500	2,600	3,000	
K_{12}	300–800	2,500	12,000	13,800	4,800	5,500	Built, operated, and tripped as the K_2 switch.
	1,200–2,000	2,500	11,000	......	5,000	....	
	300–800	4,500	11,000	12,600	4,400	5,000	
	1,200–2,000	4,500	11,000	......	5,000	....	
	300–800	7,500	9,500	11,000	3,800	4,400	
	1,200	7,500	11,000	......	5,000	....	
	300–800	15,000	6,500	7,500	2,600	3,000	
K_6	100	22,000	....	4,570	......	1,840	Built: Single-pole and single-throw. Operated electrically. Tripping coil in connection with series transformers.
	100	33,000	....	2,740	......	940	
	100	33,000	....	7,410	......	2,970	
	100	45,000	....	4,500	......	1,890	
	300	22,000	....	6,170	......	2,480	
K_{10}	300	22,000–45,000	....	20,000	......	20,000	Built: Single-pole and single-throw. Operated manually, electrically, or pneumatically. Tripping coil in connection with series transformers or with series relay. Mounted separately without cell and barriers.
	150	70,000	....	20,000	......	20,000	
	100	110,000		20,000		20,000	
K_{16}	150	70,000	....	50,000	......	50,000	Built, operated, tripped and mounted as K_{10} switch.
	100	110,000	....	50,000	......	50,000	
K_{22}				Kva.[2]		Kva.	Built: Single-pole, and single-throw. Operated manually, electically or pneumatically Tripping coil with series transformers. Mounted separately in open air.
	300–800	22.000	....	28,000	......	17,500	
	300–800	35,000	....	40,000	......	20.000	
	150–500	70,000	....	50,000	......	25,000	
	100–300	110;000	....	50,000	......	25,000	
H_3	For all voltages, amperages, and kilowatts. Built: In single-pole, single-throw units in cells. Operated: Electrically, pneumatically, or manually special. Tripping coil in connection with series transformers.						

[1] The number of poles and throws built in one oil vessel.
[2] Rating in kva. with time limit not less than two seconds.

Every oil switch is built for normal voltage and amperage, but must nevertheless be capable of interrupting the entire power which all the generators in parallel are capable of developing. In the event of short-circuit, a source of electrical energy may deliver power greatly in excess of its normal capacity, and the switch may therefore be required to interrupt not merely the normal energy delivered to the circuit but the entire power which may be developed under short-circuit conditions by all the generators which are connected to the system in parallel. The various types of oil switches are therefore constructed for given voltage and maximum breaking capacity without regard to their method of operation and action. The breaking capacity of the various types as made by the General Electric Company and Westinghouse Company are tabulated on pages 129, 130, 133 and 134.

For single-phase multiply the above figures for ultimate kw. breaking capacity, in the G. E. table, by 0.75 and for two-phase by 1.5. For oil switches used with a lower line voltage than given in the table use kw. rating of the nearest voltage given. Maximum power rating of switches for voltage of intermediate values can be found by interpolation.

The most definite and clear way of rating oil switches is to give the maximum "instantaneous" rupturing capacity, meaning by "instantaneous" the elimination of time-limit relays in tripping. The following assumptions are ordinarily made by manufacturers of oil switches as a basis for the ultimate kilovolt-ampere breaking capacity rating of the oil switches:

1. The generators on the system have an inherent reactance of approximately 8 per cent., or, in other words, give a maximum current on the first wave of short-circuit of 12.5 times the mean full-load current; the sustained short-circuit has a maximum value of three times the mean full-load current.

2. The current at the time the circuit is opened is approximately 50 per cent. of that occurring on the initial wave; or, in other words, 6.25 times the mean full-load current.

3. The oil switch is connected direct and close to source of power, so that the circuit reactance from the generators through the oil switch to the point of short-circuit is practically negligible.

4. The operation of the oil switch is normal for the particular type, and the method of tripping "instantaneous," under which condition the mechanism should be expected to operate in 0.1 to 0.2 second.

5. The normal operating voltage which the oil switch is called upon to open is that on which its rating is based.

6. Where the oil switch is used at a lower voltage than the maximum for which it is designed it is assumed that a 1 per cent. increase in kilovolt-ampere rupturing capacity is permissible for every 1 per cent. decrease in operating voltage.

Any departure from the above conditions will change the relative rating. All reactances added by lines, transformers, or otherwise will increase the rupturing capacity in direct proportion.

The location of the oil switch in the system the speed and the operating characteristics of the switch mechanism, and the method of tripping the switch are other factors to be taken into consideration in the choice of the oil switch suited for the particular requirements.

For instantaneous automatic direct connection to generator busbars, an oil switch should be selected that has an ultimate breaking capacity equal to the total capacity in kilovolt-amperes of the generators which are directly connected to the busbars, and this regardless of whether the oil switches are placed between the generators and busbars or between the busbars and other apparatus. This rule is based on the aggregate short-circuit currents of all the generators feeding through the oil switch to the short-circuit which has occurred at some point beyond the switch.

For instantaneous automatic direct connection to transformer secondary busbars, there should be selected an oil switch that has an ultimate breaking capacity equal to twice the kilovolt-ampere capacity of all the transformers feeding the busbars. This rule is based on the amount of current that the transformers will permit to flow under short-circuit conditions. In case the transformer secondaries are not connected together on the bus, or feed separate circuits, the capacity of the breaker should, of course, be twice that of the transformer feeding the circuit. Time-limit relays (see Chap. XVI), when set for an absolute minimum of two seconds on any current setting of the relay, may be regarded as doubling the capacity of any automatic oil switch to which they are applied. Non-automatic oil switches may be applied to circuits of three times the capacity to which similar automatic oil switches would be applied.

An oil switch should never be installed on any circuit where it will be subjected, under any conditions, to more than 10 per cent. in excess of its rated voltage.

Ultimate Breaking Capacity of Oil Switches, in Kw.
Westinghouse Electric & Manufacturing Co.

Type or form	Rating, amp.	Line voltage	Kw. rating			Remarks: Number of poles and throws.[1] Operation. Installation
			Single-phase	Two-phase	Three-phase	
I(A)	50 50	3,300 6,600	150 150	300 300	275 275	2, 3, 4 poles, single-throw. Operation: Manual, non-automatic. Mounted on panel or wall.
D	100–1,000 100–300	3,300 6,600	2,000 2,000	4,000 4,000	3,500 3,500	2, 3, 4 poles, single- and double-throw. Operation: Manual, non-automatic. Mounted on panel or wall.
H	10–300 10–300	600 2,500	1,500 1,500	3,000 3,000	2,600 2,600	2, 3, 4 poles, single-throw. Operation: Manual, automatic Mounted on panel or wall. Tripping coil in series with main circuit.
J	10–300	3,300	1,000	2,000	1,800	2, 3, 4 poles, double-throw. Operation: Manual, automatic. Mounted on panel. Tripping coil in series with main circuit.
F	300 10–200	3,000 6,600	1,500 1,500	3,000 3,000	2,600 2,600	2, 3, 4 poles, single-throw. Operation: Manual or electrical, automatic, non-automatic. Tripping coil in series with main circuit or in connection with series transformers. Mounted on panel, on wall, or in manholes.
B	2,000 1,200 600 300 100	600 3,300 6,600 11,000 22,000	5,000 5,000 5,000 5,000 5,000	10,000 10,000 10,000 10,000 10,000	8,500 8,500 8,500 8,500 8,500	2, 3, 4 poles, single-throw. Operation: Manual or electrical, automatic or non-automatic. Tripping coil in connection with series transformers. Mounted on panel, wall, or framework.
			Kva.	Kva.	Kva.	
C	3,000 600–1,000 600–1,200 2,000	2,500 15,000 25,000 15,000	42,000 28,000 42,000 42,000	84,000 56,000 84,000 84,000	60,000 40,000 60,000 60,000	Single-pole, single-throw Operation: Electrical, automatic or non-automatic. Tripping coil in connection with series transformers. Mounted in cells.

[1] The number of throws and poles built in one oil vessel.

ULTIMATE BREAKING CAPACITY OF OIL SWITCHES, IN KW. (*Continued*)
Westinghouse Electric & Manufacturing Co.

Type or form	Rating, amp.	Line voltage	Kw. rating			Remarks: Number of poles and throws.[1] Operation. Installation
			Single-phase	Two-phase	Three-phase	
E	1,200 600 300 100	3,500 7,500 16,500 25,000	6,000 6,000 6,000 6,000	12,000 12,000 12,000 12,000	10,400 10,400 10,400 10,400	Single-pole, single-throw. Operation: Manual or electrical, automatic, non-automatic. Tripping coil in connection with series transformers. Mounted in cells.
G	100 200	60,000 120,000	115,000 115,000	230,000 230,000	200,000 200,000	Single-pole, single-throw. Operation: Electrical, automatic or non-automatic. Mounted separately or in cells. Tripping coil with series transformers.
GA	300 600	44,000 66,000 88,000 110,000 165,000	....	....	60,000 80,000 100,000 120,000	Single-pole, single-throw. Operation: Manual or electrical, automatic or non-automatic. Mounted separately or in cells. Tripping coil with series transformers or relay with series coil.
L	50–200 50–200	60,000 88,000	12,000 12,000	....	20,000 20,000	Single-pole, single-throw. Operation: Electrical or manual, automatic or non-automatic. Mounted in cells or separate .Tripping coil with series transformers

[1] The number of poles and throws built in one oil vessel.

It is recommended that the voltage of oil switches mounted on the board shall not exceed 2,500. The above switches are constructed for operation in plants where the normal full load of the generators in the whole system or section does not exceed the maximum given kw. rating.

Oil switches are constructed so that single, double, triple, or four-pole switches are contained in one oil vessel. Moreover, sets of two, three, or four single-pole switches in separate vessels can be operated as double, triple, or four-pole switches by operating them through a common mechanism. The former are used for tensions up to 6,600 volts and the latter with isolated phases for all higher and extra-high voltages. If these switches are to be used as double-throw switches, their number must be doubled, with the exception of type K3 (G. E. Co.), which contains the necessary number of contacts and studs in one vessel. Two sets of operating mechanisms are necessary for each set of double-throw switches, which are interlocked mechanically or electrically, so that it is rendered impossible to close both throws at the same time. Each switch consists of three parts:

1. A frame holding the studs, contact pieces, and porcelain bushings.
2. A removable oil vessel mounted on the frame.
3. Movable contact bridges with operating devices.

The contact pieces with the movable contact bridges are the parts which carry the current and which when separated by the action of the mechanism open the circuit. They must be of ample cross section to carry, without undue rise in temperature, the current for which the oil switch is designed. The area of the contact surfaces must also be sufficient to carry safely the required current. It is often possible to increase the capacity of the contact surfaces of an oil switch by increasing the pressure with which these surfaces are held in contact. The design of the contacts must be such that the carrying capacity of the contact surfaces will not be impaired by the arcing which is likely to occur between these faces when opening the circuit. This end may be accomplished either by the use of auxiliary contacts which remain closed until the main current-carrying surfaces have separated and which take the arc due to the final opening of the circuit (like the auxiliary carbon contacts of a circuit breaker), or by making the contacts of a form similar to that of controller fingers where the main contact surfaces open first and the final break occurs on the points of the fingers. An ideal contact would be one which is self-adjusting, both in regard to alignment of surfaces and in compensating for wear due to continued operation or severe arcing. Freezing or sticking of contacts due to arcing at the instant of operation must be guarded against, and it must be possible to actuate an electrically operated switch manually with a reasonable degree of safety. It is, of course, highly desirable that the contact surfaces should be easily accessible for inspection and repair.

The insulating parts of an oil switch which separate the line contacts and current-carrying parts from each other and from ground must be of sufficient mechanical strength to withstand considerable rough usage and capable of withstanding some degree of heat. The material which affords the most permanent insulation is glazed porcelain. It can be used equally well in air or in oil. Porcelain has only the mechanical disadvantages that it is somewhat easily broken and that it gradually develops cracks when placed under strain. Glazed porcelain is not affected

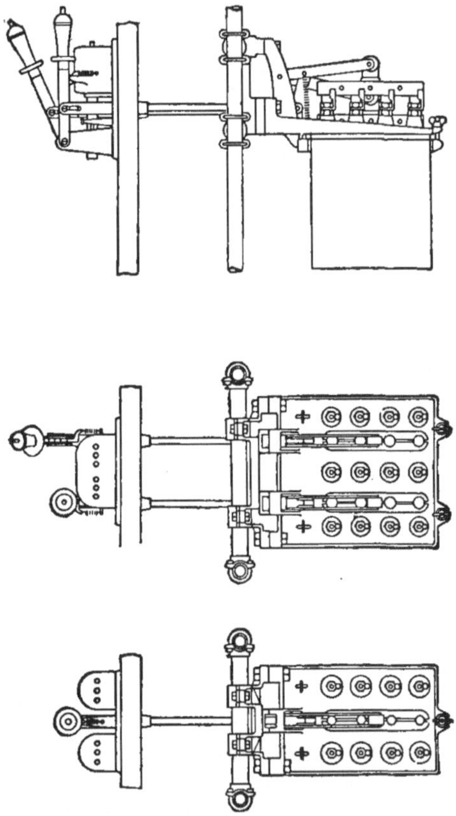

FIG. 88.—K3 oil switch mounted on pipe supports back of the panel.

by atmospheric action, will not absorb moisture even though tainted with chemical vapors, and is easily cleaned off when fouled, leaving a surface as good as before. Neither is glazed porcelain softened by heat.

Hard rubber is also a good insulator for use in air, as it is strong mechanically and has permanent insulating qualities under ordinary conditions, but it cannot be used under oil or where it is possible for oil to reach it owing to its very rapid deterioration under such circumstances. On the other hand, fiber is mechanically tougher than porcelain and can be used under oil with a good degree of safety and permanence, but it warps and disintegrates under the action of the atmosphere. A molded composition is used with very good success by the Hartman Circuit Breaker Company both for oil tanks and for insulating material on which live contact parts are mounted. Protection against atmospheric action—both from disintegration of the material and from the absorption of moisture from the air by the material—is secured by a surface skin of glaze.

Another material which is extensively used as an insulating material is wood. Dry, well-seasoned wood is an excellent insulator, is easily worked into almost any shape, is strong mechanically, and is permanent in form and insulating qualities. However, wood must be prepared against the absorption of moisture before it is placed in actual service, by closing the pores with some insulating material. In moderate-voltage service soapstone makes a good material for bases which support contacts where good insulating qualities are required.

All mechanism should be as simple and reliable in action as possible. The fewer special auxiliary attachments, the more reliable will be the operation of the apparatus. The action of the oil switch in making and interrupting circuits must be positive and rapid, because the effectiveness of the device is largely due to the quickness with which the circuit is opened.

Figure 88 shows a K3 four-pole single- and double-throw switch which release mechanism, mounted on pipe supports back of the panel.

Cables or bar connections are led to the outside terminal of the poles at safe distances from all metallic parts. The wedge-shaped copper bridges are fastened to wooden rods connected to the operating mechanism so that they can move through the frame. They are operated rapidly and simultaneously under oil. The oil vessel is lined throughout with laminated wood and is furnished with barriers of the same material, which are held securely in position between the poles of the switch. When the switch is to be tripped automatically, the tripping coil operates the linkage of the mechanism without moving the handle on the front of the board from its "closed" position, which might otherwise injure the attendant. This action is made possible by the slot in the horizontal member of the mechanism on the front of the board. If, for instance, the operator closes a switch

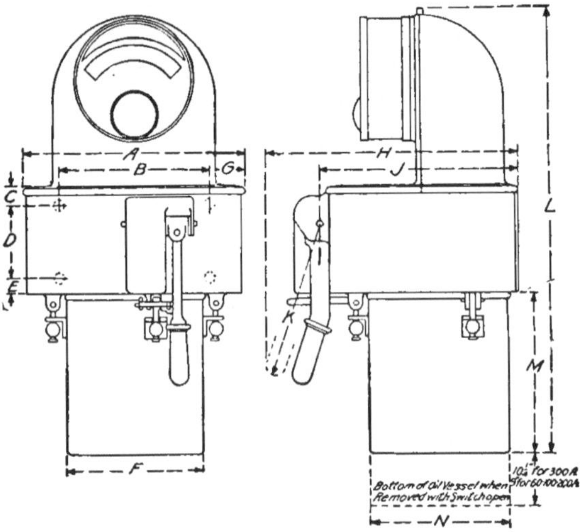

FIG. 89.—K20 oil switch with series ammeter.

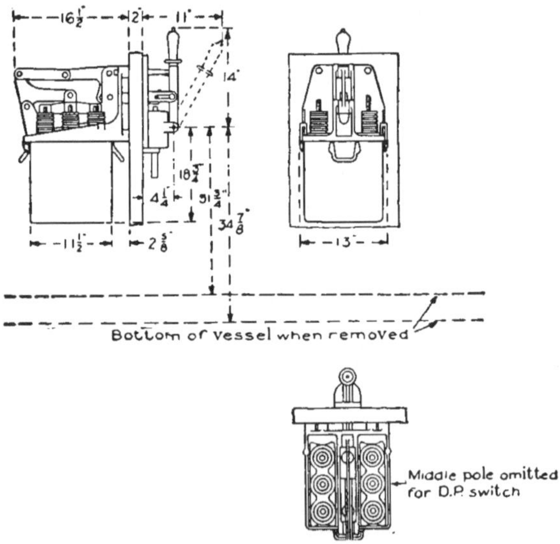

FIG. 90.—K2 oil switch.

on a short-circuit or overloaded line, the tripping coil will immediately throw open the oil switch without throwing out the handle in the hand of the operator. Such switches are termed non-closable on overload. The position of mounting thus requires a base casting and extension link. The K20 oil switch is adapted for use as a controlling device for induction motors in any industry where inflammable materials or explosive gases are

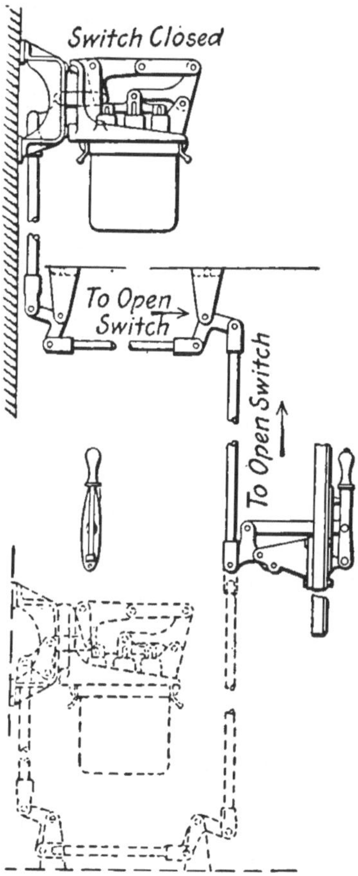

Fig. 91.—Pipe mechanism for operation of K and K2 oil switches.

present. This switch is entirely enclosed, has no live parts exposed, and therefore minimizes the fire hazard. When non-automatic it is generally employed with induction motors when fuses are installed. The automatic switch has usually an inverse time-limit adjustment of fifteen seconds at the starting current of the motor or approximately five times the running current.

This gives a time delay of about three minutes at 25 per cent. overload. Figure 89 shows the K20 switch with a series ammeter on the top. The leads are brought out through the bottom of the cover frame. In Fig. 90 we have one triple-pole, single-throw K2 switch with electrical trip (one coil) for use with series transformer. For the double-pole, single-throw switches the middle pole of the three-pole switch is omitted. The contact fingers of this switch are flared and made of drop-forged copper, fastened to the studs by heavy springs which insure good contact with the moving blades. The shape of the blade is such that the final interruption of the circuit occurs on the flared portion of the contact fingers and above the working surface of the blade. The oil vessels are of sheet metal lined with maple. Figure 91 shows the method of mounting double, triple, or four-pole K2 non-automatic oil switches. They are mounted on the wall or in cells removed from the board. The mounting also applies to automatic switches. The method of mounting shown allows the placing of the oil switch in any position relative to the board. Care must be taken, however, that the length of compression members of the operating mechanism be not too great, as otherwise bending will take place, so that the switch is not completely opened or closed. Such members may be strengthened by guides or through increase in cross section. Figure 92 is a photograph of three single-pole, single-throw K2 switches for 15,000 volts and 300 amp., operated from the same mechanism by means of a common shaft. Each oil switch is mounted in a separate cell of brick or other fireproof material (omitted in the cut). The mounting of the cells and the corresponding switches relative to their control panel is such that the rod from the board operates the shaft between its bearings or immediately outside of one of the bearings.

Figure 93 shows arrangements for an electrically operated three-pole K2 oil switch. The switch and d.c. solenoid are mounted on pipe supports. Their relative positions may be varied as desired, provided the necessary changes in operating mechanism are made. Sets of two, three, or four single-pole switches in cells similar to those shown in Fig. 92 may be operated with common solenoids in place of mechanical devices. Figure 94 shows a triple-pole oil switch, type K4 with electric trip. For double-pole, the middle pole of the triple-pole switch is omitted. The oil switches are mounted on the board and are

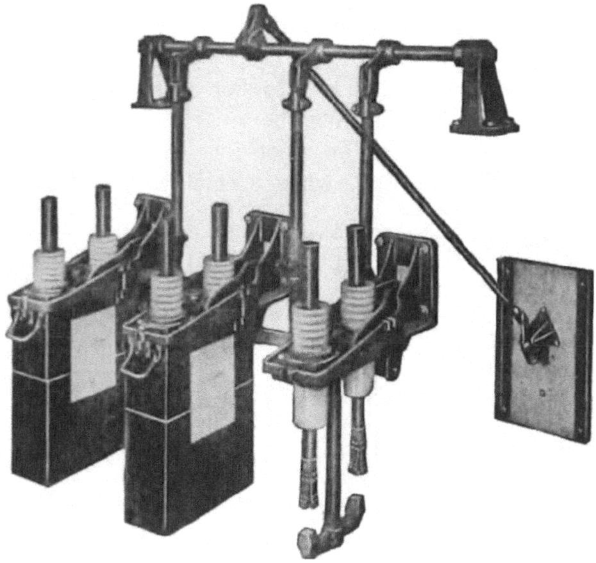

FIG. 92.—Triple-pole, single-throw 15,000-volt K2 oil switches for remote control. (Each pole is installed in a separate brick cell, not shown.)

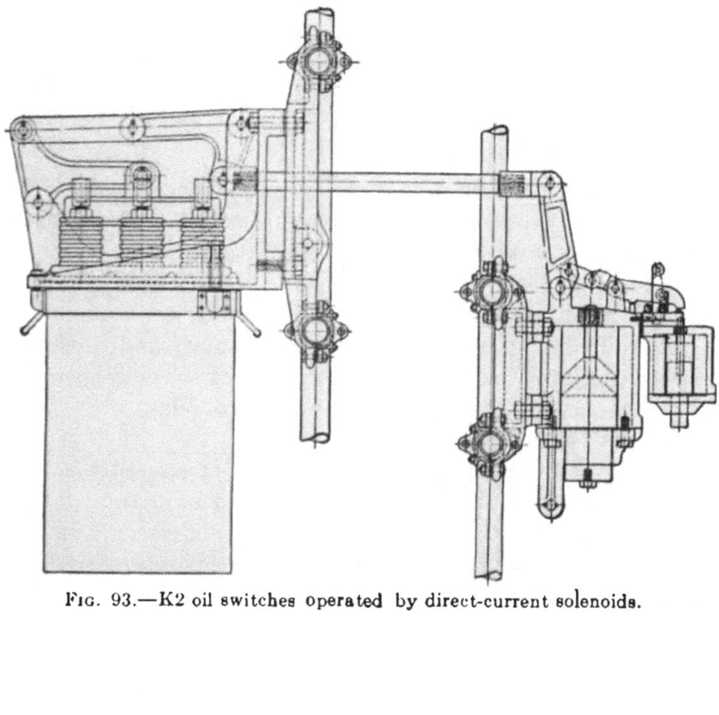

FIG. 93.—K2 oil switches operated by direct-current solenoids.

braced by two pipe supports apiece because of the greater weight and lever arm.

When mounting oil switches, provision must be made allowing removal of the oil vessel when the switch is open. In exceptional cases allowance can be made for removal of the vessel when the switch is closed, when special care in handling is

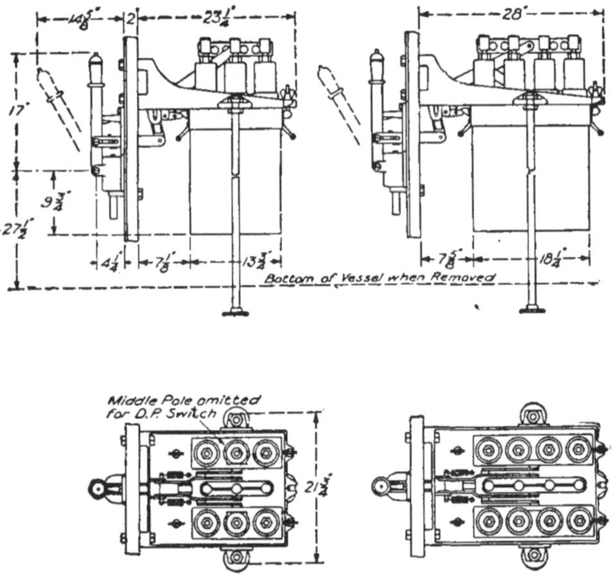

FIG. 94.—K4 oil switch.

necessary. This condition is fixed by the distance between the bottom of the vessel and the floor or cell bottom. The method of mounting a three-pole type K4 oil switch on pipe supports is given in Fig. 95.

These switches may also be furnished with a low-voltage coil or with different relays (see following chapter) and auxiliary switches (for interlocking electrically). When solenoid-operated switches are used, control switches and control relays with signal lamps must be provided.

Recently the General Electric Company introduced the K12 oil switch, which supersedes the K2 and K4 switches. It has many features like the last mentioned and differs only in the mechanical construction of the operating mechanism.

Figure 96 is a wiring diagram for a d.c. solenoid for operating an oil switch. It has two windings (see Fig. 93), the larger one, *A*, being the closing coil and *B* the tripping coil. A small double-pole controlling switch is mounted on the board, which controls the current for the opening and closing coils. A red bull's-eye lamp shows when the switch is closed, and a green one when it is open. On the solenoid there are also two small auxiliary switches *1* and *2*. Switch *1* is open when the oil switch is closed, that is,

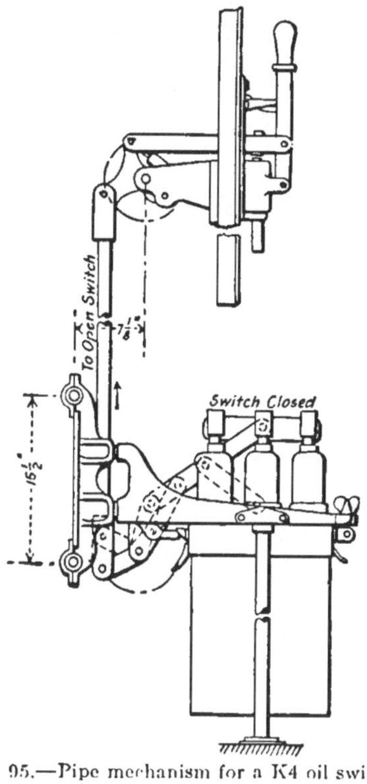

FIG. 95.—Pipe mechanism for a K4 oil switch.

after coil *A* has operated. It is closed when the oil switch is open. The action of switch *2* is the reverse of that of switch *1*. The object of each of the two auxiliary switches is to disconnect the coil with which it is in series, after this coil has operated, and at the same time to throw into circuit the coil of the other switch, thus preparing the solenoid for reversing, which is made possible by operation of the controlling switch. The auxiliary switches also operate the signal lamps. Resistance *R* is used only when the d.c. circuit for the operation of the solenoids has an e.m.f. over 125 volts. The solenoids and lamps are protected by the fuses *F* and *f*. The opening coil *B* is smaller because the weight of the contact bridge helps to open the oil switch. The diagram is for non-automatic operation.

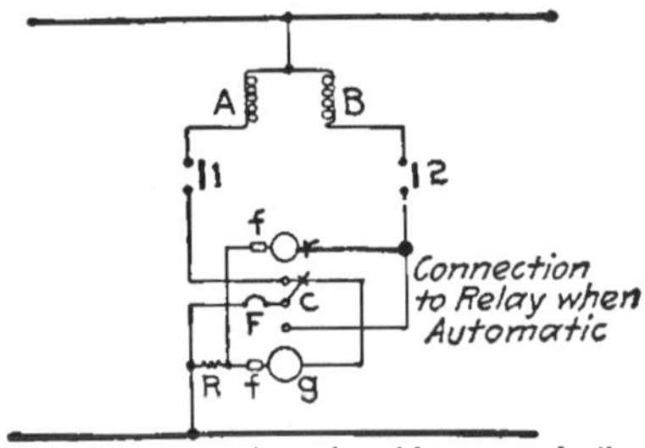

FIG. 96.—Diagram of a solenoid-operated oil switch.

When oil switches are to be used as double-throw switches, as is the case when transferring connections between busbar sets, the simultaneous closing of both throws must be prevented. For mechanically or electrically operated switches this is provided for in the mechanical or electrical interlocking devices.

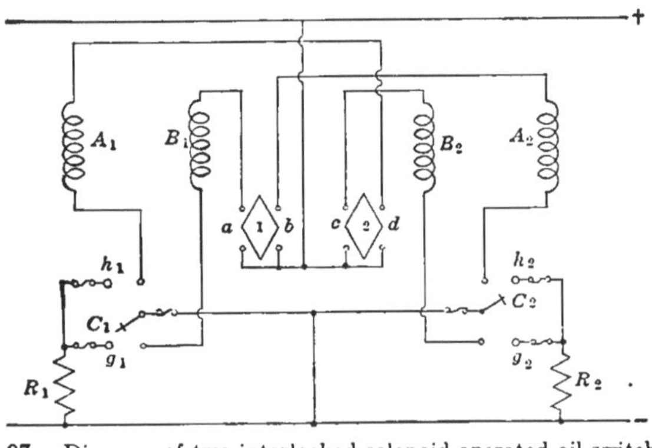

FIG. 97.—Diagram of two interlocked solenoid-operated oil switches.

Figure 97 is a wiring diagram for two electrically interlocked solenoid mechanisms for the operation of two oil switches. A_1 and A_2 are the closing and B_1 and B_2 the opening coils. Auxiliary switch *1* closes the side *a* when the oil switch U_1 is closed. Similarly switch *2* closes side *c* when switch U_2 is closed. Switch *1* closes side *b* when U_1 is open, and switch *2* closes side *d* when U_2 is open. Consider, for example, that U_2 is closed, as indicated by the red lamp h_2, and that C_2 is in the upper position. Then switch *2* will close side *c*. If we now attempt to close switch U_1, we throw C_1 into the upper position in order to excite coil A_1. But *d* of switch U_2 is open. Therefore A_1 is not in position to close U_1, and therefore U_1 cannot be closed. This shows that both switches cannot be closed at the same time, but that they can be opened together. In Fig. 97 both switches are non-automatic.

As the table shows, the K10 and K15 oil switches are adapted for potentials between 22,000 and 110,000 volts. Each pole is contained in a boiler tank of such dimensions that no barriers between the leads and no cells for the tank are necessary. The switches are especially suited for overhead and also open wiring. The bushings of the outgoing leads are of the built-up type having porcelain ends with intermediate rings of compound and threaded fiber washers between joints. As to the operation of these switches, attention is called to the attempt made to use compressed air for remote control when direct current for the actuation of a solenoid mechanism is not available. For that purpose a small a.c. air-compressor equipment with an air-diaphragm mechanism in connection with remote-control electrically operated valves is to be used.

The overload protection of these switches sometimes permits the adoption of a series relay trip. (See following chapter.) By means of hand-operated switches this trip acts directly on the toggle mechanism, while by means of the solenoid operation of the oil switch it influences the tripping coil of the solenoid to close an auxiliary small switch of an electrical circuit.

Figure 98 shows a single element of a K15 110,000-volt oil switch solenoid-operated.

For outdoor service and very high voltages up to 110,000 the General Electric Company, recommends the use of its oil switch K22 of the tank type. Figure 99 shows a triple-pole oil switch consisting of three units. Each pole of the switch is contained and supported by an oil tank of boiler iron double-lapped, riveted and calked. The cover is provided with vents to release the gases formed by the arc set up by opening and closing the switch. The cover supports the switch mechanism and bushings. The bushings are made for the higher voltages (70,000 to 110,000 volts), of porcelain petticoat tubes outside the cover, while inside the cover they are built up with rings of compound and treated fiber washers, forming a shell of high dielectric strength. Through the center of the bushings passes a rod which is joined to the contact stud of the switch. The contacts

FIG. 98.—K15 oil switch, single-pole, solenoid-operated.

are of the sliding-wedge construction, each phase containing two of these contacts in series. The stationary contacts consist of widely flared fingers and extra-long arcing trips. The movable contacts are of wedge-shaped copper blades, being drawn up in vertical plane by the switch mechanism. The operating rods of each element are connected together and may be operated by hand lever, direct-current solenoid, or air diaphragms. The automatic operation is actuated by trip coils acting directly on the switch toggle and operated by current transformers. The transformers are of the bushing type having for its primary the metal rod of the switch bushing and for its secondary a copper ribbon wound on an iron core which is placed around the bushing.

The General Electric Company recommends types H3 and H4 oil switches for the largest stations with any voltage. (See Fig. 100.) These switches are distinguished from all other types by three characteristics. Each terminal of a pole pair is contained in a separate oil vessel, the outside terminals are on the bottoms

FIG. 99.—Triple-pole oil switch comprising three units.

of the vessels instead of the top or side, as in other types, and finally the switch is operated by a d.c. motor. Each pole pair is mounted in a separate cell, three such cells constituting a three-pole switch. Since each pole is enclosed in a separate vessel the arcs at the points of interruption are separated, which increases the safety of the device. Since the cable connections are made at the bottom of the vessels they are separated from all movable parts, affording better insulation and easier access. The metal plungers which project through the top of the insulators are connected to a metallic cross arm, which in turn is joined to the operating mechanism by wooden rods. The operating height of the plunger is 12 inches for 6,000 volts and 17 inches for 12,000 volts. All poles open simultaneously, six arcs being formed in the three-pole switch. The cells enclosing the switches are made of brick. The top and bottom are slate, the bottom containing the insulators for the cable connections.

The motor operation has an advantage over the solenoid method in that it operates the switches more rapidly than the solenoid under lower service voltage. Figure 102 shows an H3 switch designed for 15,000 volts, good for 1,200 amp. The oil vessels are made of brass or sheet steel lined with insulating material. For 60,000 volts and over, the vessels are made barrel-shaped

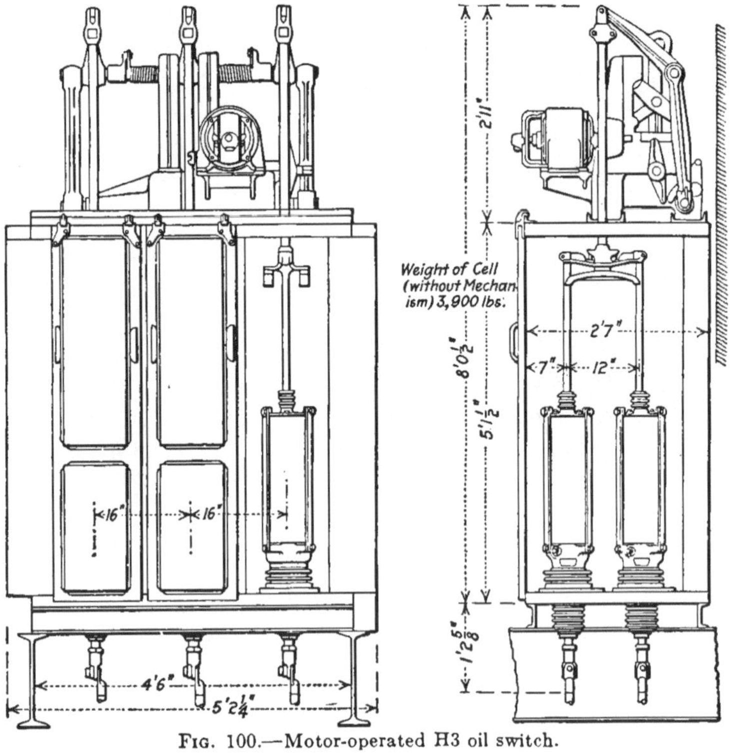

Fig. 100.—Motor-operated H3 oil switch.

and are constructed of wood, held together by rope windings, being supported in the cell on four insulated legs. (See Fig. 101.)

Figure 102 is the wiring diagram for the motor of an H3 oil switch. Its operation consists essentially in winding up two spiral springs after each throw of the switch. Opening of the oil switch is entirely independent of the motor. A control switch or relay on the board controls the circuit of an electric magnet operating a toggle. This toggle releases the spring which operates the switch, simultaneously starting the motor to wind up the springs.

The floor where the cells are set up requires special construction on account of the cell weight and the cable connections on the bottom of the cell.

Westinghouse oil switches may be classified in two groups, the first comprising those in which the interrupting device is similar to a knife switch, so that each pole pair is interrupted on one side

FIG. 101.—Motor-operated H3 oil switch for 60,000 Volts.

only, the second group including forms where there is interruption at both terminals of each pole pair. In the first group are the several knife blades connected together by a specially treated wooden piece which is itself connected by a rod of the same material to the operating lever. Knife-blade contacts are used for type I and D switches as they insure the best contact for low-temperature rise. Each jaw has an arcing piece attached which takes the final break, thus preventing any possibility of arcing between the jaws and the blades. Insulating barriers are fastened to the cover between the pole pairs. In type B each pole has a separate tank lined with insulation. All tanks, terminal insulators, and operating mechanisms are carried on a cast-iron frame fastened to the switchboard or wall for remote control. The contacts are of brush type and are held in open position by gravity. (See Fig. 103.) The tanks and rods carrying the contacts serve as barriers between the points

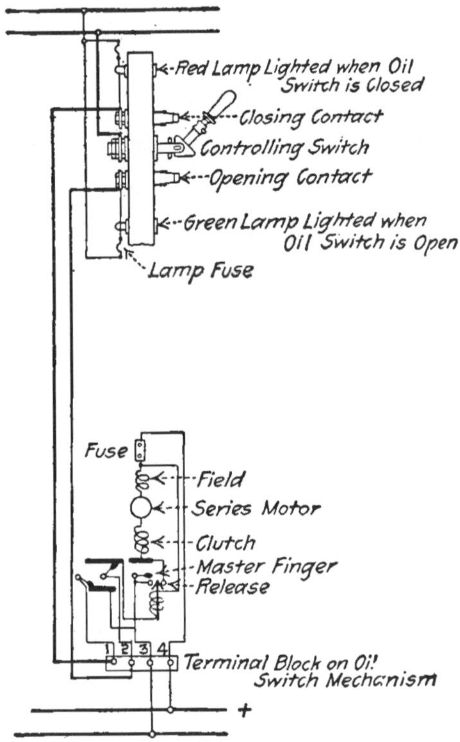

FIG. 102.—Connections of controlling circuits for H3 oil switch with magnetic release. (Using double-throw controlling switch, normally open with double-throw contact fingers.)

where arcing may occur. At overload or short-circuit the tripping coil which opens the oil switch releases a trigger, so that the handle is not thrown open. For voltages above 6,600 it is advisable to ground the metallic framework of the breakers, and in cases where the power is more than 4,000 kw. per circuit it is the best practice to operate the oil switch apart from the switchboard. For higher voltages and larger power the Westinghouse Company makes use of type C switches electrically operated and mounted in fireproof cells. They are used under conditions similar to those where types K6 or H3 of G. E. make would be used. Figures 104 and 105 show the outlines of the cells and the operating mechanism and a view of the switch proper. Each cell contains one pole pair in one oil tank, which is easily removable.

The main moving contacts are of the laminated butt-brush type. The form of lamination is such as to provide an end-on

Fig. 103.—Contact mechanism and terminals, 2,000-amp. oil switch, type B, open position.

contact at minimum pressure. The brushes have a wiping or self-cleaning action during the motion of closing or opening. The movable contacts are connected to the mechanism by a rod. The insulating portion of this rod is of treated wood, but the part of the rod that passes through the top of the expansion chamber is made of steel of smaller diameter to make the area exposed to the internal gas pressure as small as possible. Arcing contacts are provided making contact with the head of a removable stud on the stationary contact. The leads are brought out at the rear of the switch and may pass directly into a masonry conduit. Fireproof insulating barriers are provided between phases on the back of the cell, and smaller barriers of slate or asbestos are placed between cables of the same phase. The oil tanks are of elliptical shape, constructed of heavy sheet metal

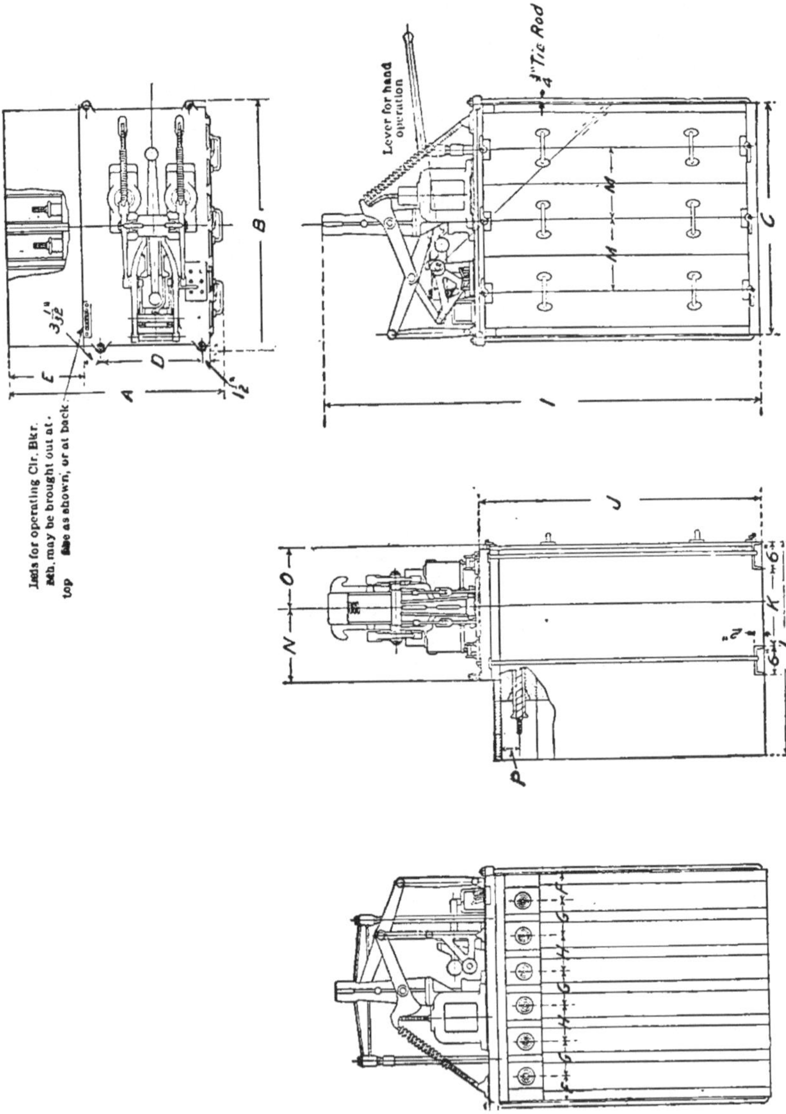

FIG. 104.—Type C oil switch, electrically operated, distant control.

with lap-welded joints. The interior of the tanks is provided with seamless micarta linings as an additional protection against arcing from contacts to the tank. Over the tank is situated an expansion chamber which provides a large air space above the oil level for the formed gas. The chamber is made of bronze

FIG. 105.—Type C oil switch, solenoid-operated.

alloy and braced to the tank. Each chamber has a vent to provide an exit for gases, but baffled so as to prevent the throwing of oil at times of excessive disturbances. The operating mechanism consists of a system of levers actuated by a direct-current shunt-tripping solenoid requiring a current of 125 or 250 volts. To make the opening of the breaker rapid and positive, accelerating springs are used in order to force it to the open position. Dash-pots absorb the momentum of the mechanism in closing and opening.

A control switch governs the operation. Tell-tale indicators and red and green signal lamps are used to indicate the open and closed positions of the oil switch. The break in these switches occurs near the surface of the oil instead of in the lower portion of the tank as in type H3.

Figure 106 is a manually operated type E switch, mounted in cells. It is provided with an automatic tripping coil. A toggle

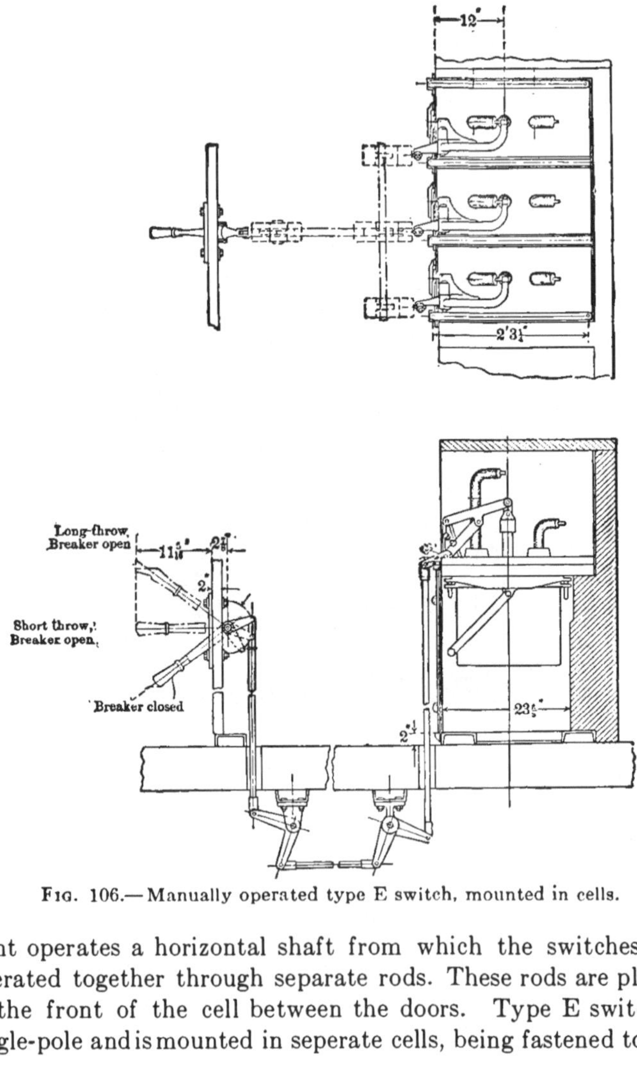

Fig. 106.—Manually operated type E switch, mounted in cells.

joint operates a horizontal shaft from which the switches are operated together through separate rods. These rods are placed in the front of the cell between the doors. Type E switch is single-pole and is mounted in seperate cells, being fastened to the slate or soapstone cell cover. The same type of switch with electrical operating mechanism is shown in Fig. 107. The single-pole switches are similar to the manually operated type E switch, and are similarly mounted on the covers of the cells in which they are contained. Each switch has its own solenoid mounted on the cover, but they are all simultaneously operated from a single control switch on the board. Under certain conditions of installation an interlocking mechanism for coupling together the poles of a multipolar, electrically operated oil switch may be used.

In Fig. 108 we have a set of three single-pole type G oil switches, built for 60,000 volts. They are operated by a common solenoid mounted at the foot of the cell, the tanks of which (one for each switch) are made of boiler steel lined with insulation. Poles in the same tank are separated by insulating barriers. The heavily insulated cable terminals project about 4 ft. 6 in. above the top of the cell. The three vessels may be mounted in cells or on iron-frame supports. The crosspiece operated by the solenoid is connected to three sets of toggles which operate the wooden rods of the U-shaped contact bridges.

Figure 109 is a set of three single-pole type G oil switches for 120,000 volts. The boiler-steel tanks are supported on iron framework. All three switches are operated by one solenoid through a cross arm and rod. The tanks are separated from each other by a certain distance. A new oil switch, termed type GA has recently, come into use for high-potential systems where circuits carrying large amounts must be broken. This type partly supersedes the G type oil switch. Each pole of the switch is a separate, distinct unit enclosed in an oil tank composed of boiler sheet steel. The walls of the tank are lined with rope cement paper.

Figure 110 is a sectional view of the GA oil switch in its closing position.

This type is suitable for use on high-potential circuits where large amounts of energy are to be handled. They are adapted for both indoor and outdoor service. Each pole is a self-contained unit consisting of an oil tank of heavy sheet steel. On the top of the tank is mounted the operating toggle mechanism. In the outdoor type the mechanism is entirely covered by a hood with proper gaskets for weatherproofing. The lower or movable contacts are carried by a metallic cross bar and consist of pieces of cylindrical brass rod—butt type—and backed by compression springs which render the contacts self-aligning. The terminal bushings are of the condenser type and consist of concentric layers of insulating material separating layers of tinfoil. The conductor, consisting of metal tube or rod, is the core of the bushing. The sections of the condensers are so proportioned that

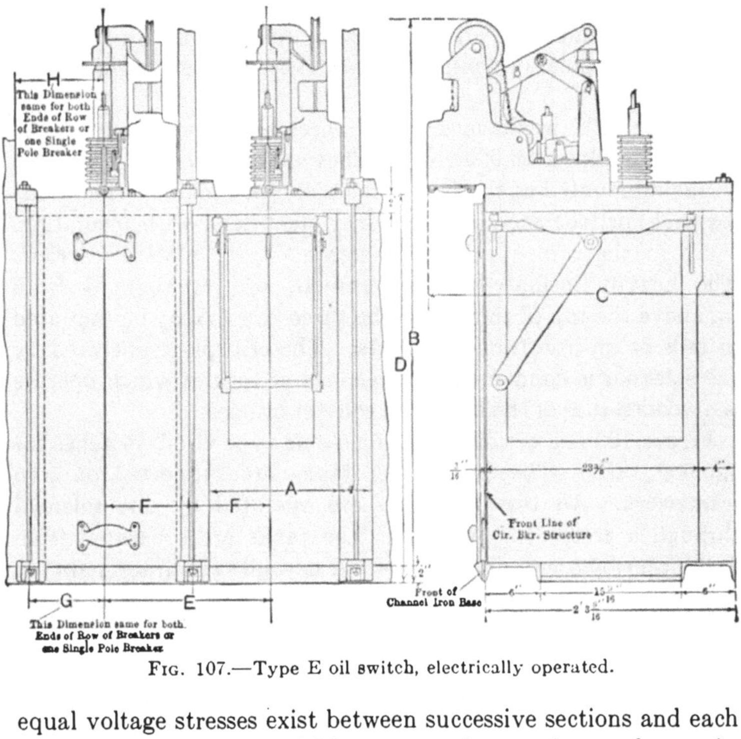

FIG. 107.—Type E oil switch, electrically operated.

equal voltage stresses exist between successive sections and each layer of insulating material has an equal stress imposed upon it. In the outdoor type the condenser terminal bushing is protected by a series of porcelain rain shields cemented together to make a weatherproof unit. In the indoor style the condenser bushings are protected against the ordinary moisture frequently present, particularly in hydro-electric power plants, by a covering of waterproof insulating tubing sealed with moisture-resisting compound. The GA oil switch may be operated either by hand or electrically. The operating mechanism of the manually operated switch consists of a single system of levers, bell cranks and links actuated by a toggle mounted on a panel. The distance of the panel from the switch should, however, not exceed 30 feet. The electrically operated switch, as well as the hand-operated mechanism, is equipped with an accelerating and cushioning device.

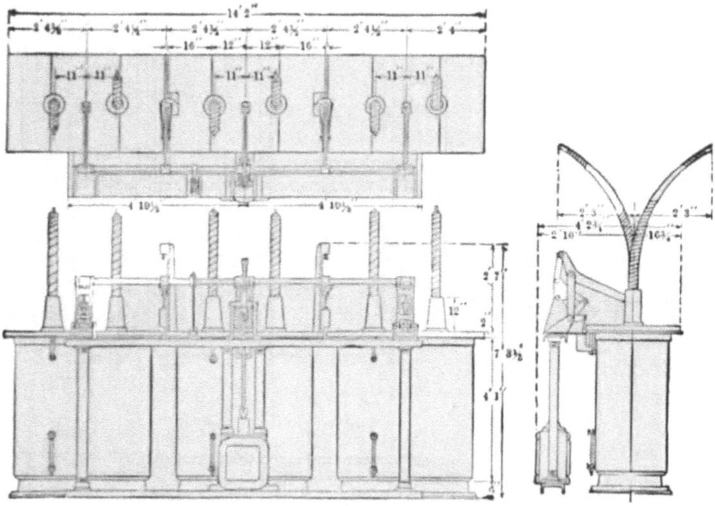

FIG. 108.—Type G oil switch, electrically operated for 60,000 volts.

The breaking distances, or distances between the contacts when the oil switch is open, are given in the following table.

VOLTS	DISTANCE PER BREAK, INCHES
44,000	11.5
66,000	16.5
88,000	20.0
110,000	23.5

Figure 111 shows the outlines of a type L oil switch for 88,000 volts, which is operated either manually with toggles or electrically. The distance of contacts above the cell when the switch is open is 17 in. for a 60,000-volt switch, and 20 in. for 88,000 volts. The tanks are made of hard wood, covered with metal sheeting. In each tank there is a double barrier between the poles, and the contacts are fastened to bushings in the middle of each of the two resulting chambers. A movable rod between the barriers operates the contact bridge through the slots cut into the barriers. The solenoids are mounted on the lower part and

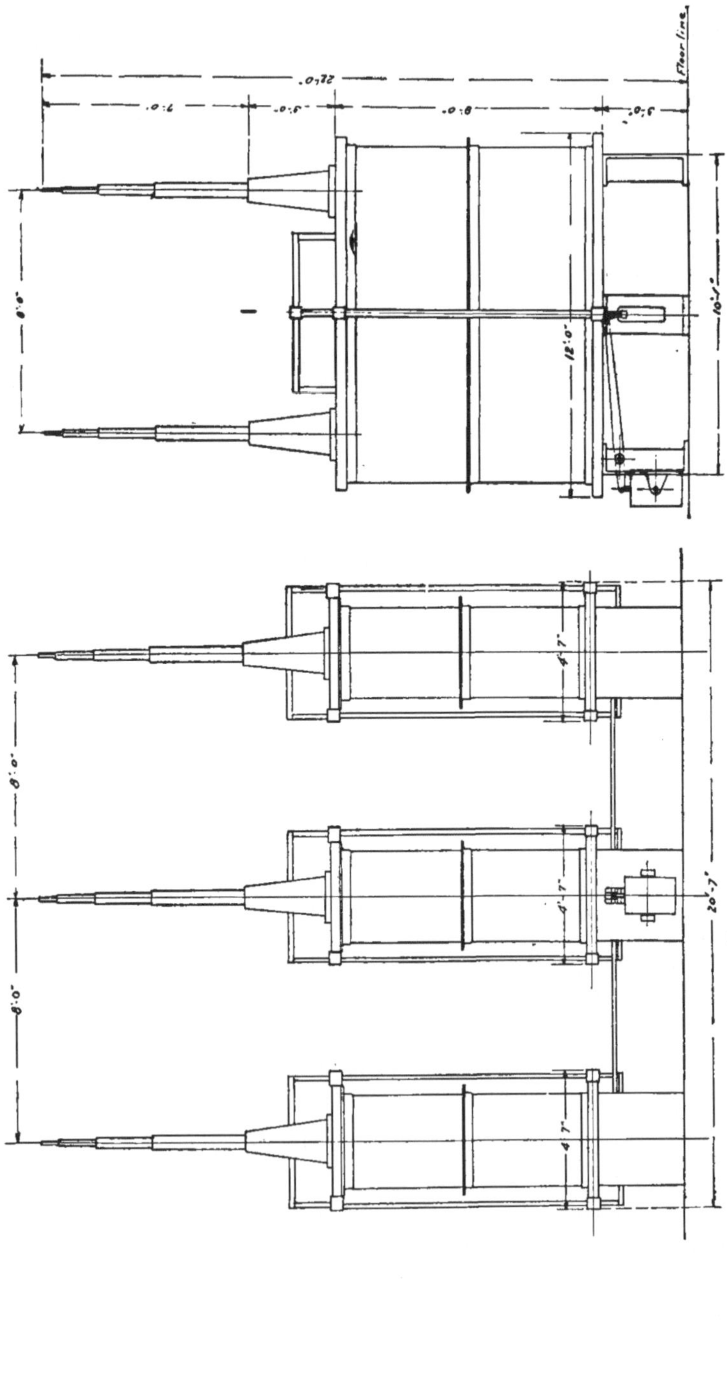

FIG. 109.—Type G oil switch electrically operated for 120,000 volts.

operate the rods of the contact bridges through the cross arm and toggle mechanism. In all of the last-mentioned switches heavily insulated cables project from the bushings in the cover, to which the circuit connections are made. These switches can

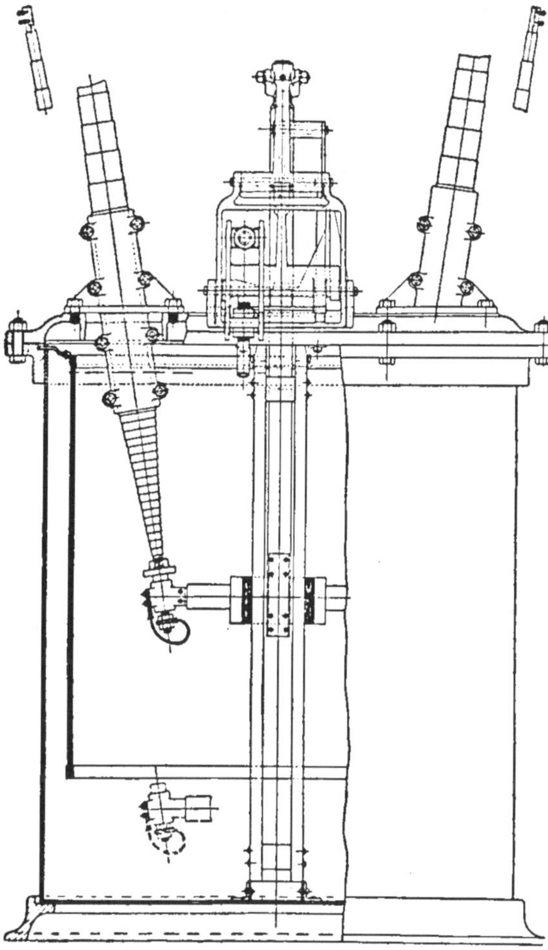

FIG. 110.—Type GA oil switch.

be opened automatically or non-automatically, as shown by the table for Westinghouse switches. Special appliances for automatic interruption will be discussed in the chapter on relays.

Considering the gradual increase in the rating of oil switches, it is noted that as the current-carrying capacity grows the weight of the current-carrying parts becomes greater, and also the mechanism for operating these parts. In other words, the inertia of the oil switch is increased and also the friction of the mechanism. A combined accelerating and dash-pot device is adopted for accelerating any mass, as may be necessary, and in turn for

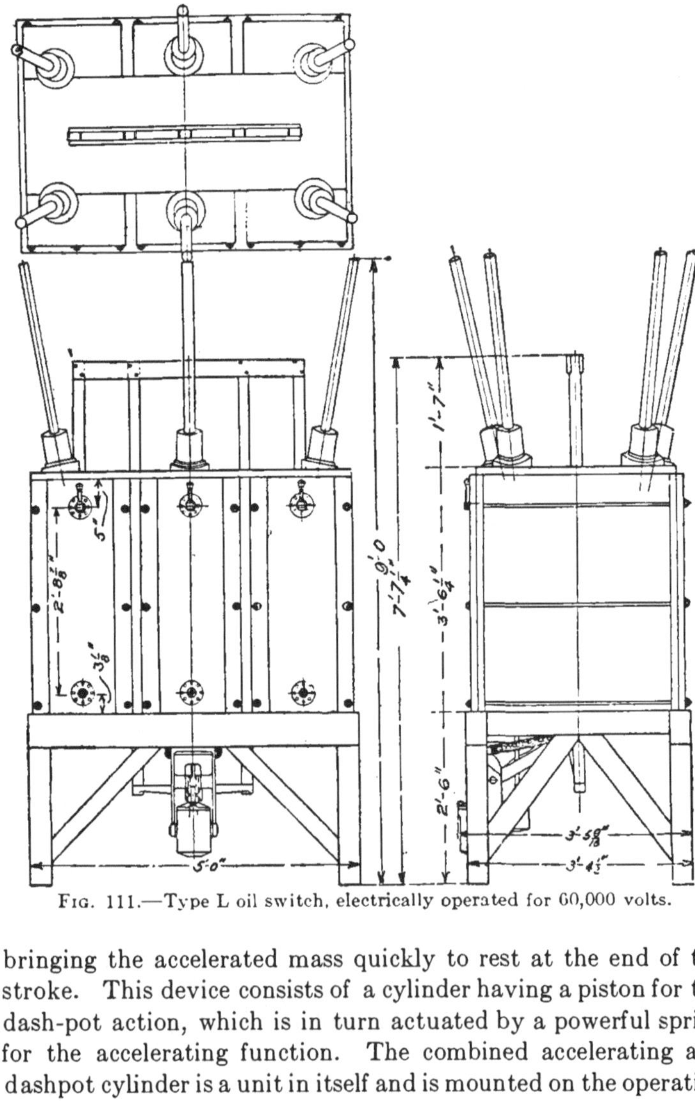

FIG. 111.—Type L oil switch, electrically operated for 60,000 volts.

bringing the accelerated mass quickly to rest at the end of the stroke. This device consists of a cylinder having a piston for the dash-pot action, which is in turn actuated by a powerful spring for the accelerating function. The combined accelerating and dashpot cylinder is a unit in itself and is mounted on the operating lever of the magnet mechanism in the electrically operated types, and on one of the bell-crank brackets in the mechanical remote-control types. The use of this mechanism is necessary, especially in high-powered oil switches, in order to secure positive action under conditions of short-circuit which cause high pressure in the tanks. Such a pressure acting on the contact-carrying rods without this device has been known to prevent the movable contacts from traveling more than a short distance away from the

FIG. 112.—15,000-Volt type C non-automatic oil switch.

stationary contacts, when opening very slowly, and has caused the arc to persist in the oil for a length of time sufficient to generate a pressure in the oil switch strong enough to disable it.

The scope of this treatise does not make it possible to discuss the products of all manufacturers of oil switches. Only those have been selected which are most used in service, and their special characteristics have been emphasized. Some of the products of the Hartman Circuit Breaker Company also deserve mention. The oil tanks of switches of this company are made of molded fiber, so that a good separation of both poles is attained with the aid of the rods carrying the contact bridge.

The pole pairs of a type A switch are in separate oil tanks, and the whole switch, including operating mechanism and automatic tripping coil, is carried on two pipes fastened to the switchboard For low voltages and small capacity, all three-pole pairs are placed in one tank. Effective barriers against arcing are found in the extra wooden rods between the phases and the operating rods for each pole pair. Up-and-down motion of the rods is brought about by a crank-like motion of the operating mechanism. Figure 112 shows type C, for 15,000 volts, with tanks removed. The covers of the tanks, together with the bushings, are made of treated fiber. The illustration shows the movable laminated contacts of jaw type and the wedge-shaped fixed contacts which fit into them.

For high tensions type C is used, which is operated electrically from the board or by a wire-rope arrangement. Figure 113 illustrates a three-pole switch with two series transformers enclosed in tanks similar to oil-switch tanks. In the illustration one of the pole-pair tanks and one of the transformer vessels are removed. The outer vessels are the transformer tanks. The entire apparatus is carried on two pipe supports. Arrangements for wire-rope operation and electrical manipulation are shown in Figs. 114 and 115. In both cases the switches are mounted on a distant wall.

The type H switch is built for 60,000 volts. Instead of a double break, a quadruple break is used. (See Fig. 116.) Two auxiliary poles are supplied for each pole pair, and the two contact bridges are operated mechanically by two rods. The treated fiber tanks are fitted with barriers and projections, which, with the movable rods, afford separation of the poles and auxiliary poles. This oil switch, like the type C, is operated electrically or by wire rope, and can be opened automatically or non-automatically. The fiber sleeves carrying the contacts are high and quite thick.

The Pacific Electric & Manufacturing Co. produces an oil switch for high tension in which each pole pair is actuated by a double-contact arm revolving about a vertical axis. Figure 117 shows a section and outline of one of these switches for 110,000 volts. The contact arm is mounted on a bushing on the lower end of a vertical operating rod. The outer ends of the contact arm connect with the poles, which are joined to the circuit leads. Two or three of these switches may be manually operated as two- or three-pole switches. If the angle of rotation of the arms is made 90°, an extra long interruption of phases is obtained.

To limit the disturbance in the oil switch and on the line, where extremely large powers at high voltages must be controlled with certainty under all conditions, a type of oil switch has been developed in which a reactance is cut into the circuit before the main contacts open.

These switches have two sets of contacts per pole included in and operated by the same mechanism. One main heavy set of

FIG. 113.—Three-pole type C automatic oil switch for wall mounting and remote control. (Shown with one switch and one transformer tank removed.)

contacts regularly carry the load current, while the auxiliary second set only carry the reduced current after the main contacts have opened and thus introduced the reactance into the circuit. The auxiliary contacts rupture the final current when they separate and are also first to make contact when closing the switch.

The reactance switch clears the circuit in two steps, and although the first step produces the greatest reduction of current, it really makes a very small disturbance inasmuch as the rate of change is comparatively low. The second step will therefore have but a small proportion of what the single-step form of oil switch would have. The single-step switch makes an infinite rate of change in cutting the maximum flow instantly to nothing,

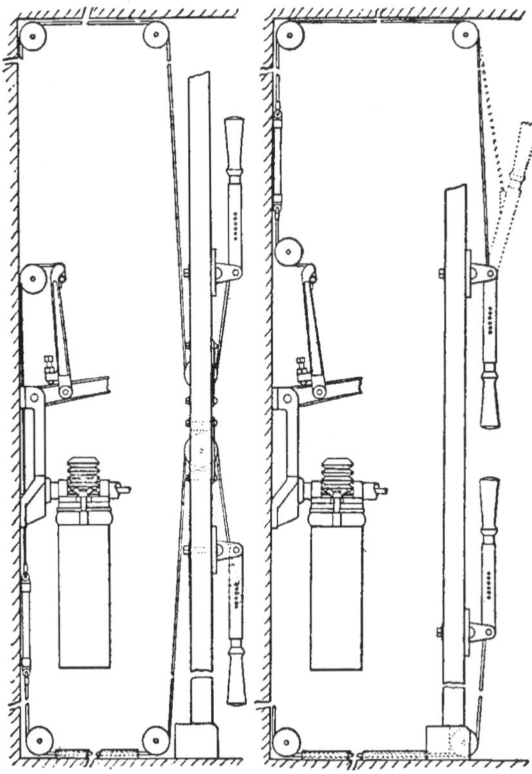

FIG. 114.—Rope operation of types C and H oil switches.

and therefore produces the maximum of potential surge. When the first step of a two-step switch acts, it still leaves a closed circuit in which the stored energy of the apparatus and system will dissipate at a relatively slow rate, whereas in the single-step switch this energy has to be absorbed instantly or is dissipated in destructive arcing under oil.

The design of a suitable reactive coil involves a knowledge of the maximum current which the switch will be required to rupture, the potential and frequency of the circuit, and the time which the

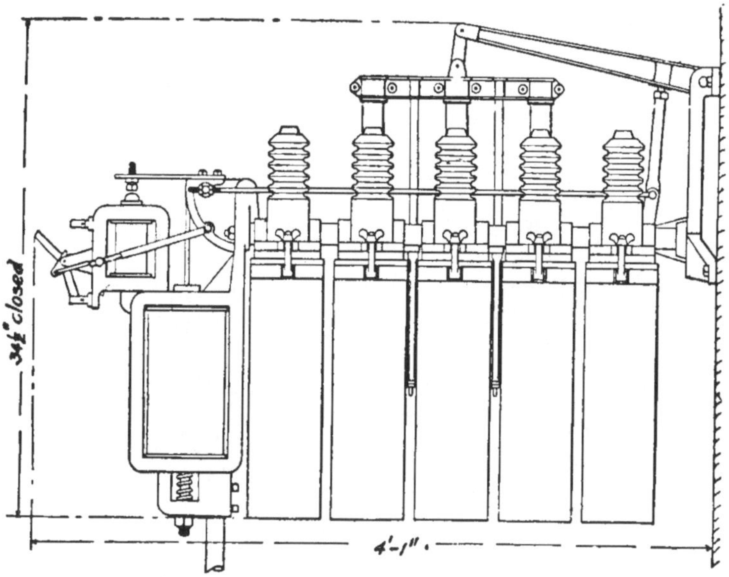

FIG. 115.—Type C oil switch, electrically operated for 22,000 volts.

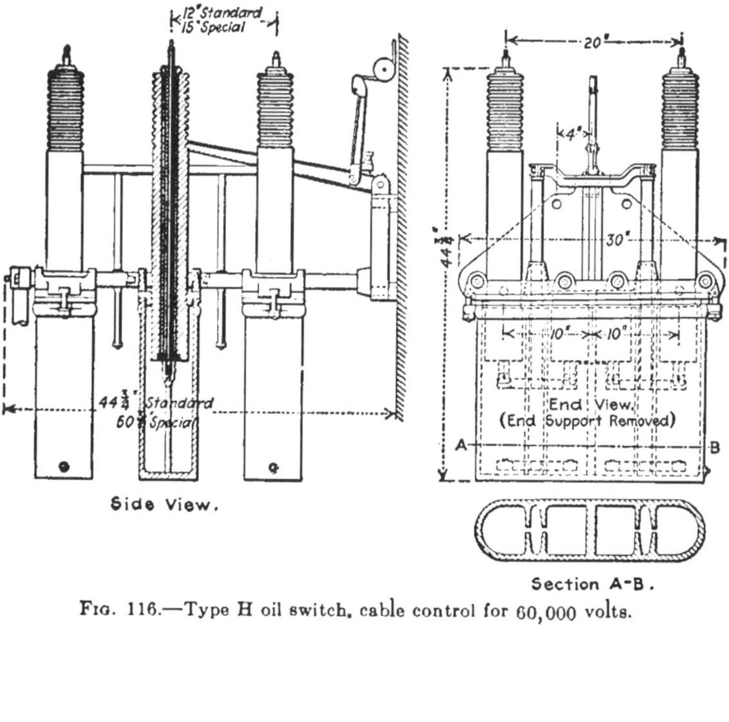

FIG. 116.—Type H oil switch, cable control for 60,000 volts.

reactance coil may be required to carry the maximum short-circuit current which the system can deliver through the coil. A coil properly designed according to these data will limit the currents to desirable values and carry, with a good margin, the maximum short-circuit currents for periods which will never be

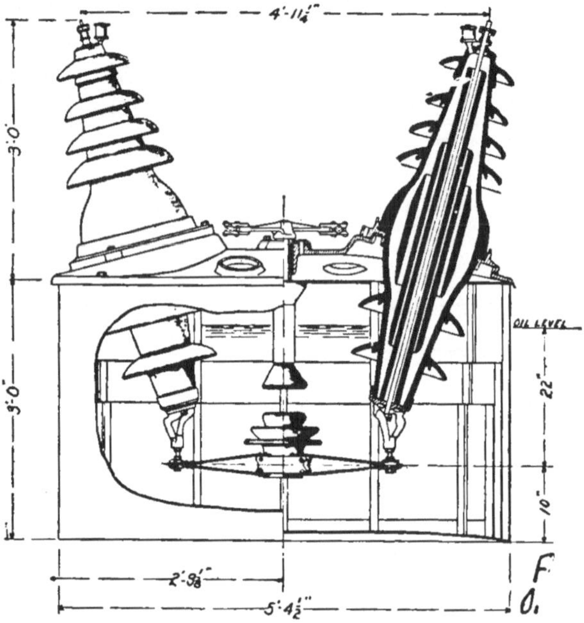

FIG. 117.—Oil switch for 110,000 volts.

exceeded in the normal operation of the switch. The size of conductor in such reactance coils is small, compared with that normally required for the currents involved, because of the short periods during which the reactance is in service.

The reactance-coil units for oil switches up to 22,000 volts consist of one well-insulated coil for each pole mounted in the expansion chamber of each pole on substantial treated wood strain insulators. For switches from 25,000 to 110,000 volts the construction of the reactance coils is similar to that of high-tension core-type transformer coils. These coils are substantially mounted and are completely submerged in oil.

Analysis of test results and theoretical considerations indicate that the reactance oil switch properly designed is adapted to any possible power and class of service. This is due to the readiness with which the current-limiting reactance can be introduced into a circuit carrying enormous currents, which are thereby reduced to values easily within the rupturing capacity of the switch.

An idea of the size of this unit may be obtained from the following data: Height to tip of terminals, 16 feet; floor space, three poles, 24 by 18 feet.

Filling the tank of an oil switch with oil should be possible without taking the switch apart or disconnecting it from service. It is, however, advisable to fill the tank when the apparatus is not in use on account of the danger incurred in handling a live switch. In order that repairs may be undertaken, provision must be made to disconnect the switch from all live parts. This is usually provided for by a disconnecting switch between the oil switch and the line. All live parts, such as studs, cable terminals, or copper rods projecting from the vessel, must be wound with good insulation, or must be screened, in order to obviate the dangers due to contact, fire, and short-circuit. The cells should have fireproof doors.

CHAPTER XVI

RELAYS

In large plants with expensive apparatus and where the service is continuous it is necessary to protect the apparatus and machines by making provision for automatic current interruption. Such provision is also necessary to protect stations or a series of sub-stations against serious shut-downs. The desired result is attained by equipping the apparatus, like oil switches, with tripping mechanisms which are actuated by relays at certain critical moments.. The relay itself is operated by the current which it is to interrupt, while the solenoids of tripping devices are energized from the secondary windings of series or shunt transformers which are on the main circuit, or from an independent source. They are adjusted to a predetermined condition of operation. Their action is the opposite of that of telegraph relays. With the latter a strong current is required to operate the receiving apparatus, while the former open their switches with a relatively weak current, as it would be difficult, if not impossible, to operate the relays directly with the main current when this current assumes large proportions. A classification of relays may be made on the basis of current influence and kind of action:

Current Influences	Action Time Element
1. A.c. overload relay	Instantaneous Definite time limit Inverse time limit
2. (a) A.c. reverse-current relay (b) A.c. reverse-current selective relay	Instantaneous Inverse time limit
(c) A.c. reverse-current differential relay	Instantaneous
3. A.c. reverse-phase relay 4. A.c. underload relay 5. A.c. low-voltage relay 6. A.c. overvoltage relay	Instantaneous
7. D.c. overload relay	Instantaneous Inverse time limit
8. D.c. underload relay	Instantaneous
9. D.c. reverse-current relay	Instantaneous Inverse time-limit

Current Influence	Action Time Element
10. D.c. low-voltage relay	Instantaneous
11. D.c. overvoltage relay	Instantaneous
12. D.c. differential (three-wire) relay	Instantaneous
13. D.c. solenoid-control relay	Instantaneous
14. Auxiliary relay (bell alarm, station signal, time limit.	Instantaneous Definite time limit

1. An overload relay, as the name indicates, serves to protect the line apparatus, or machines, against a load in excess of a given maximum. Such relays are made to act instantaneously at the points of energy consumption, especially when fire risk is great, in feeders which will deliver an excessive current under short-circuit or overload. In this case an instantaneous interruption is preferable to a momentary disturbance, and this action at the place of consumption relieves the other time-limit or inverse-time-limit relays. Instantaneous overload relays are often used to prevent the current from exceeding the maximum current rating of instruments. A time-limit relay maintains the service of the line in which it is connected, without regard to any danger, for a certain limited period. A device of this kind allows all the less important lines to be cut out by the instantaneous or inverse-time-limit relays before it itself interrupts the main circuit. When the main current is interrupted it shows that the cause of interruption was not momentary or subject to recovery after a short time, such as strong currents caused by load variations or a burning-out short-circuit, and that the cutting out of the less important lines has not sufficiently relieved the main circuit. Synchronous converters are protected on the a.c. side by inserting time-limit relays on the high-tension side of the transformers. It is recommended by some engineers to insert them in the feeders on the sub-station side, their action being made independent of the current direction. A special type of relay is that mentioned above, in which the time before action is inversely proportional to the amount of overload, so that the greater the overload the quicker will be the action of the relay, being instantaneous at short-circuit. This relay may be adjusted for time variation from ½ second to about 30 seconds, delay depending on the time setting and the magnitude of the overload current.

Figure 118 is the ampere-time curve for a bellows-type overload relay. This curve was obtained by adjusting the device to operate with 5 amp. in 10 seconds. The current was then increased step by step, and the times for the relay to operate were noted. This type of device has the advantage of disconnecting the feeders consecutively, so that the feeder nearest the source of disturbance receiving the greatest amount of current disconnects first, thus relieving the other feeders and relays. If the relief is not sufficient, the next relay disconnects, etc. Another method of securing consecutive operation is obtained by adjusting the definite time-limit relays to different time elements, in such a way that the farther a relay is from the source of power, the seorter is the predetermined time element. Figure 119 shows the

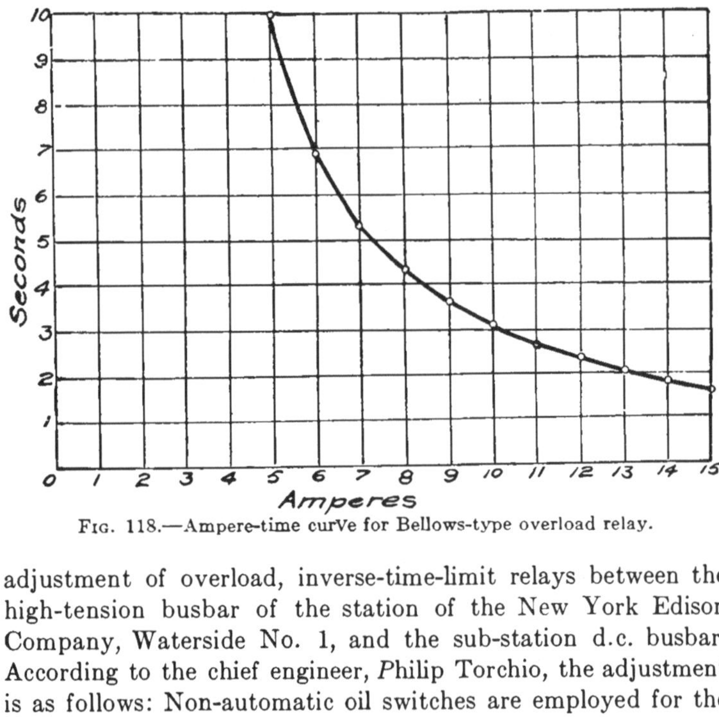

FIG. 118.—Ampere-time curve for Bellows-type overload relay.

adjustment of overload, inverse-time-limit relays between the high-tension busbar of the station of the New York Edison Company, Waterside No. 1, and the sub-station d.c. busbar. According to the chief engineer, Philip Torchio, the adjustment is as follows: Non-automatic oil switches are employed for the generators, as reliance is put upon the attendant to disconnect the generators by hand operation of the oil switches whenever he finds it necessary to do so. To guide him, an overload relay, operating signal lamps, is mounted on each generator. This relay is without time limit. On each high-tension feeder in the generating station is mounted an overload relay with a variable time limit in inverse proportion to the value of current. Curve 1 shows the characteristic curve of this type of relay, the periods at which it is adjusted, and its relation in time and load to other similar relays at the sub-station end of the feeder. The feeder switch in the sub-station is automatic and is controlled by a relay similar to the one at the Waterside end of the feeder. The adjustment is shown in curve 2. The high-tension side of the synchronous converters is equipped with a relay of the same kind as previously discussed, the calibration being shown in curve 3.

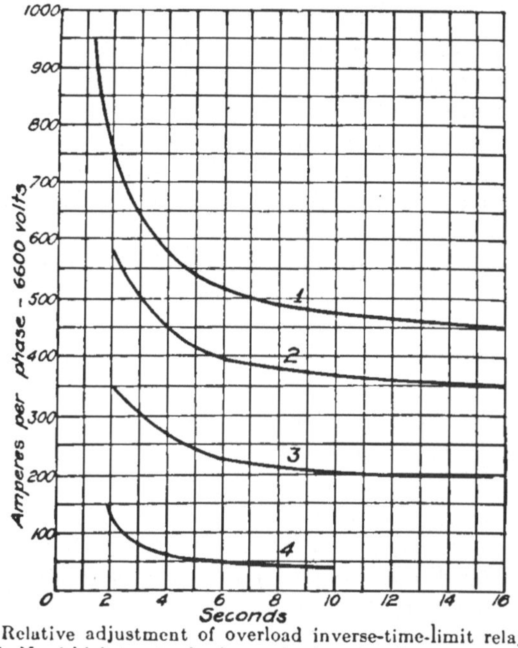

Fig. 119.—Relative adjustment of overload inverse-time-limit relays between Waterside No. 1 high-tension busbar and substation direct-current busbar.

The d.c. side of the synchronous converter is equipped with a reverse-current, inverse-time-element, direct-current relay, the adjustment of which is shown in curve 4, which is figured on the basis of primary amperes for the purpose of comparison with the curves above it. The consecutive operation of the relays is plainly evident from a comparison of the four curves.

Another use of the definite time-element relay is in relieving oil switches. It is well known that the ultimate breaking capacity of an oil switch depends on time of breaking. If a ground or short-circuit can persist for 2 seconds before the circuit is opened, the ultimate breaking capacity is approximately doubled, that is, the strain of the switch is reduced to half owing to the drop in voltage and current occurring during the interval.

2. A reverse-current relay is one which acts on reversal of energy flow. It consists of two windings, in series and shunt with the line respectively, so that one depends on the current and the other on the voltage of the line. Under normal conditions of the line only the difference in the magnetomotive force of the windings acts. On reversal of energy, however, the sum of the m.m.f. of the two windings comes into action, and the relay operates the switch. The ideal device of this type should possess the following characteristics: It should operate on overload at normal pressure, on short-circuit at zero or reduced pressure, or when the direction of the flow of energy is reversed.

It should have the additional feature of an inverse-time-element action so as to exercise a selective influence and trip out the circuit in trouble without interrupting those circuits which are not affected, and also to prevent interruption of service due to the flow of synchronizing currents between machines or to overload currents due to temporary grounds or short-circuits.

As a matter of fact it meets only some of the above conditions, and these only to a certain degree. The application of such relays therefore becomes limited. They are used here and there in the operation of generator oil switches to protect against damage due to reversal of power when accidents happen to generators or prime movers or through mistakes in synchronizing, or they are used at the sub-station ends of feeders, but they are not reliable. The disadvantage is that, being dependent upon the e.m.f. supplied from the line, a condition will arise at a time of severe short-circuits when the e.m.f. will drop to a very low value, in which case the relay will lose its reverse feature and operate as an overload relay with a high setting. This brings about the action of all overload relays on parallel circuits, causing a shut-down. This disadvantage was sufficient to cause the Interborough Rapid Transit Company, of New York City, to replace its instantaneous "differential" relays in the sub-stations of the Manhattan division by straight overload time-limit relays. In other cases inverse time-limit relays have been recommended to take their place.

In systems where a number of power houses feed a common high-tension underground network of feeders, or where the sub-stations are interconnected by means of an underground ring network, a protective arrangement called the differential or Merz-Price system is applied. In this arrangement a differential relay with a corresponding oil switch is used at each end of each cable. A series transformer is inserted in each phase on both ends of the cable. The secondaries of the transformers on both sides of the cable are so interconnected by means of an auxiliary cable that no current flows through this auxiliary cable as long as the main cable remains intact. As soon as a short-circuit occurs on the main cable, current flows in the auxiliary line and energizes the differential relays built in the line.

The relays actuate the corresponding oil switches to cut out the faulty cable on both ends at the same moment.

In plants in which two or more banks of transformers are operated in parallel between high and low-voltage buses it is desirable that each transformer bank have an autocratic oil-switch equipment which will act selectively and disconnect only the banks in which trouble may occur. With an oil switch on each side of each transformer bank, selective action may be secured by means of an instantaneous differential relay connected to the secondaries of series transformers installed on both the high- and low-voltage sides of each transformer bank, as described before in connection with the Merz-Price system.

The relay has to operate instantly on a low-current reversal on either side of the bank.

It is recommended that there should be used for installations with only two outgoing lines in parallel, and having the same load conditions in both lines, a selective reverse-current relay acting instantaneously under short-circuit conditions, thus obtaining a selective tripping of the oil switches. This relay is similar to the previously mentioned reverse-current relay and is connected to the secondaries of series transformers in each high-voltage line and potential transformers in the low-voltage bus.

A relay has been devised for use in connection with automatic TA Tirrill regulators (see Chap. XVII) for minimizing the effect of short-circuits and possible voltage rises in the transmission system. If a voltage regulator is weak and a short-circuit should occur somewhere on the system, the action of the regulator would naturally be to deliver the maximum excitation so as to keep up the voltage of the system. This in turn necessitates that the governors of the prime movers be wide open, and if the short-circuit should be suddenly relieved, the voltage often rises to very high values owing to the time element involved in closing the governors and in demagnetizing the field. The relay is provided with a current coil and a potential coil and will automatically reduce the excitation in case of excessive loads, high voltages, or any other cause tending to increase the voltage.

3. Reverse-phase relays are used to open motor switches not properly connected. They are useful for the protection of elevator motors or in any case where change of phase rotation is objectionable.

4. The object of the underload relay is to throw out one or more machines for load values under an economic load factor at which the remaining machines can run economically.

Underload relays are often used to trip the automatic oil switch that is placed in the primary of arc-lighting circuits to prevent an abnormal rise of secondary voltage in case of a breakdown in the secondary circuit. This relay acts on a decrease of current.

5. Low-voltage relays are used for motor switches, to insure proper connections in starting. They close the tripping circuit of the oil switch whenever the voltage decreases to approximately 50 per cent. below normal. Low-voltage relays are often used with motor-booster sets to prevent the disastrous speed of the booster which might result from the loss of alternating-current power.

6. Over-voltage relays are applied in connection with storage batteries. They are also used sometimes to protect induction motors against injury from high-voltage conditions.

7. The d.c. overload relays are employed when several circuits are controlled by one shunt-trip or overload and shunt-trip carbon-break circuit breaker. Each relay is selected to suit its particular circuit, and any relay will trip the main breaker. These relays are built in the main circuit.

8. The d.c. underload relay closes the secondary tripping circuit at 25 per cent. of the normal rated load.

9. Automatic d.c. circuit breakers are often equipped with a reverse-current relay inserted directly on the circuit breaker or in the feeders. Its object is to prevent the synchronous converter from running away when its field circuit is opened, in which case it is used in connection with a speed-limit device. It is advisable to have the relays independently energized by a battery or exciter for, under short-circuit or overload, it may occur that the relay refuses to act on account of the low voltage. These relays are especially adapted to protection against current reversal in

an installation where the line is fed by storage batteries and converters.

When synchronizing machines to a system operating rotary converters, momentary and harmless corrective currents are liable to flow toward the rotaries on the d.c. side. In order to prevent interruption of the circuit by such flow, where reverse-current relays are present, it is necessary that the latter have a time element. This element may be of the inverse order to give quick interruption on overloads and short-circuits and to give a selective action so as to cut off affected areas.

The d.c. reverse-current relay may also be used to protect generators, storage batteries or main station buses from damage on reversal of current due to short-circuits, or from the grounding of machines or connections.

In general, it is built as a horseshoe-magnet frame with a shunt-wound armature pivoted between its poles. The magnet is mounted on the current-carrying stud of the circuit breaker and is excited by the current flowing through the stud, and the armature is connected across the line in series with suitable resistance.

Rotation of the armature in the normal direction is prevented by a stop. Reversal of the current flowing through the stud changes the direction in which the armature tends to rotate, thereby causing it to move away from the stop, close the circuit through an auxiliary trip coil, and trip the circuit breaker.

10. Low-voltage relays for direct current are similar to the same type for alternating current. They release at approximately one-half rated voltage.

11. Overvoltage relays are similar in construction to a.c. overload relays and may be applied when automatic protection against high voltage is desired.

12. In many power plants d.c. three-wire power service is furnished by high-voltage two-wire generators operating in connection with balancer sets consisting of two low-voltage machines on a common shaft. With this combination of machines a short-circuit or heavy overload on one side of the system will shift the neutral considerably, and the lamps on the opposite side may burn out. To protect the lamps, a differential relay which operates on 15 volts unbalancing is commonly recommended. This relay is connected to trip either the generator circuit breaker or a circuit breaker connected in the

bus between the balancer set and the other generators. (See also Chap. VI, case 3.)

13. Solenoids for operating large oil switches, circuit breakers, field switches, etc., frequently require large operating currents in the closing coils. This necessitates the use of relatively heavy

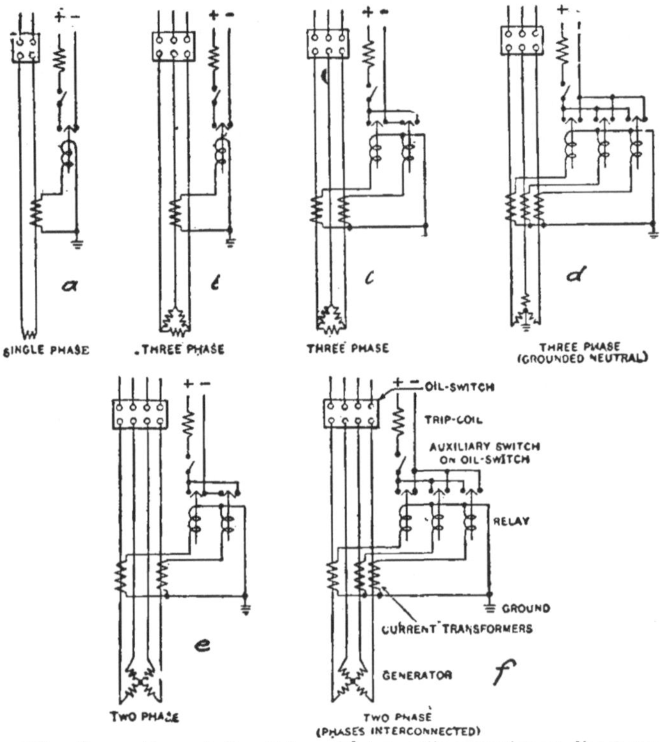

FIG. 120.—Connections of oil switches with trip coils operating on direct-current circuit using circuit-closing relays.

leads between the control switch and the solenoid. Severe arcing at the control switch may therefore be caused. The use of a control relay eliminates these objectionable features.

14. To auxiliary relays we may add the time-limit relay which is used with an instantaneous overload relay or reverse-current relay to introduce a definite time element between the occurrence of overload or reverse current and the opening of the circuit. This relay is operated by the d.c. auxiliary circuit.

A relay switch is introduced when the current required to open an oil switch is greater than the current-carrying capacity of the primary relay used. This relay switch has ample carrying capacity and is energized from a d.c. circuit.

A bell relay indicates when an oil switch has operated by energizing a gong circuit and maintaining that circuit until the same is opened by the switchboard attendant. It does not operate, however, when the oil switch is opened intentionally either by hand or by electrical remote control as the windings of this relay are in series with the automatic trip circuit. Many large power companies require more or less elaborate signaling systems between the switchboard and the engines. Sometimes lamps are used for the different signals and bells or whistles are added to call the attention of the operator to the fact that signals are being made.

In regard to the action of a.c. relays, they may either close a d.c. auxiliary circuit for energizing the trip coils or they may open an a.c. auxiliary circuit which under normal conditions is kept short-circuited by the relay. In the first case the auxiliary d.c. is independent of the main current, its source being a battery, exciter, or d.c. generator, of 125–250 volts. In the second the current is supplied by the secondary winding of a series or potential transformer, inserted in the main line, which at the same time operates the relay.

In Fig. 120 there are shown the connections of oil switches with trip coils operating on a d.c. circuit using circuit-closing relays. The number of poles corresponds to the number of transformers in the line, which is in turn dependent upon the number of phases of the system, upon the balanced condition of the load, and upon the kind of armature winding connections (with or without grounded neutral), as has already been mentioned under Fig. 87. Figure 121 is the same as the corresponding diagram of Fig. 87 with the exception of the relay installation, which under normal conditions keeps the secondary transformer winding short-circuited, opening the short-circuit only at overload, when it excites the trip coil.

Figures 120 and 121 show the switching arrangements of automatic oil switches operated manually.

Single-pole relays are generally used on single-phase and on balanced three-phase circuits. Double-pole relays may be used on ungrounded three-phase and on quarter-phase. Triple-pole relays are sometimes applied on three-phase grounded neutral and interconnected quarter-phase.

Polyphase maximum relays operated by the current in two or three of the phases should, however, be used as seldom as possible, and instead single-phase relays connected entirely separately in two or three of the phases on three- or four-wire systems respec-

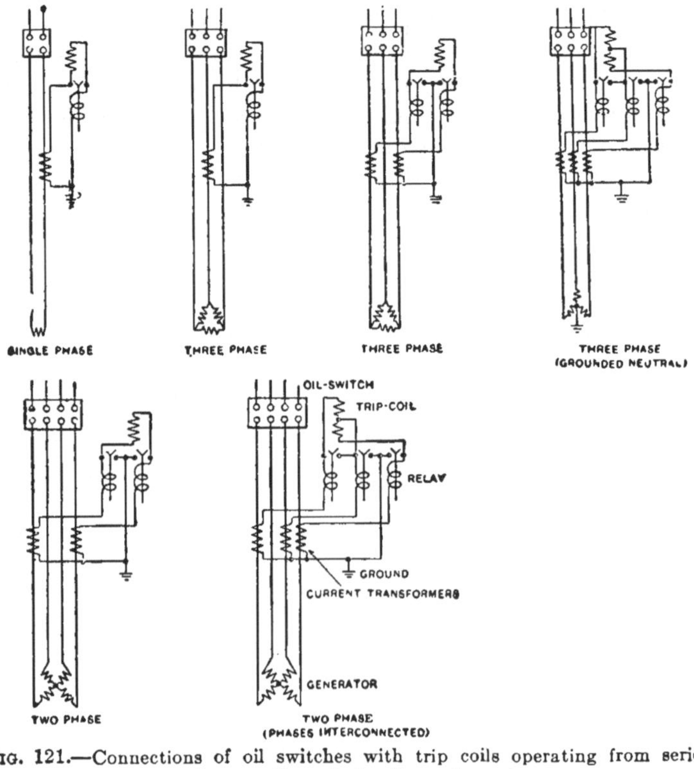

FIG. 121.—Connections of oil switches with trip coils operating from series transformers through circuit-opening relays.

tively should be used. With the former very large dangerous overloads can continue to exist in any one phase without causing the relay to operate and open the circuit. The same objections apply to polyphase reverse-current relays.

One of the wrong connections often met is that where the so-called "resultant scheme" of relay connection is in use, namely, one single-phase relay connected to two series transformers so that current in the relay is the vectorial resultant of the current in each of the phases in which the series transformers are situated, assuming a three-wire, three-phase system. Assume the normal full-load secondary current of the transformer to be 10 amp. and that the relay be set to operate at 200 per cent. of full-load current; that is, it will operate at 34.6 amp. If overload should occur on both phases connected through the transformers to the relays and the phase displacement in each phase is the same, then the circuit is opened when there is 20 amp. secondary current in each

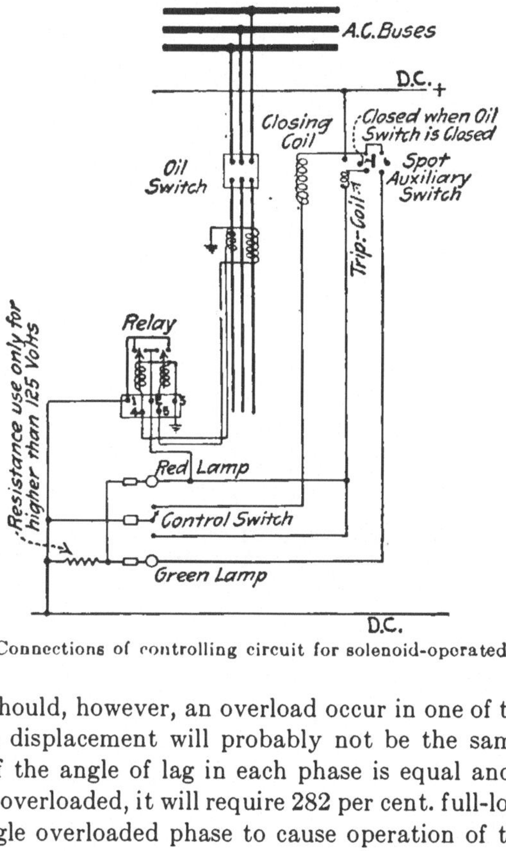

FIG. 122.—Connections of controlling circuit for solenoid-operated oil switches.

phase. Should, however, an overload occur in one of the phases, the phase displacement will probably not be the same in each phase. If the angle of lag in each phase is equal and only one of them is overloaded, it will require 282 per cent. full-load current in the single overloaded phase to cause operation of the switch. Should the current in the single overloaded phase lag 60° more than the other phase, then the current in this overloaded phase must increase to 386 per cent. of full-load current before the relay actuates the switch. The protection therefore is quite uncertain. Similar conditions may occur in a four-wire, three-phase system when three series transformers with only two single-phase relays are used. It is quite possible for one phase to be overloaded, since it has a large return path through the fourth wire. If this overload has a large phase displacement relative to the currents in other phases, then the condition may arise of very large overloads in the one phase and the switch not opening; consequently the same number of single-phase relays as of series transformers should be used, with four wires, three transformers and three relays, and with three wires, two transformers and two relays.

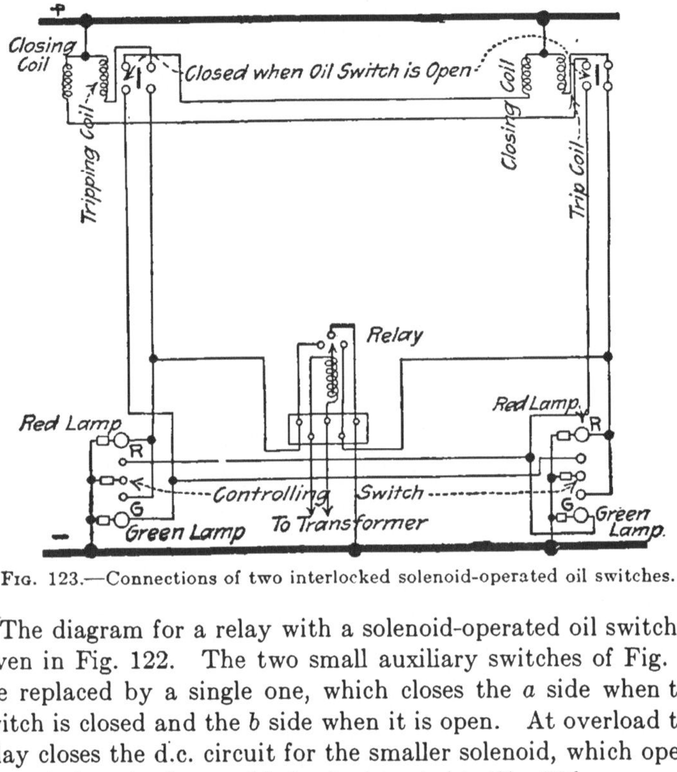

FIG. 123.—Connections of two interlocked solenoid-operated oil switches.

The diagram for a relay with a solenoid-operated oil switch is given in Fig. 122. The two small auxiliary switches of Fig. 96 are replaced by a single one, which closes the *a* side when the switch is closed and the *b* side when it is open. At overload the relay closes the d.c. circuit for the smaller solenoid, which opens the switch. A relay could also be inserted in Fig. 96 by connecting terminal *1* of the relay to any point between the red-lamp terminals and switch *2*, and terminal *2* of the relay to the negative side of the d.c. source.

Figure 123 is a diagram for a single-pole relay with two electrically interlocked solenoid-operated oil switches. Figure 124 shows the connections of a d.c. motor with a relay for the H3 and H4 oil switches. These switches are opened by a spring released by a trip coil, which is connected to the circuit by the controlling switch on the board, or which may be excited automatically by the relay.

Figure 125 shows a three-pole circuit-opening reverse-current relay which opens the short-circuited secondary winding of the series transformer at the moment of current reversal and admits current to the trip coil. P_1, P_2, and P_3 are connected to the

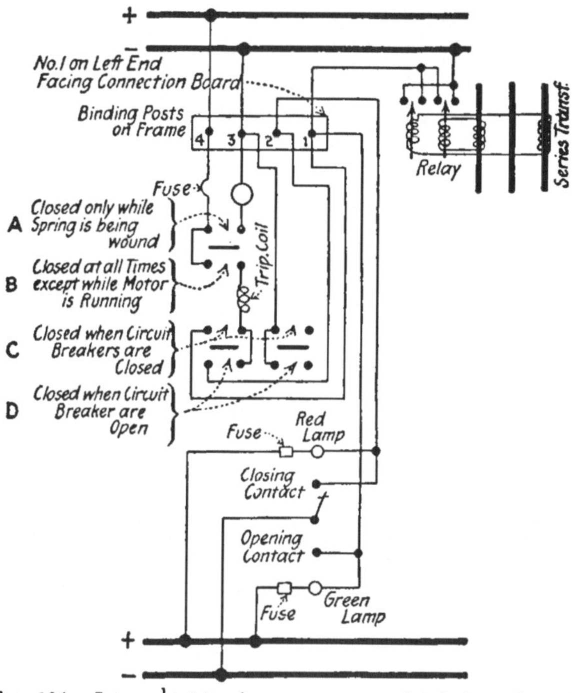

FIG. 124.—Internal wiring for motor-operated switch mechanism.

secondary windings of shunt transformers, C_1, C_2, and C_3 to the secondary windings of series transformers, and A_1 and A_2 to the trip coil. Figure 126 is a wiring diagram for a double-pole circuit-closing reverse-current relay with a separate d.c. source for the trip coil. The connections of A, C, and P are the same as those in Fig. 125. A d.c. differential relay is shown in Fig. 127.

M. C. Ripinski in the *Electrical World* for Nov. 2, 1907, gives the following diagram for the connection or relays for a three-phase system: Three generators with Y armature windings and neutrals grounded through resistances supply a set of bus-bars from which four feeders lead to the sub-station, each two feeders supplying one set of sub-station busbars. One set of busbars supplies a low-tension set of busbars, through step-down transformers, which in turn supply a.c. feeders. The

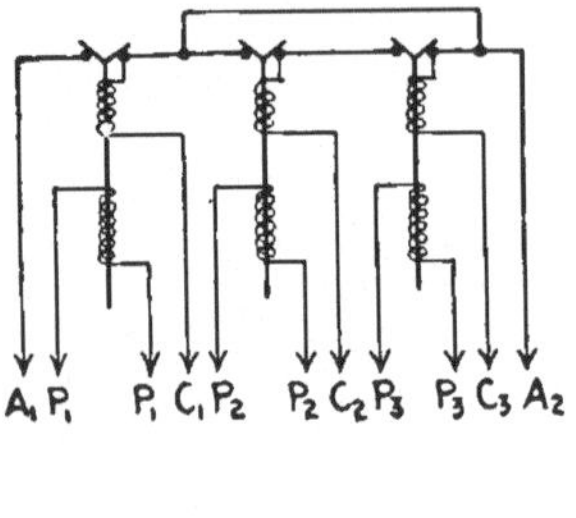

FIG. 125.—Three-pole circuit-opening relay.

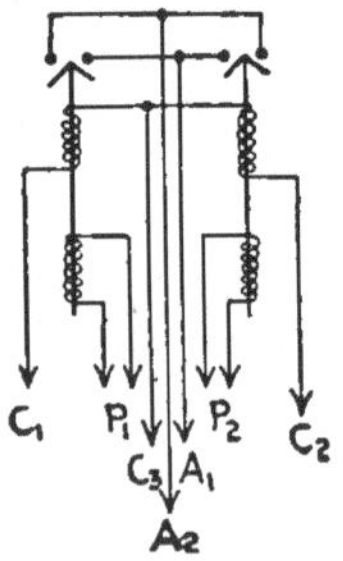

FIG. 126.—Two-pole circuit-closing relay.

current from the other busbar set feeds converters through transformer sets which deliver d.c. for power distribution. (See Fig. 128.)

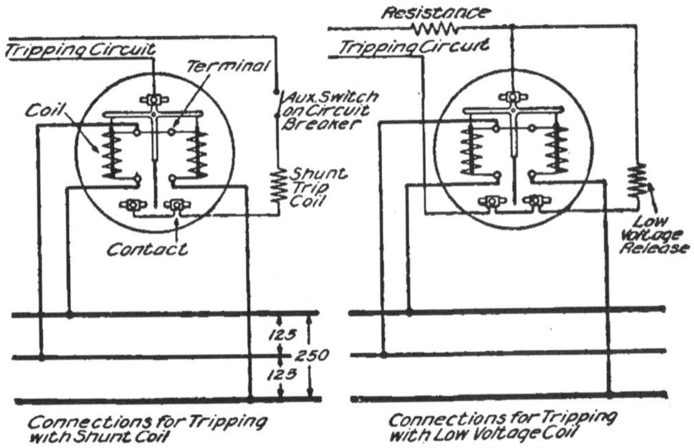

FIG. 127.—Direct-current differential relays.

Overload reverse-power inverse-time-limit relays operate at the moment of overload or short-circuit for the normal direction of energy flow. In case of reversal of energy in a feeder this relay operates as a reverse-current relay for a current strength from one-third to one-eighth of the overload current for which it is set. Let us consider three feeders leading from central station A to sub-station B. (See Fig. 129.) In case of short-circuit at point X there will be more current through switch D than through

E or *F*, because the short-circuit is nearest to *D*. Therefore, the overload current will open this switch by means of its overload relay. At the same instant, however, there is a rush of current from the sub-station toward X, which operates the reverse-current relay *P* and opens its corresponding switch. In *Q* and *R*, and also in *E* and *F*, very large currents exist in the normal

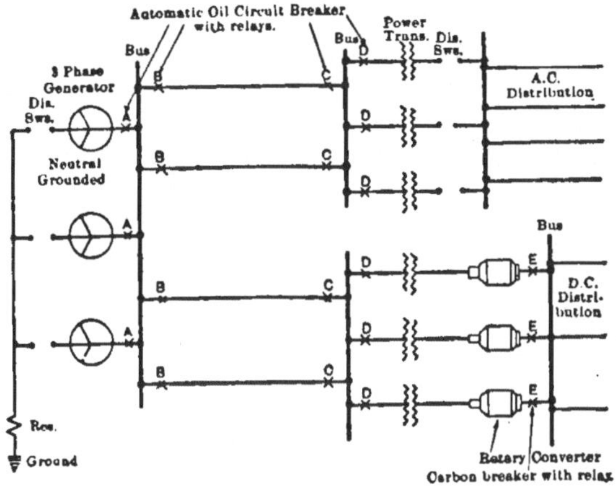

FIG. 128.—Relays in a four-wire, three-phase system. *A* and *C*, alternating-current overload reverse-power inverse-time-limit relays; *B* and *D*, alternating-current overload inverse-time-limit relays; *E*, direct-current reverse-power inverse-time-limit relays.

direction, but *D* and *P* are opened more quickly because of the still greater current through *D*, and the action of the reverse current through *P*, which suffices to open *P* at one-third the value of the overload current. This action relieves the other relays quickly enough to prevent their being opened. In this case the devices act as inverse-time-limit relays.

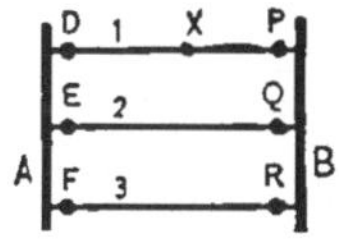

FIG. 129.—Connection of relays for a three-phase system.

In Fig. 130 we have a three-phase generator supplying energy to four outgoing feeders which feed four synchronous converters in the sub-station. In the sub-station the a.c. busbars are equipped with bus-sectionalizing switches, so that eventually the machines and feeders can be worked as independent units.

The generator oil switch is automatic and is operated by means of a differential relay. Should a short-circuit occur at the point X on the B feeder, there would be a rush of current from the

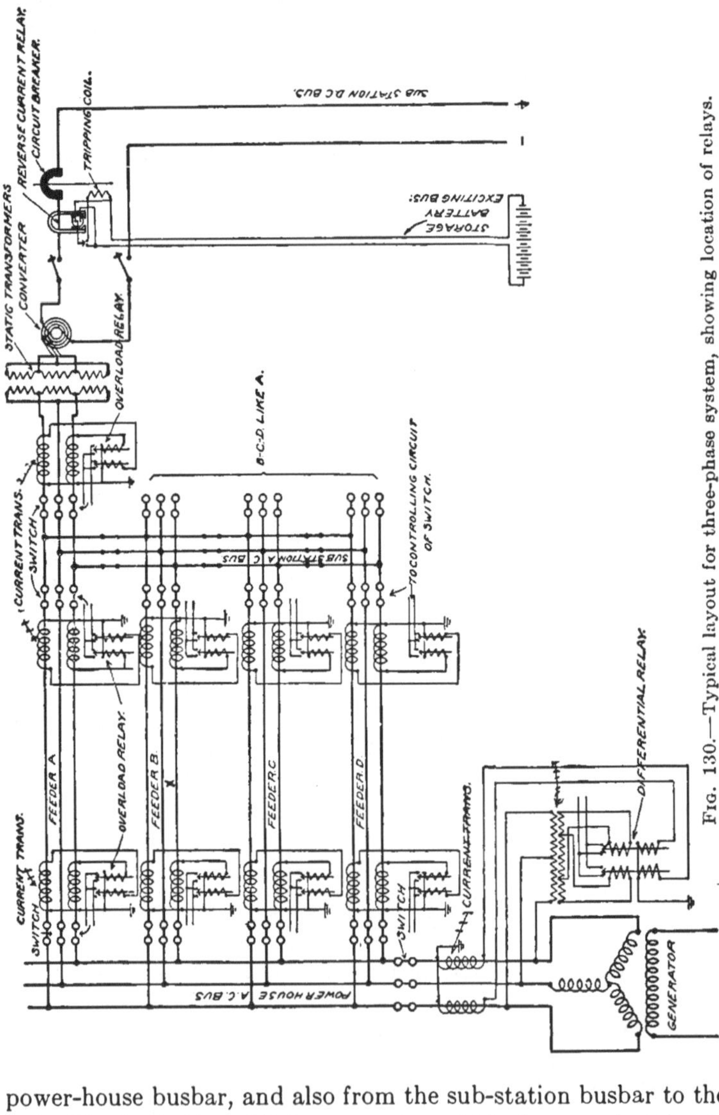

FIG. 130.—Typical layout for three-phase system, showing location of relays.

power-house busbar, and also from the sub-station busbar to the point X. In A, C, and D the current would be increased, owing to the short-circuit being fed by these feeders through the sub-station busbar. Since a synchronous converter generates an a.c. electromotive force corresponding to rated pressure at synchronous speed, in case the point X is near the power house, it is possible that the direction of the flow of energy in feeders A, C, and D would be reversed, due to the synchronous converters feeding the short-circuit back to the power-house busbar as well

FIG. 131.—High-tension series overload relay.

as directly over the short-circuited feeder. For this reason the switches on the sub-station side of the feeders should be protected by an instantaneous reverse-current relay, while the power house should be guarded by an overload time-limit relay. Since, as has already been mentioned, the reverse-current relays are dependent upon the pressure of the system, they would become practically inoperative at zero or low pressure, which exists at the moment of short-circuit. They therefore lose their reverse action and become worthless at the very instant when they are most necessary.

Since the relative value of the time element of relays is determined by the various conditions of service and load, it is impossible to formulate any empirical rules applicable to all cases.

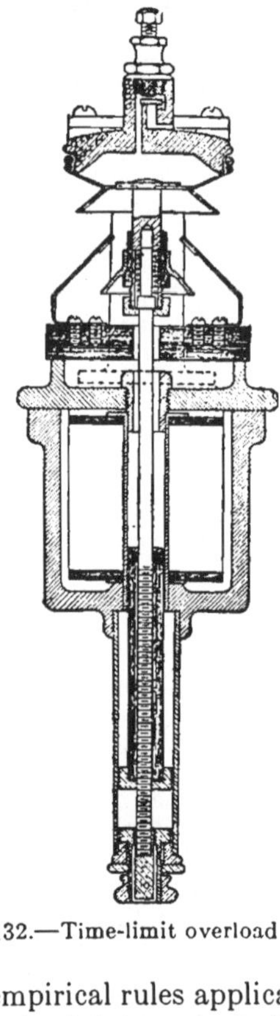

FIG. 132.—Time-limit overload relay.

The choice of the proper device is left to the judgment of the engineer, who must take into consideration the reliability of the various types available. In some instances the relays are used simply to operate tell-tales or signal lamps, by which the attention of the operator is called to the action of the corresponding switches or other apparatus.

A relay consists of a mechanism operated by a solenoid which is excited directly or indirectly by the current of the line to be guarded. The mechanism carries a set of contacts which open or close a corresponding set of fixed contacts as the case may be. A relay adjusted to a certain time element possesses a special mechanism to accomplish this adjustment. The exciting

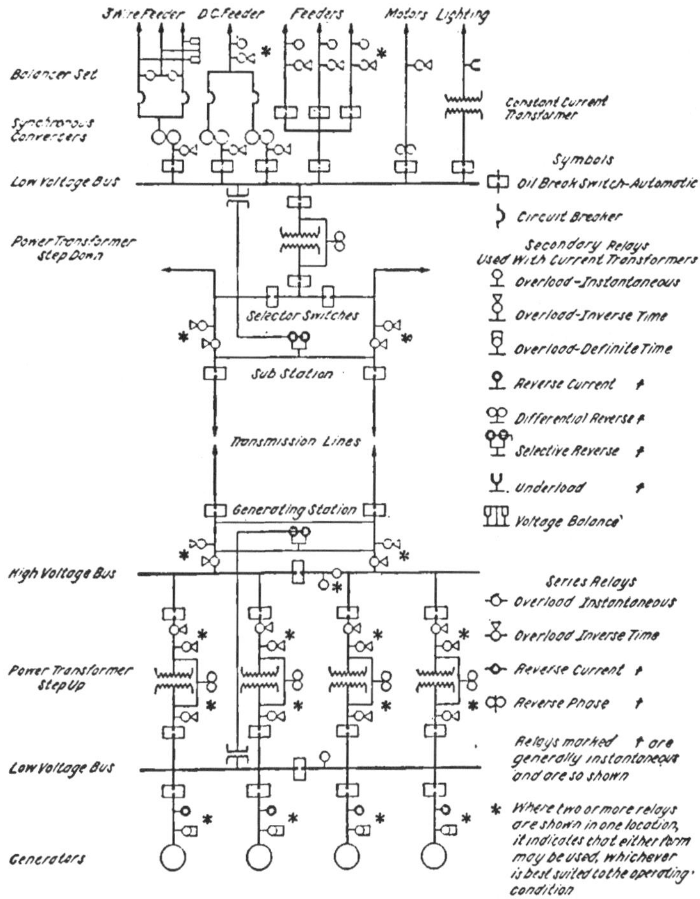

FIG. 133.—Typical system of connections for a large power house.

mechanism generally consists of a solenoid with a core, while in other cases it is a movable armature winding similar to that on a motor, or it may be constructed as a wattmeter. Still others are operated by clockwork. There are two pairs of metal contacts, one movable and the other fixed. The fixed contacts are either in shunt with the operating current, being closed under normal conditions, as is the case with an a.c. excited trip coil, or they are in series with the operating current, being open under normal conditions, as is the case with a trip coil excited from an independent d.c. source. The fixed contacts are bridged by the movable set.

Series relays are connected directly in series with the line and are used chiefly with high-tension oil switches for overload protection. By means of a specially treated wooden rod the relay operates a tripping switch, closing a separate tripping circuit. Figure 131 shows the arrangement of a high-tension series overload relay.

Figure 132 is a section through a single-pole time-limit overload circuit-closing relay, as made by the General Electric Company. The bellows on top are used to adjust the time element, by regulating with small set-screws the air passage from the bellows through a small channel at the moment of the upward motion of the solenoid core. By combining two or three single-pole relays we may obtain a double- or triple-pole apparatus. Instantaneous relays are similarly constructed with omission of the bellows. An arrangement of oil switches with various types of relays for a large power station is given in Fig. 133.

CHAPTER XVII

POTENTIAL REGULATORS

A potential regulator is a device intended to maintain a constant e.m.f. of the generators, distributing feeders or transmission lines, independently of load or in accordance with the respective variations of load, but without disturbing the e.m.f. of other parts of the system. Three kinds of regulators are to be distinguished, the Tirrill and Thury regulators, which maintain a constant e.m.f. of the generators; the feeder regulators, which maintain a constant feeder e.m.f., and the synchronous condenser to regulate the e.m.f. of transmission lines.

Tirrill Regulators.—This apparatus regulates the e.m.f. of the generator by varying the field strength with the load. There are two types, of which the first regulates the voltage of shunt- and compound-wound machines, while the second regulates the pressure of machines requiring a separate exciter, the latter being applicable to both d.c. and a.c. machines. In the alternating-current systems the Tirrill regulator may also be used to regulate the power factor of the transmission line. Figure 134 shows the elementary connections of a regulator (type TD) for a compound-wound machine. The regulation is accomplished by automatically throwing the field rheostat on or off the circuit. This eliminates the variations in the machine voltage due to variations in load by decreasing or increasing the field strength. The mechanism consists of a main control magnet energized by current from the main busbars or from the center of distribution. In the first case it is desired to regulate the voltage on the busbars, while in the second case such regulation depends upon a given value of the pressure at the load center, whence we have direct or indirect voltage regulation. The magnet actuates a lever carrying a contact stud at its free end, opposite to a fixed contact. A spring opposes the action of the magnet, which is adjusted so that at the required voltage the lever will vibrate under the action of both magnet and spring. This vibration of the lever opens and closes the main contacts, which causes an opening and closing of a part of the relay contacts of a differential relay. The relay consists of two equivalent windings, wound opposed to each other, both being in shunt with the main current, but one being energized permanently, and the other only at the closing of the main contacts. When the main contacts are open, there is current through only one of the relay windings, the relay is magnetized, attracts an armature, and opens the relay contacts. With the closing of the main contacts, there is also current through the second relay winding, and since the windings are opposed

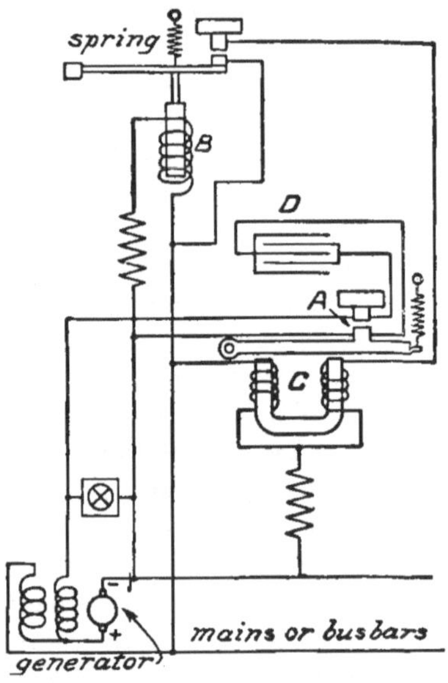

FIG. 134.—Diagram of connections of a Tirrill regulator for direct current.

to each other the relay becomes demagnetized and the armature closes the contacts under action of the spring. The opening and closing of the relay contacts causes the field rheostat to be thrown on or off, the time element of action for either closing or opening being dependent upon the momentary fall or rise of the pressure on the busbar or feeder as the case may be. In order to prevent arcing between the contacts, these are connected to a condenser set.

Figure 135 is a wiring diagram for a Tirrill regulator connected to two generators in parallel, which supply a three-feeder system. The size of the control resistance in the magnetic circuit is varied according to the required voltage. If a 125-volt generator, for instance, is to generate 110 volts, binding post 14 on the resistance box is joined to binding post 1 on the regulator. Tables made for each regulator give the binding posts to be used for the required voltage. An adjustment of the spring opposing the magnet must be made for the required voltage, so that when this pressure is reached the main contacts will vibrate properly. If the voltage at the load center or busbar drops, due to increase in load, the spring overcomes the magnetic pull and closes the main contacts, which results in demagnetizing the relay and closing the relay contacts that short-circuit the field resistance. The voltage is then allowed to rise until normal conditions again prevail. These fluctuations are quite rapid, so that as a matter of fact the voltage is kept at a constant value.

A Tirrill regulator (type TA), for regulating the voltage of separately excited machines, is similar in operation to type TD. in that it regulates the impressed exciter voltage with fluctuations

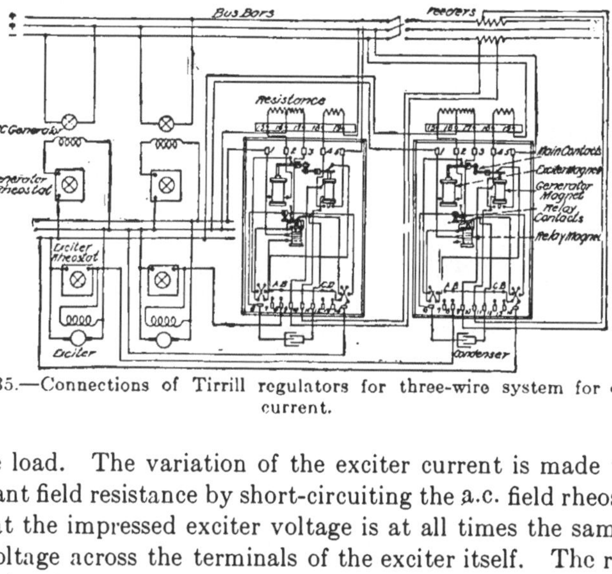

FIG. 135.—Connections of Tirrill regulators for three-wire system for direct current.

of the load. The variation of the exciter current is made with constant field resistance by short-circuiting the a.c. field rheostat, so that the impressed exciter voltage is at all times the same as the voltage across the terminals of the exciter itself. The regulator therefore influences the generator field indirectly, and the shunt field of the exciter directly, the last similar to the action of type *TD* on the generator field. A further similarity between this type and type *TD* is that it throws the shunt rheostat of the exciter on or off, by means of a main control magnet, main contacts, differential relay, and relay contacts. The two types differ in that the second main contact piece, instead of being fixed as before, is fastened in the TA type to a lever actuated from an a.c. solenoid core. Two windings, one a potential and the other a current winding, make up the a.c. magnet, which receives current from transformers joined to the busbars or feeders. (See Fig. 136.) With d.c. generators the transformers are of course omitted, the potential windings being connected directly to the busbars or load center, while the current winding is joined to the line through a resistance. Current is supplied to the main magnet by the exciter, and the magnet works on a lever on whose further arm one of the main contacts is fixed. Four springs, fastened to different points of the lever on the side of the main contact, oppose the action of the magnet. They come into action consecutively according to the position of the lever, that is, according to the value of the exciter voltage.

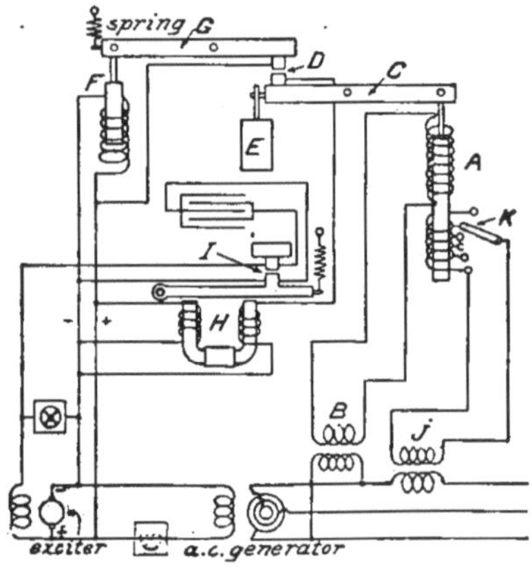

FIG. 136.—Diagram of connections of a Tirrill regulator for alternating current.

The adjustment and action of the springs and magnet is such that the rise of the contact piece is directly proportional to the change in the exciter voltage. Since the latter is in fact a fluctuating e.m.f. due to the variation of load on the generator, the lever with the main contact is set vibrating under this influence. The maximum amplitude of vibration is $\frac{1}{16}$ inch. The position of the upper contact is, therefore, determined by the exciter voltage. The winding of the a.c. magnet is supplied with a current whose value is at all times proportional to the generator voltage. The solenoid pulls the laminated core upward. This core is attached to a lever on whose outer arm a counterweight is applied, and the second stud of the main contact pair is also fastened on this end. The motion of the core is damped by an oil brake.

For a given generator voltage the position of the second contact is determined by the position of the core in the solenoid and the size of the counterweight. Since a different electromotive force is applied for each different position of the core in the solenoid, under the same generator voltage, a position of the core is chosen which will call the maximum electromotive force into action at the moment when the contacts are closed. The upper contact is kept in position by the exciter voltage, which corresponds under normal conditions to the required generator voltage. The counterweight is so chosen that it will exactly balance the weight of the core and the electromagnetic pull on the core at the moment the contacts are closed. The sum of the moments acting on the lever must be zero for the required generator voltage, so that the system is in equilibrium. As soon as the lever is changed from this position, the electromagnetic pull acting on the core is changed. It, in fact, becomes less, so that the moment due to the weight of the core predominates in all other positions of the core. The lever, therefore, is in unstable equilibrium for the voltage in question. If the load on the generator is changed from zero to maximum, and the exciter voltage varies at the same time from minimum to maximum, the upper contact is shifted about 1/16 inch. Since the solenoid lever remains in contact with the upper stud, and since the shifting of this stud is very slight, the lever may still be considered to be in equilibrium as long as the generator voltage remains constant. Should the generator voltage drop, the moment due to the weight of the core predominates, producing clockwise motion. If the generator voltage rises above normal value, the magnetic pull increases and the lever is under the influence of a counter-clockwise moment. It must be borne in mind that the solenoid lever was adjusted to maintain equilibrium for a given generator voltage.

The following example may serve to illustrate the action of the complete apparatus. Let it be assumed that the constant voltage of the system, as expressed in the voltage of the secondaries of the potential transformer, is 110 volts, and that the load and speed of the generator are constant, so that the vibration of the contacts is uniform. The core of the solenoid is set for the given voltage. Let us start with the time element during which the exciter field rheostat is short-circuited, as represented by Δt_1 in Fig. 137, that is, when the main contacts, and hence the relay contacts, are closed. During this interval the current D of the shunt-field winding of the exciter rises, causing the exciter voltage C to increase, which in turn increases the exciter current B and the effective generator voltage A. The values are changed, but are displaced in phase on account of self-inductance. The upper contact moves upward on account of the rise in exciter voltage. As long as the generator voltage remains below 110 volts the solenoid lever is subjected to a clockwise movement, causing the lower contact to rise also and press against the upper one. When the generator voltage reaches

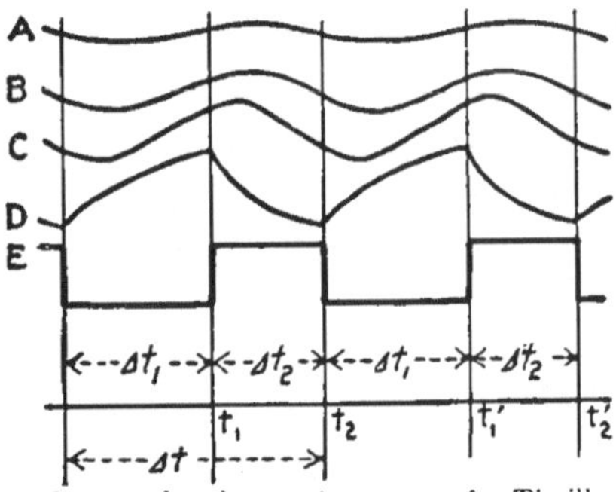

FIG. 137.—Curves showing performance of a Tirrill regulator.

110 volts or more, the solenoid lever commences to turn in an opposite direction, thereby opening the main contacts. This occurs at the point t_1, so that the exciter field rheostat is again thrown in. The upper contact continues to rise under the growing exciter voltage until the influence of the exciter field rheostat, now in circuit, becomes noticeable. The exciter voltage begins to fall off, and the upper contact reverses its movement. During this time the generator voltage continues to rise above 110 volts. The lower contact is lowered until the generator voltage reaches a maximum value and commences to fall under the influence of the falling exciter voltage. When this falling voltage reaches 110 volts and less, the motion of the solenoid lever is reversed, causing the lower contact to rise. The two contacts approach each other until they meet at the point t_2, when the exciter field resistance is short-circuited. The tendency of the contacts to continue their motion is counteracted by a flat spring to which the upper contact is attached. The spring is bent, and both contacts follow the same constrained path until they are again

separated. The motion of the mechanism remains constant as long as the load and speed of the generator remain unvaried. All of these motions are very rapid. The larger the maximum resistance of the exciter field circuit on throwing in the field rheostat and the smaller its minimum resistance on short-circuiting the field rheostat, the more rapid will be the motions. The more rapid are the fluctuations in the generator field and in the magnetic pull on the solenoid core, the more rapid will be the motion of the mechanism and the opening and closing of the

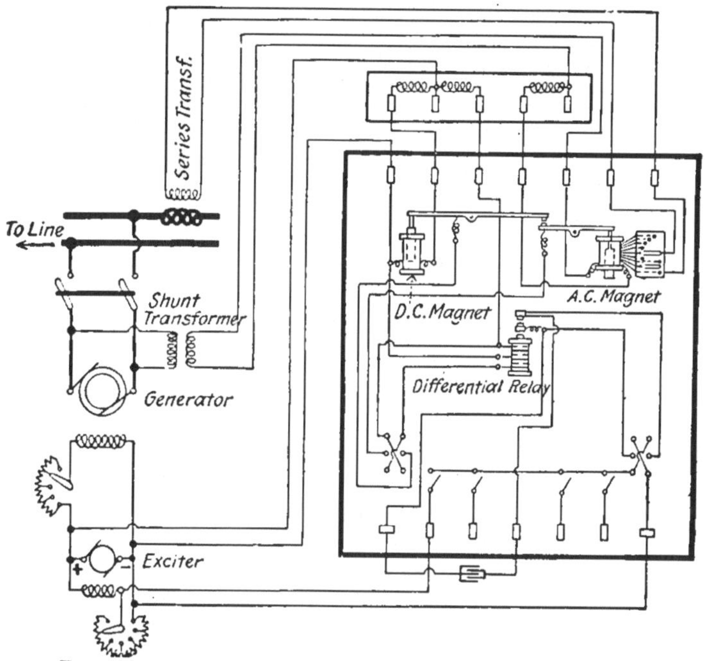

FIG. 138.—Connections of Tirrill regulator with one alternator.

contacts. The greater the frequency of the contact vibration and the less the time allowed for changes in the exciter excitation, however, the less will become the fluctuations of the exciter and generator voltages. And finally, the greater the frequency of the generator excitation, the more rapidly will the voltage fluctuations due to the load variations be removed. These conditions, therefore, call for as high a maximum resistance and as low a minimum resistance as possible in the exciter field. The minimum resistance is kept to a low value by short-circuiting the a.c. field rheostat. The maximum resistance, on the other hand, is obtained by setting the exciter field rheostat to a value which, when permanently connected to the exciter, will reduce the generator voltage to from 40 to 65 per cent. of its normal value. Any change in the load or speed of the generator will result in a disturbance of the uniformity of the contact vibration, which will last as long as the voltage fluctuation continues. For instance, under rise of load and fall of voltage, the solenoid

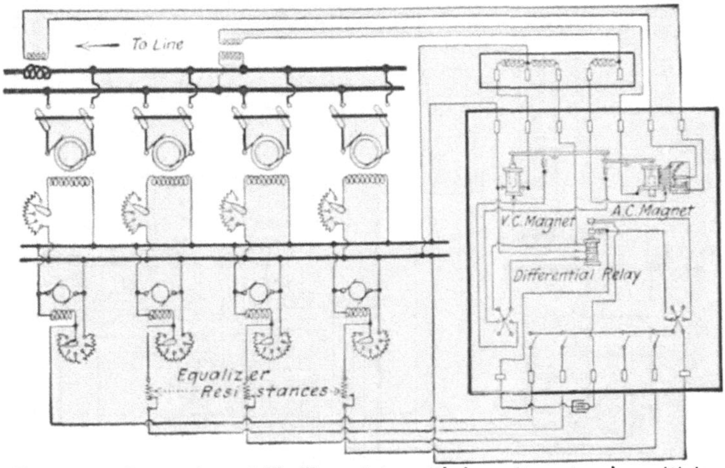

FIG. 139.—Connections of Tirrill regulator with four alternators in multiple.

lever moves in a clockwise direction, which closes the main contacts and short-circuits the exciter field rheostat. It is seen therefore that the regulator brings about an immediate rise in the exciter voltage so that its generator voltage reaches its normal value in the shortest possible time. As soon as the e.m.f. rises above 110 volts the contacts are opened and the normal vibration commences. The voltage fluctuation under normal conditions is so rapid and so small that the normal voltage of 110 may be considered practically constant.

Figure 138 shows a diagram for connecting a TA regulator to a generator, the voltage of the busbars being kept constant.

If the generator voltage is to be increased by a certain amount, although the solenoid lever is set for 110 volts, the series coils of the solenoid may be connected in series with the potential coils, so that the two windings are opposed to each other. This does away with the series transformer. Both windings are excited by the same current which before supplied the potential coils alone, with the result that the effective electromagnetic pull becomes less by an amount corresponding to the decrease in the number of effective ampere-turns. The decrease depends upon the number of series turns opposing shunt turns. The action of the decreased magnetic pull on the core and lever is the same as would be that of a decreased generator voltage. The result is an increase in this voltage of as great a percentage as the percentage decrease in the effective windings of the solenoid.

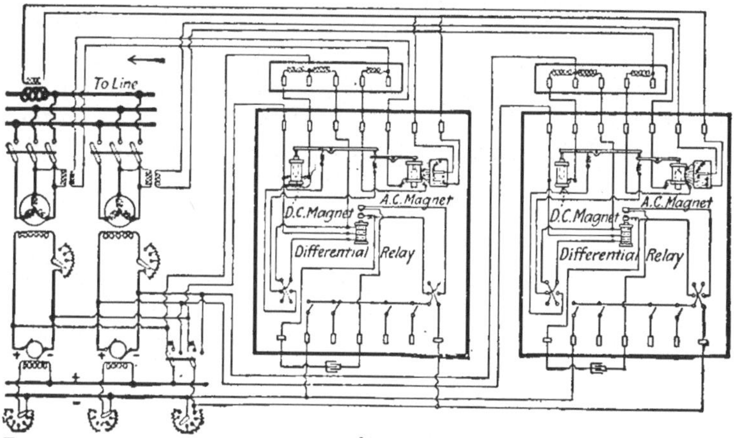

Fig. 140.—Connections of Tirrill regulator with two alternators and separate exciters.

If the voltage of the system or of a single feeder is to be kept constant, the current winding is joined to the secondary of a series transformer connected to the feeder in question. When the generators are in parallel and are excited by the same exciter, their voltage can be kept constant by a regulator joined to the exciter busbar, and also through transformers to the main bus.

Figure 139 shows a system of connections for a regulator with several generators and exciters. All of the control rheostats are connected in parallel with the relay contacts and are short-circuited simultaneously. Any one of the exciters can be disconnected from the regulator at will. Resistances are introduced into the regulator leads in order to prevent equalizing currents. If it is not desired to connect the exciters in parallel, but that each exciter work directly on the field of its generator, then the field windings of the exciters are connected in parallel with each other, and also in series with a common resistance, which is influenced by a Tirrill regulator. (See Fig. 140.) Only one of the two regulators is in use at any given time, the other serving as a reserve.

For exciters with greater shunt-field currents Tirrill regulators are used with a number of relays and corresponding switches, to avoid sparking at the relay contacts. The exciter rheostat is then subdivided and the subdivision points are joined to the contacts of the different relays.

On long transmission lines where synchronous motor-generator sets are used, the generator supplying direct current for a railway load, the Tirrill regulator may be employed as a power-factor regulator, if it is desired to hold a leading power factor on the motors at all times irrespective of what the load may be on the generator. This regulator is designed with two stationary coils and one moving current coil on the a.c. solenoid.

The current of the moving coil is 120° out of phase with the potential at unity power factor. Should the power factor tend to change in either direction, the excitation of the synchronous motor is increased or decreased proportionally, so that the power factor on the motor or line will be held constant at any desired value. For holding a leading power factor the adjustable coil is placed at a suitable angle to the potential coil, so that with this regulator a power factor anywhere from 80 per cent. leading to 80 per cent. lagging can be maintained.

The Thury Regulators.—In Europe the Thury automatic regulator has come into use. In this device the power of an independent motor actuates the field rheostat of a generator and thus controls the voltage of the machine, constituting practically an electrically operated mechanical relay. By means of an electromagnetic balance it communicates to the regulating switch a rotary motion in one direction or the other, from a rocking arm which is kept in motion by an independent motor.

The a.c. type (Fig. 141) consists of an electromagnet *F* with a laminated core and of a moving coil *B* attached to the balance lever *E*. The lever is pivoted at its center and carries at one end the adjusting spring *A* and the compensating spring *R* and at the other a tempered steel step *C*. The current in the coils tends to lift the moving parts.

A non-inductive resistance in series with the magnet-winding is generally used to avoid irregularities due to variation of frequency. A dash-pot over the lever E is used for damping vibrations set up by the current.

The rocking motion from which the power is derived for operating the regulating switch is produced by an eccentric on a shaft driven by an independent motor. On the rocking arm D are mounted the tappets K and K' and the pawls I and I', by which the electromagnetic balance utilizes the oscillation of the rocking arm to move the toothed wheel H, which is coupled to the regulating device.

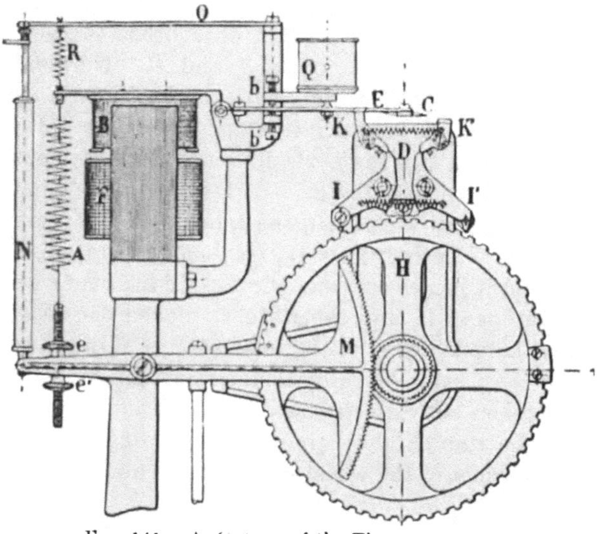

FIG. 141.—A. C. type of the Thury regulator.

For controlling the output of electrical machines the apparatus drives, either directly or through transmission gear, a switch arm passing over contacts on a dial between which are connected suitable resistances to regulate the excitation of the machine.

When a current of normal strength flows through the magnet coil, and also through the moving coil which is in series with it, the latter is in the midway position and so is the stop C. The balance is in equilibrium between the magnetic force exerted and the pull of the spring A. When the current increases, the magnetic force tends to lift the moving coil against the pull of the spring until it is in its highest position and the stop C is in its lowest. On the other hand, a diminution of current strength allows the spring A to pull the coil into its lowest position, raising stop C to its highest. The moving of the lever is limited by a couple of adjusting screws.

In its lowest position the stop C comes opposite the tappet K', and in its highest position opposite the other tappet K; while in the normal position of equilibrium it misses both tappets. These several positions of the stop bring about the adjustment of the regulating switch by making use of the motion of the rocking arm to rotate the wheel H slightly one way or the other.

The rocking arm, which is kept in motion by an eccentric driven by a motor, communicates this motion as an impulse or series of impulses in one direction or another to the toothed wheel H by means of the pawls I and I'. Normally these pawls are held up out of contact with the teeth of the wheel H; if one of the pawls is released, it drops into a gap between the teeth of the wheel, and the next motion of the rocker pushes the wheel round to the next tooth.

When the current is normal, the tappets on the rocking arm miss the stop C entirely; but if the current increases, the stop C drops until it comes opposite the tappet K', which strikes it, releasing the pawl I′, and the wheel is moved one tooth to the right with the next swing of the rocking arm. On the return of the swing, the pawl is lifted by a stationary pin fixed below the spring stretched between the pawls. If the increase of current still holds the stop C down, the following swing brings about a further movement of the wheel H, and so on until the regulation is complete.

The travel of the toothed wheel is limited by clamps fixed on the side of the rim which engage with rollers on the pawls and hold them up as soon as that point is reached, so preventing them from pushing the wheel round any farther.

To throw the regulator out of service, it is sufficient to stop the rocking motion of the pawls or to move a small pivot fork which is fixed on the regulator bracket in front of the screws b and b'. This fork holds the lever E in its middle position.

This regulator is fully competent to deal with non-inductive circuits, accumulators and lighting feeders, etc.

For regulation of electrical machinery and circuits containing self-induction, a corrective device is necessary to avoid the possibility of hunting due to the lag which results from the self-induction. The movement of the regulating switch is accordingly transmitted to the electromagnetic balance by means of the toothed sector *M*, which works a lever pivoted at a point on the supporting column. This lever is connected to the flat spring *O*, and, as a result, every time the switch moves it tends to oppose the further movement of the balance.

This flat spring is deflected or released as the regulating device is moved one way or the other. When it is deflected it slackens the compensating spring *R*, allowing the spring *A* to exert more pull on the balance to bring it back into the position of equilibrium and *vice versa*. The result of this is that whenever the regulating switch moves it modifies the conditions governing the balance and tends to prevent the movement being prolonged beyond what is required. The end of the flat spring *O* is not connected rigidly to the sector lever *M*, but only through the long dash-pot *N*, which allows the flat spring gradually to regain its normal position after being deflected either way. The piston of the dash-pot has an adjustable opening to allow the more or less rapid flow of the oil.

The rocking arm makes 150 complete oscillations a minute, thus turning the toothed wheel *H* by one tooth in 0.4 second. The regulating switch possesses as many contacts as the case on hand requires, for example, 40 contacts for lighting or power work. As the contacts are spread over 240°, the switch arm turns through its complete range in 16 seconds.

The sensitiveness of regulation is limited only by the subdivision of the rheostat. It depends on the degree of elasticity of the springs *A* and *R*. To render the apparatus less sensitive the number of turns of *A* can be diminished and *vice versa.*

The Thury regulator permits very close regulation, independently of the force exerted in obtaining it. This regulation is operated mechanically throughout, no electrical contacts such as those of the Tirrill regulator being required.

The regulator can maintain the voltage of a system constant:

1. By varying the resistance of the exciting circuit to the voltage of the machine which supplies the exciting current.
2. By raising or lowering the pressure at the central station to compensate for drop in feeders.
3. By keeping the voltage of a motor-generator constant independently of variations in the supply pressure.
4. By altering the transforming ratio of a.c. transformers.
5. By varying the number of cells in a secondary battery.
6. By keeping the voltage constant in a battery booster circuit.

This type of regulator is also used to maintain constant current in arc lamps or series motor circuits. It can automatically regulate electric furnaces, etc.

Figure 142 shows the diagram of connections for a Thury regulator in a d.c. circuit. The d.c. regulator varies only in insignificant details from the a.c. regulator. The resistance *RA* is used to absorb the excess pressure of the system to be regulated, the current then passing through the magnet winding *F* and the moving coil *B*, which are connected all in series across the generator terminals.

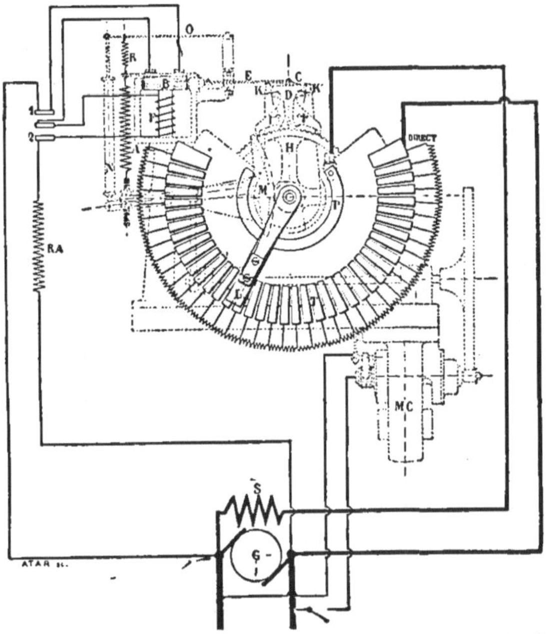

FIG. 142.—Diagram of connections for a Thury regulator in a D. C. circuit.

If the regulation is required at the center of the distribution system, the magnet coil is fed by a voltmeter pilot wire from the point in question. When the distance between the generator and the center of the system is too great to permit the expense of pilot wires, a special magnet winding can be applied to compensate at any moment for the loss in the feeder. This superimposed winding placed in a reverse direction on the magnet is fed from the terminals of a resistance shunt inserted in the feeder. The more the superimposed winding opposes the field of the electromagnet, as the load on the generator increases, the more voltage will the moving coil require in order to raise the balance lever.

If the current instead of the voltage is to be kept constant, the magnet winding of the regulator must either be connected in the main circuit or it can be shunted by a suitable resistance.

Figure 143 shows the diagram of connections of a Thury a.c. regulator. The regulation is generally affected by varying the resistance in the field circuit *S* of the alternator; also by putting the rheostat in the field circuit of the exciter. In the case of

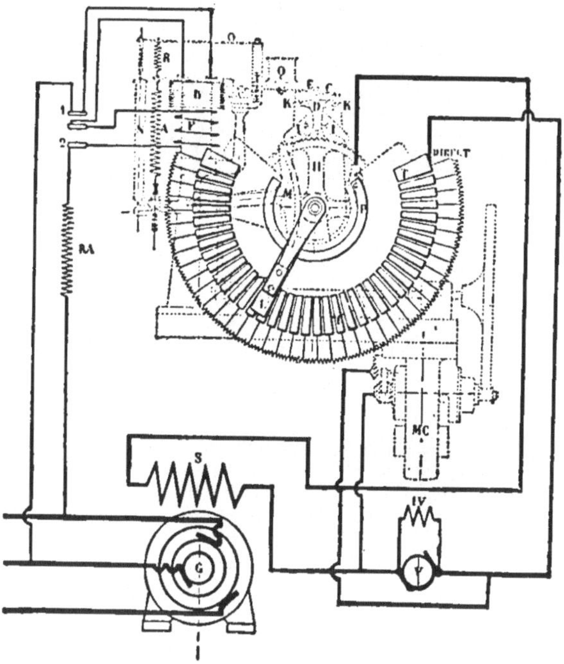

FIG. 143.—Diagram of connections for a Thury regulator in an A. C. circuit.

high-tension alternators the regulator is connected through a small transformer. One phase is usually regulated and this phase is selected for electric lighting. Voltage regulation at the end of a feeder is obtained, as in the case of the d.c. type, by providing a compound winding on the magnet coil.

In case a traction system also supplies lighting to a small town or private estate, the Thury regulator in connection with a motor-generator can be used to reduce the pressure to 110 or 220 volts and to deaden the speed changes produced by pressure fluctuations. Slight momentary variations are thus eliminated.

With several generators in parallel, automatic regulation can be effected either on each machine or simultaneously on the group. In the first case, each generator is provided with its own

Fig. 144.—D. C. Thury regulator mounted on a rheostat.

regulator and all work simultaneously to correct variations. The simultaneous regulation of all machines in parallel is done preferably by a single controlling mechanism, all rheostats being coupled to the said regulator.

These regulators are mounted either on the switchboard or on resistance boxes according to conditions. Figure 144 shows a d.c. Thury regulator mounted on a rheostat.

When a group of alternators excited from a common exciter only the current of the exciter is regulated. If the exciting current is furnished by a storage battery, the regulator can be used in conjunction with the battery booster to vary the voltage as required.

Feeder Regulators.—One of the most important considerations in an a.c. lighting system is to keep the voltage in the main feeders constant. Since these feeders are subjected to a constantly varying load, an efficient regulation of the individual feeders gives an effective regulation of the entire system. The use of feeder regulators is of particular value for incandescent lighting systems, where it is essential that all the lamps should burn with equal brilliancy. The candle-power and life of light-

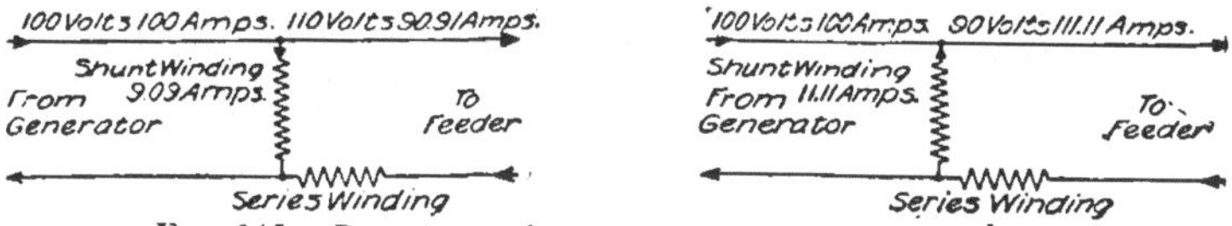

FIG. 145.—Boosting and lowering by a potential regulator.

ing units, the speed and torque of motors, etc., are directly affected by any change of the voltage from normal. When the generator supplies only a single circuit, the line voltage can be kept constant by regulating the generator voltage for the given line loss. This would be impossible where there are a number of circuits to deal with unless all of the circuits have the same line loss. Under such conditions the best that can be done is to regulate the generator pressure for the average line drop, so that the pressure in some of the feeders will be higher than in others. An independent voltage regulation of the circuits is made possible, however, by inserting feeder regulators in each feeder. Potential regulators are recommended for the generators so as to maintain a constant voltage at the bus or at the sub-station independent of the changes of the load, and the various feeders should each be controlled by feeder regulators.

The action of the feeder regulator is similar to that of a transformer, or rather compensator with two distinct windings. The primary is connected in shunt and the secondary in series with the feeder to be controlled. The product of the voltage and current on the busbar side is equal to the corresponding product on the feeder side of the regulator, with a slight correction due to the loss in the feeder. Figure 145 shows two diagrams of connections for boosting and reducing the feeder pressure for a single-phase line, given a 100-volt, 100-amp. circuit, and a regulator whose range is 20 per cent., that is, 10 per cent. boosting and 10 per cent. reducing, neglecting the losses in the regulator itself. With a rise of feeder voltage a part of the current goes back to the line through the primary winding, and this amount is deducted from the useful current in the feeder, while an increase of current results from a drop of pressure. It is important to distinguish between regulators and control resistances or reactances. The latter reduce the voltage by absorbing it. For since their windings are in series with the line, the current at both ends must be the same, so that there is a loss in voltage. The product of amperes by the difference in pressure across the ends of the winding represents a loss, and since the regulation of the current in the feeder is made to depend on this loss the result is a low efficiency.

Feeder regulators may be divided into switch or control regulators and induction regulators, the latter being single-phase or polyphase. They are all subject to manual or automatic operation.

Switch or Control Regulators.—A switch or control regulator is a transformer with both windings on the same core. The

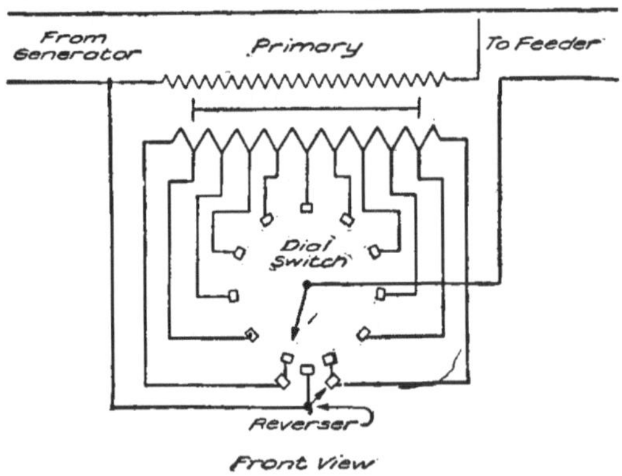

FIG. 146.—Internal connections of dial-switch feeder regulator.

primary winding is connected across the feeder, and the secondary is in series with it. The secondary can be gradually thrown on or off the line by a dial switch, upon which fact the operation of this type of regulator is based. Figure 146 shows the internal connections of a type CR regulator (G. E. Co.) In the position of maximum boost the dial switch is in the extreme left-hand position, where all secondary windings are thrown into circuit. By turning the switch from left to right the windings are thrown out, one by one, until they are all short-circuited in the extreme right-hand position. At this point the line voltage is the same as that on the generator side. With a continued motion of the switch in the same clockwise direction, a reversing switch is automatically

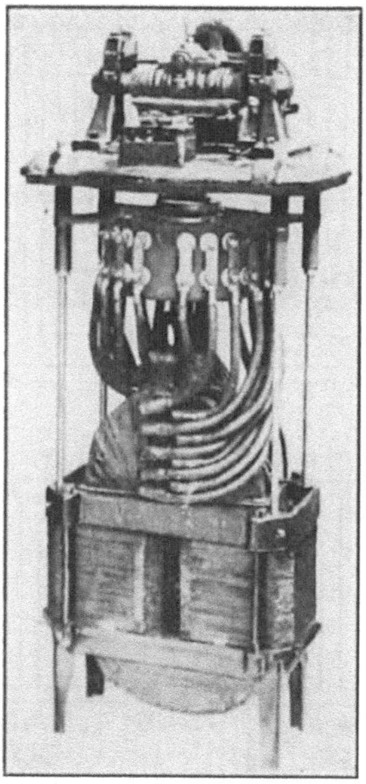

FIG. 147.—Controlling-type regulator (without casing).

thrown, causing the dial switch to cut the secondary winding in again step by step, and so lower the generator voltage. Two complete rotations of the dial switch are therefore possible, one for boosting and the other for lowering the feeder pressure. At the positions of maximum or minimum voltage the direction of rotation must be reversed. In the extreme positions the switch is automatically arrested to prevent it from turning too far in the same direction. The change in feeder voltage is therefore made step by step. Precision of regulation depends upon the number of contacts or taps on the secondary. Great precaution is necessary in breaking the circuit, while the dial switch is moving from one contact to another, as adjacent coils must not be short-circuited, or it can also be controlled automatically by a Thury regulator. The regulating switch has to be suited for one, two, or three phases according to the system in use, the contacts of the switch having connection with the ends of the sections of the secondary winding of the transformer. The levers of the switch are con-

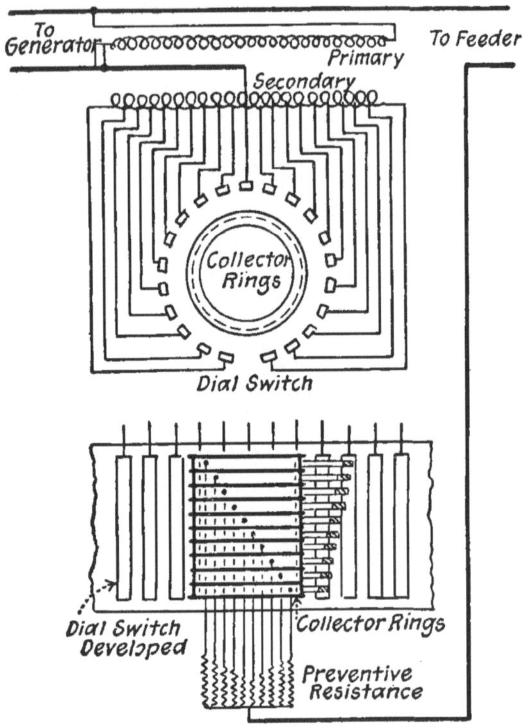

FIG. 148.—Internal connections of controller-type regulator.

trolled by the Thury regulator. The regulator, called also a boosting transformer, may also be driven by a motor, being controlled from the switchboard by a small reversing switch. It is undesirable to open the circuit each time in moving from one contact to the next. Arcing tips are therefore placed on the contact arm so that the contacts are not short-circuited but always have a gap of about 1/16 inch in the circuit during the time of the change from one contact to the next. This air gap acts as a resistance. The handle has a snap movement as quick action is obtained by compressing a spring which carries the blade to the next contact. This type is air-cooled.

Another form of switch regulator is given in Figs. 147 and 148. This is an oil-cooled transformer with both windings on one core. The coils are suspended from the cover. The switch resembles a controller in form. The device consists of a stationary drum on which are fastened the contacts for the various coils of the secondary winding. Inside the stationary drum is a second one on which is mounted a row of collector rings insulated from each other. A set of metallic fingers of different lengths make permanent contact with the collector rings so that when the vertical

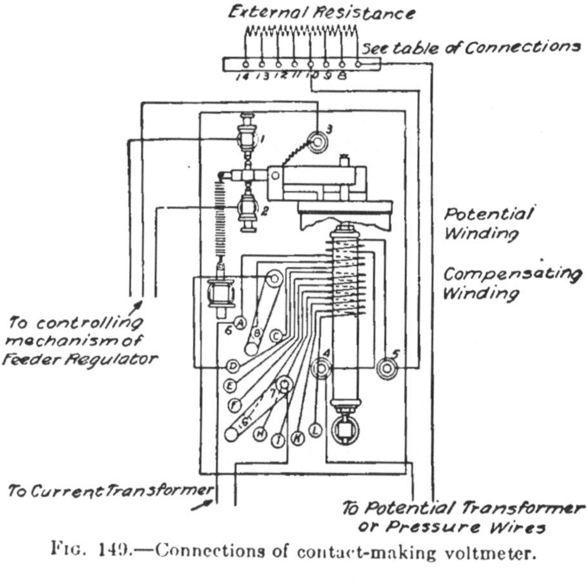

FIG. 149.—Connections of contact-making voltmeter.

axis of the regulator is rotated the fingers move on the upper faces of the contact pieces. The rings are connected in parallel through protective resistances and are joined in series with one end of the feeder. The other end of the same feeder is permanently connected to the middle point of the secondary. Regulation is accomplished by connecting the finger switch to a rotating flywheel by a horizontal bevel gear and two vertical pinions. A magnetic clutch keys one or the other of the pinions to the flywheel shaft, on which they run loose, according as boosting or depressing is required of the regulator. The flywheel is actuated from a rotating shaft which generally serves for operating a series of regulators. The magnetic clutch excitation is controlled by a contact-making voltmeter, quite similar to the a.c. solenoid and lever discussed under Tirrill regulator, type TA.

Figure 149 shows the connections of the contact-making voltmeter. The apparatus consists of a solenoid whose core actuates a lever. A spring at the farther end of the lever opposes the magnetic pull of the core. The lever is connected with one side of a low-tension a.c. circuit (see Fig. 150) and vibrates between two contacts which are connected to the armatures of the magnetic clutches. A potential and a current winding, fed by shunt and series transformers in the feeder to be controlled, make up the solenoid. The current windings can be partly cut in or out by a dial switch, thus limiting the regula-

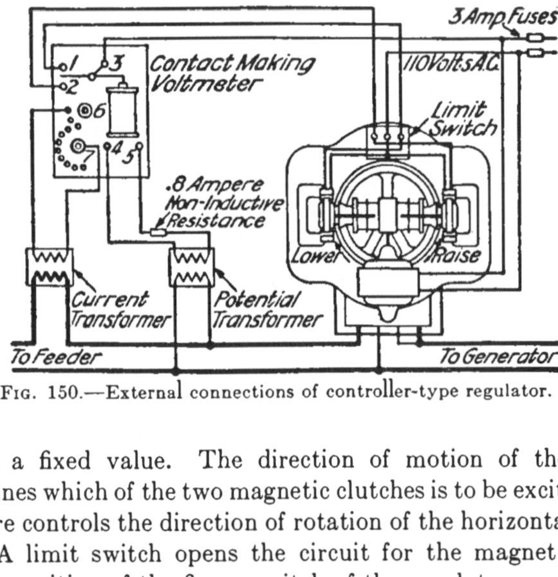

FIG. 150.—External connections of controller-type regulator.

tion to a fixed value. The direction of motion of the lever determines which of the two magnetic clutches is to be excited and therefore controls the direction of rotation of the horizontal bevel gear. A limit switch opens the circuit for the magnet in the extreme position of the finger switch of the regulator.

The efficiency of this regulator is very high, reaching almost 100 per cent. Since the motion is imparted to the small finger switch the moment of inertia and turning moments of the movable parts are small, so that the time element of the apparatus is very short. Moreover, quite precise regulation is obtainable because of the large number of contacts. Switch regulators are used for single-phase leads rated at less than 200 amp. at 2,300 volts, 60 cycles, and with regulating range of 10 per cent. either way, making a total of 20 per cent.

The Westinghouse dial-type "step by step" regulator has the dial switch mounted on the board away from the transformer to which it belongs, and the individual taps of the transformer secondary are joined by leads to the contact studs of the switch. These dial switches can be used directly for currents of 100 or 200 amp. To obtain regulation on a low-voltage circuit in which the current is larger than can be handled by the dial switch, a series transformer and a shunt or auto-transformer are used. The auto-transformer has a number of taps connected across the line by means of the dial switch; the series transformer is placed in series with one side of the line and two leads of the secondary are connected to the dial.

If it is desired to obtain voltage regulation on circuits from 2,500 to 6,600 volts, this can be done by using the dial with shunt and series transformers similar to those previously described.

The Westinghouse drum-type "step by step" potential regulator likewise consists of a separate control and a regulator transformer. The controller is similar to the kind used for electric street railway service. It consists of a cast-iron top and base held together by steel bars. Two of the bars carry the insulated contacts of the controller to which the transformer leads are connected. Inside the casing there are two revolving drums with their respective sets of fingers, one of which accomplishes the switching on or off to the transformer secondaries, while the other serves to reverse the regulation; i.e., according as a boosting or lowering of the voltage is called for. A so-called "floating coil" is used for the large number of steps of this regulator. It is simply an independent part of the secondary winding having a large number of well-insulated taps. The main portion of the winding is also subdivided by taps into a number of equal sections, each one being equivalent to the whole of the floating coil.

In the operation of the regulator the floating coil and main winding are first connected in series with each other and with the line to be regulated. The floating coil is then cut out of the circuit step by step. When entirely cut out it is transferred to the next lower tap on the main winding, after which it is again cut out of the circuit step by step. When again entirely cut out it is transferred to the next lower tap of the main winding, after which it is further cut out step by step and then transferred again. Precise regulation of the voltage is therefore possible with a maximum degree of freedom from arcing when cutting the windings in and out.

Induction Regulators.—The operation of these regulators is based upon a principle of regulation somewhat different from those discussed till now. The primary and secondary windings are respectively shunted across and in series with the main line as before, but they are wound on separate cores. The secondary is wound on the inner surface of a stationary core, while the primary is wound on the outside of a movable core. Both windings are polar windings and are so mounted that a given pole of one winding lies opposite to the same pole of the other. The regulation depends upon the displacement of one of the poles relative to the stationary pole. As long as two similar poles lie opposite each other the regulation is at maximum boost, but as soon as they become displaced, so that dissimilar poles are adjacent, the regulation drops until it reaches its lower value. The regulation with this type of apparatus is gradual. Two types of induction regulators may be distinguished, namely, single-phase and polyphase.

Single-phase Induction Regulators.—Figure 151 shows the arrangements of primary and secondary cores and windings in single-phase induction feeder regulators. The primary flux, and consequently the flux through the secondary coils, has a constant direction with respect to the movable core, namely, perpendicular to the plane of the primary winding. As the core is rotated gradually the relative direction of the primary flux to the fixed secondary winding is varied and produces a gradually varying e.m.f. in the secondary, from the maximum positive, through zero, to the maximum negative value. In the position where the direction of the primary flux is opposed to the secondary flux, the voltage generated by the primary in the secondary is added directly to the line voltage, but is subtracted when the direction of the flux is the same. In any intermediate position of the primary winding, the primary flux, and consequently the flux in the secondary coils, is proportional to the angular position of the core. The generated e.m.f. is, however, always in phase with the excitation and is therefore added directly to or subtracted directly from the line voltage. As shown in Fig. 151, the rotary core contains a short-circuited winding arranged at right angles to the shunt windings. Its object is as follows: In the positions of maximum boost or lowering, the primary flux neutralizes that of the secondary, for the fluxes are then in the same or opposite directions. With the armature in the neutral or no-boost, no-lower position the flux produced by the current in the secondary passes equally on either side of the primary coils, which cannot therefore neutralize the flux due to the secondary. The secondary flux sets up a self-inductance in the windings which reaches its maximum with the neutral position of the primary coils. This self-induc-

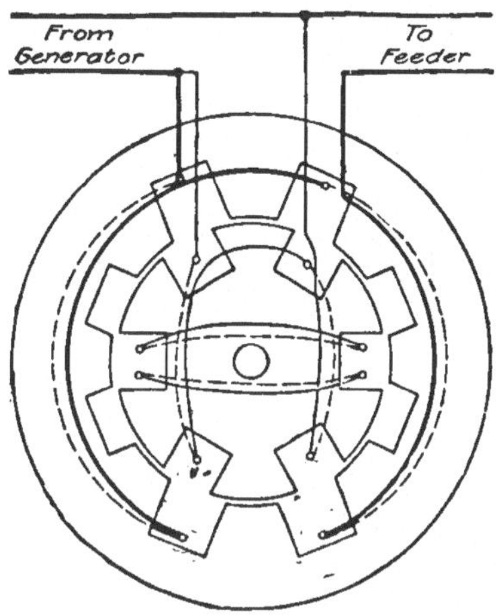

Fig. 151.—Arrangement of primary (armature) and secondary (field) cores and windings in single-phase feeder induction-regulator. Armature in maximum lowering position.

tance of the secondary, if the line current is constant, requires a gradually increasing e.m.f. to maintain the current through the series windings. The voltage so absorbed would be at right angles to the fine voltage, and the result would be a poor power factor in the feeder. The short-circuited coil on the armature, however, which is in direct inductive relation to the series coils when the armature is in the neutral position, acts as a short-circuit on the secondary winding and therefore reduces the voltage necessary to force full-load current through this winding to only a trifle more than that represented by the resistance drop across the secondary and short-circuited windings. This short-circuiting of the secondary is gradual from zero in the maximum boosting position of the regulator to the maximum short-circuiting in the neutral position, so that by the combined effect of the primary and short-circuited coils the reactance of the secondary is kept within reasonable limits. The total ampere-turns of the primary plus the ampere-turns of the short-circuited windings are always approximately equal to the ampere-turns of the secondary. Figure 152 is a graphical representation of boosting and lowering of feeder voltage by induction at no load and full load. The exterior of the regulator is shown in Fig. 153. The device is operated either directly by hand or by chain and sprocket, by hand-controlled motor, or automatically. If automatically operated, the actuating motor should be of the polyphase type. The motor is controlled by means of a small double-pole, double-throw switch which is automatically thrown in by a small magnet, energized through a contact-making voltmeter.

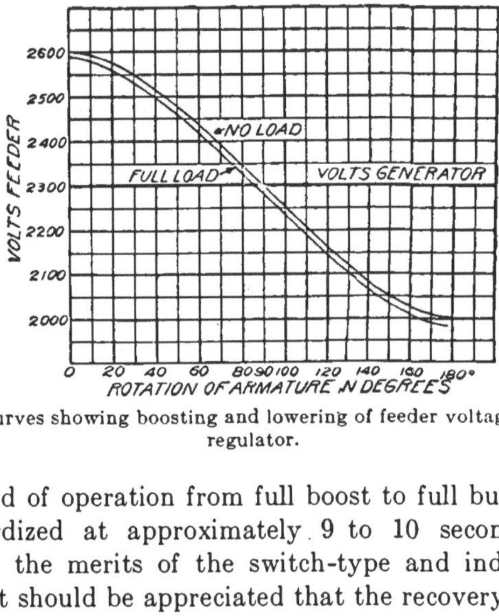

FIG. 152.—Curves showing boosting and lowering of feeder voltage by induction regulator.

The speed of operation from full boost to full buck seems to be standardized at approximately 9 to 10 seconds. When considering the merits of the switch-type and induction-type regulator, it should be appreciated that the recovery of the first type of regulator is a straight-line curve and at a fixed rate, whereas the induction regulator curve is relatively flat at the start, due to the comparatively heavy effort required to accelerate its moving parts. As the rapidity of connections is most important in all regulating apparatus, this feature of operation at the start is worthy of careful consideration.

Figure 154 shows connections for an automatically operated single-phase induction regulator. The action of the contact-making voltmeter is similar to that described under Fig. 150.

The up-and-down motion of the armature controls the excitation of the two coils of the relay switch magnet, and the regulator is operated by the motor in either direction.

Polyphase Induction Regulators.—The general construction of polyphase induction regulators is similar to that of the single-phase type of apparatus. The primary shunt windings for the different phases must be identical in every way and arranged so that the various windings magnetize a given pole of the regu-

FIG. 153.—Single-phase induction regulator.

lator in the same direction. Both windings are wound on their cores like the windings on an induction motor. With the polyphase type the field is not an alternating one as in the single-phase apparatus, but is rotary, and the speed of the rotation of the field per pole is the same as that of the generator. The value of the e.m.f. generated in the secondary is not changed by the motion of the primary coils, but the phase of both voltages is displaced, and this changes the value of the load voltage which is the resultant of the two. In other words, both coils set up rotating magnetic fields in the same direction. The e.m.f. generated in the secondary is determined in value and phase by the value and phase of the resultant field. Since the absolute values of the magnetic fields do not change, their resultant value, and consequently the absolute value of the generated voltage, also remain constant. The phase position of the resultant field, does change, however, and hence the phase of the secondary voltage relative to the line pressure. The resultant line voltage is the vector sum of the generator and generated voltages. The diagram in Fig. 155 shows the change of phase relation for this type of regulator.

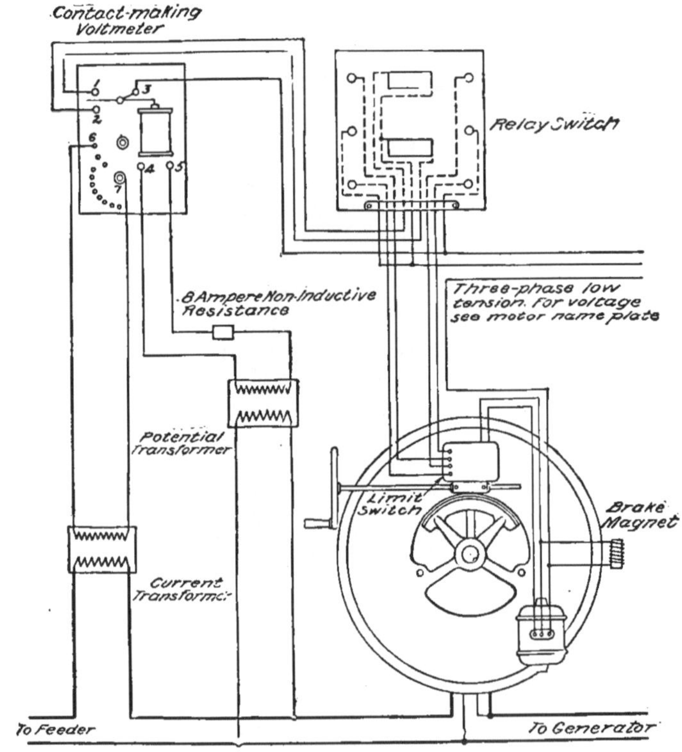

FIG. 154.—Connections of automatically operated single-phase induction regulators.

Let *EO* represent the normal e.m.f. of a certain primary phase both in value and in time-phase position; let the radius of the arc *ABC* represent the constant voltage generated in the secondary phase winding of the regulator. With the primary coil directly opposite the secondary coil of the same phase, the voltage generated will be in reverse phase with that impressed and will be represented by OD. The regulator will lower by the maximum amount, and the difference, ED, is the resultant feeder voltage. Since, however, the primary is rotated out of this position, the

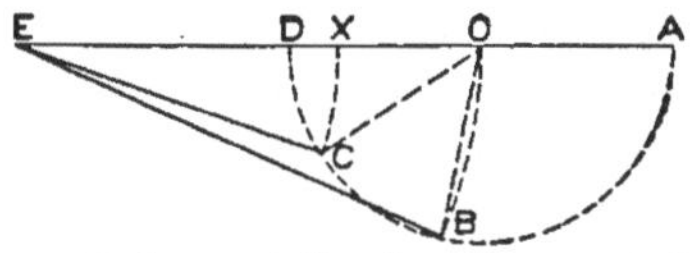

Fig. 155.—Change of phase relation in polyphase induction regulator.

secondary winding considered will be partially excited by the next winding on the armatures, so that the voltage generated is not deducted directly from OE, but at an angle, as OC, and the

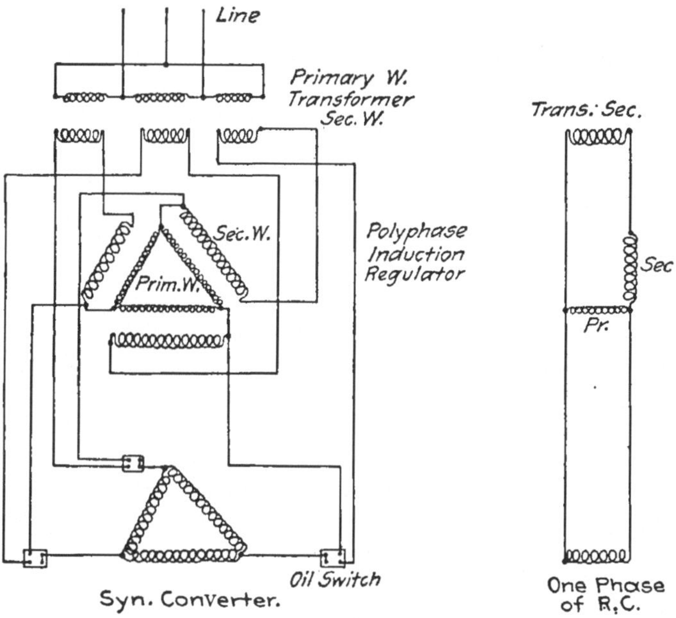

Fig. 156.—Connections of a polyphase induction regulator with synchronous converter.

feeder voltage is a resultant of EO and OC, or $EC = EX$. In a position of nearly 90° from the maximum lowest, the regulator is neutral or in no-boost, no-lower position, completing the full range of 180°; the armature is in an opposite field of the winding surrounding the dissimilar pole, so that the secondary voltage, is in phase with the primary, but in the same direction, and the regulator is boosting. The resultant *EA* represents the maximum boost voltage.

Due to the rotation of a similar field produced by the currents in the series coils, the currents in the shunt windings are constant, regardless of the position of the armature. For a given line current the currents in the shunt windings are taken from the line or are delivered back into the system as the armature is rotated from maximum boost to maximum lower in the same phase relation as represented by the secondary voltage generated. (See Fig. 145.) The arrangement and operation are the same as for single-phase regulators. They are used in lighting systems to regulate the voltage in the feeders, and also in connection with synchronous converters. (See Fig. 156.)

Polyphase regulators change the voltages of all of the phases simultaneously and by an equal amount, depending on the excitation, regardless of the load or the unbalancing of the line voltage between phases, whereas by using the single-phase type each phase can be controlled separately and a perfect balance obtained at the point of distribution on the feeder regardless of the unbalancing of the system.

The induction type of potential regulators can also be controlled by a Thury regulator. Induction regulators are oil-, air-, or water-cooled, depending upon the capacity, pressure, and frequency of the line in which they are inserted.

Synchronous Condensers.—In the older method of transmission systems the voltage of the system is controlled entirely from the generators. The chief disadvantage is the large variation in voltage, which is indeed the factor limiting the carrying capacity of the system. Since at no-load both the generating and receiving end of the line are at approximately the same voltage, while under load there is considerable drop at the receiver end on account of the reactance of the transmission line, a variation in voltage is unavoidable as the load increases or decreases. This variation may take place at the generator end or the receiver end, as desired, but in either case it puts a limit on the amount of power which the line can transmit. When the load on a transmission system becomes so large as to produce too great a variation in voltage for good service, it has been customary practice to install larger conductors or build additional transmission lines, or finally to adopt a still higher line voltage. With high-voltage systems the gain of doing so is very slight and other means must be resorted to for keeping the regulation within commercial limits. The effect of the inductance and condenser capacity of the line on high voltage causes the voltage to vary within very wide limits from full load to no-load. At no-load the large-capacity current may cause a rise of voltage from the generating station to the receiving end, while at full load the lagging inductive current taken by the load in general more than offsets the effect of the capacity current and causes a drop of voltage from the generating station to the receiving end.

Instead of controlling the voltage from one end only, special machinery, consisting generally of synchronous motors or synchronous condensers, is installed at the receiver end, and the voltage is controlled at that end also by adjusting the power factor. By adjusting the field current of the synchronous condensers, causing the condenser to draw a lagging current from the line at no-load and a leading current at full load, the receiving voltage of the line can be kept constant.

A synchronous condenser is practically a specially designed synchronous motor for operating without energy load from minimum current to full kva. This apparatus is also employed in connection with distribution systems to compensate for unsatisfactory conditions created by consumers' apparatus. While this is always a feasible method of improving power-factor conditions, it should be resorted to only as a last extreme after all other available schemes for improving operating conditions have been tried.

Low power factor, except in special cases, is usually the result of the prevailing tendency to install larger motors than are actually required. In certain classes of service, owing to their nature, low power factors are necessarily developed and may require special attention for improvement, but aside from these cases it appears possible in the majority of situations to materially improve the operating conditions by a more intelligent application of power-consuming devices.

If the voltage established at the receiving end of a transmission line is such that the voltage drop is small, being controlled by a synchronous condenser with a voltage regulator, there is no necessity in some cases for a voltage regulator at the generating station.

Controlling the voltage at the receiving ends of transmission lines by governing the wattless current so as to operate lines with small voltage drop not only brings about highly satisfactory service, on account of the excellent voltage regulation, but makes the operation of the system very flexible. Since the power factor in the transmission line and its transformers and generators is above 0.9 except at higher loads, the water wheels or steam units, as the case may be, can operate at all times at efficient points of their load curves without any limit being encountered due to transformer, generator, and transmission-line capacity such as may occur if the power-factor were low. Since the copper losses vary inversely as the square of the power factor for a given load, considerable net saving may be accomplished by correcting power factor, although losses occur in the synchronous machines effecting the correction. The carrying capacity of a 25-cycle line can be doubled by the use of synchronous condensers, and the capacity of a 60-cycle line can be multiplied by two or three. Another advantage that comes from operating a system in this manner is that the small voltage drop avoids to some extent the necessity for transformer taps. But most important of all seems to be that considerable reactance may be used in the generators, transformers, and transmission lines, eliminating thus the destructive effects of short-circuits. (See Chap. XX, "Reactive Coils").

In water-power plants the voltage and also the frequency are subject to sudden changes due to variations in load which are too rapid for water-wheel governors to compensate for immediately. In such case a large flywheel effect on the system tends to minimize the trouble. Synchronous motors add directly to the flywheel effect and are found very useful in this way.

The most advantageous application of constant-voltage transmission is in large transmission networks connecting all the hydro-electric plants and the cities within a radius of several hundred miles. The generators placed at various points of the network themselves partly take the place of synchronous motors for maintaining constant voltage, and thus the total capacity of synchronous motors required is somewhat less.

A desirable feature of synchronous condensers on a high-voltage transmission system is that they offer protection against those voltage surges that arise due to a sudden loss of load, which might throw the generating stations on the unloaded transmission line with their generators on heavy field excitation. Under such circumstances, due to the effect on generators and transmission lines of the leading current set up by the charging current, a destructive voltage would occur. Overvoltage devices may be applied to the generators to give protection. However, with synchronous condensers on the receiving ends of the transmission lines, and each one equipped with an overvoltage device; an ideal solution of the problem is effected. since some of the synchronous condensers could be on the line at all times with their excitation under control through the desired range.

The adjustment of the field current of synchronous condensers must be under control of the transmission-line operator in order to have the correct effect on the line voltage. An automatic regulation of the condenser-field current can also be readily accomplished by means of a Tirrill regulator. As the excitation must at times be practically zero, a special exciter arrangement must be resorted to in order to provide for the limiting voltage range of the regulator. This is accomplished by exciting the condenser from a 250-volt exciter, whose field in turn is separately excited from a 125-volt source consisting of two small exciters which are running series bucking. One of these exciters is designed for 125 volts and the other for 250 volts, the regulator being placed across the 250-volt exciter. In this manner the range of the regulator is from 125 to 250 volts, never reaching zero, while the excitation of the condenser is from zero to 250 volts.

Synchronous condensers can be designed so as to remain in step when running unloaded at 100 per cent. power factor as tenaciously as a fully loaded induction motor. If necessary, the starting torque could be sacrificed for the synchronizing power, and the condenser could be started by small starting motors. At times of very light load on the line, such condensers can operate safely with weak field and lagging power factor. Thus, in spite of the fact that the loaded synchronous motors and converters are commonly regarded as a very unstable element in transmission-line operation, it may be stated that the addition of properly designed unloaded synchronous condensers to a transmission system does not decrease the reliability of the system.

A flexible generator and receiver voltage may also be aecomplished by a combination of a synchronous condenser with a series booster, both machines arranged on the same shaft, the excitation of the synchronous condenser being controlled with a voltage regulator and that of the booster by hand. At the generating station the excitation should also be controlled by a voltage regulator. At both stations regulating rheostats should be installed in the circuit of the potential transformers of the voltage regulators. By means of these regulating rheostats the voltage settings of the busbars can be changed at will of the operator.

For a given voltage difference and an assumed power factor for the load, the kva. of the synchronous condenser will vary definitely with the kilowatt load, and by changing the voltage difference the kva. of the synchronous condenser may be changed; thus controlling the voltage difference gives a control of the kva. load on the synchronous condenser. An equivalent effect to this can be established by holding the voltage constant in both stations and compensating for voltage by means of a synchronous booster.

The function of a synchronous booster, therefore, is to effect voltage compensation, the amount of buck or boost being controlled by field excitation. When used in connection with a synchronous condenser, controlled by a voltage regulator, a means of controlling the wattless kva. is effected. Without the booster for a given voltage setting at the generating and receiving stations, the wattless kva. supplied by the synchronous condenser will depend upon the kilowatts load for a given assumed power factor. With the booster, whenever the synchronous condenser comes up to its overload limit, the field excitation of the booster can be changed to reduce its kva. load, without affecting the voltage at either the generating or receiving station, and conversely, when the voltage setting at either station is changed, tending tobring the synchronous condenser kva. to an undesirable value, the booster excitation can be changed, effecting the proper voltage compensation to bring the kva. to a suitable value. A flexible voltage can be therefore aecomplished at both ends of the line, and an independent means of power-factor control for the transmission line is thus secured. This offers a very considerable advantage, in that it is frequently necessary to have a long transmission line with one or two intermediate stations.

A number of important installations have been made with synchronous condensers. One includes two 15,000-kva. condensers for voltage regulation on a 50-cycle, 150,000-volt transmission system, where they are used not only to compensate for low power factor but for the excessive charging currents developed by such a system under light or no-load conditions. In some instances turbo-generators in reserve steam plants, auxiliary to transmission systems, are used to advantage as condensers for power-factor correction and regulating purposes.

The use of synchronous condensers for constant-voltage transmission has the following advantages and disadvantages:

Advantages

1. Better service—no variation in voltage.
2. Better protection due to permissible use of high reactance.
3. Tendency to use the frequency of 60 cycles.
4. Increasing carrying capacity of line. The limit is changed from maximum voltage variation to maximum energy loss. This allows more power to be transmitted, or the distance to be increased, without the voltage being raised.
5. Lower total cost for long lines. The saving in cost is greatest for long lines or large networks, large quantities of power, large conductors, and for the frequency of 60 cycles.
6. The method is easy to apply to existing lines. The change can be made gradually, and no change is necessary in line construction.

Disadvantages

1. Cost and attendance of additional rotating machines.
2. Higher total cost for short lines.
3. In order to obtain the greatest economy from constant-voltage operation, the losses must be increased, and the number of separate lines, which are useful as reserves, must be reduced.
4. Tendency of synchronous machinery to drop out of step and delay in putting load on the line again after shut-down.

CHAPTER XVIII

CONSTANT-CURRENT SYSTEMS

In series a.c. systems for arc and incandescent lighting it is necessary to maintain the current value of the circuit at a con-

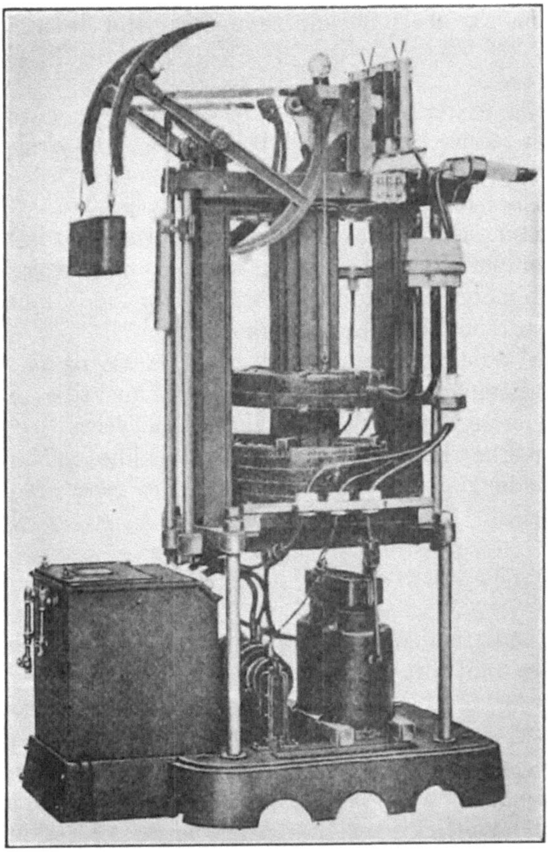

FIG. 157.—Internal arrangement of an air-cooled, constant-current transformer in combination with a mercury-rectifier outfit.

stant value regardless of the number of lamps. This maintenance is brought about by a Thury regulator, as mentioned in the previous chapter, or by a constant-current transformer.

The principle of the constant-current transformer depends upon the relative linear shifting of the secondaries with respect to the stationary primaries. Within certain limits the repulsion between the fixed and moving coils for a given position is directly proportional to the current in the coils. The transformer may be set for a given current value by adjusting a counterweight so as to balance the movable coil for a certain position.

Fig. 158.—Internal arrangement of an air-cooled, constant-current transformer with two primary and two secondary windings.

Figure 157 shows the interior connections of a 50-lamp air-cooled, constant-current transformer in combination with a mercury-rectifier outfit, and one for 100 lamps is given in Fig. 158. Apparatus rated up to and including 50 lamps are built with two flat coils enclosing the central core. The lower coil, which is the primary, is fixed in position, while the upper one is suspended from the two inner arms of a double lever and can move freely along the central core. The outside arm of the lever carries a counterweight of such value that it will exactly balance the weight of the secondary coil minus the electrical repulsion due to the normal currents in the coils. Therefore, if the weight is reduced, the current value is raised.

The 75- and 100-lamp transformers shown in Fig. 158 have four coils, two primaries and two secondaries. In the 75- and 100-lamp oil-cooled apparatus and the 75-lamp air-cooled transformers the two primary coils are fixed at the extreme upper and lower ends, while the secondary coils are free to move up and down along the central core. The 100-lamp air-cooled transformer is arranged with the secondary coil stationary and the primary movable. The two moving coils are balanced one against the other, and the counterweight serves merely to draw the coils together in opposition to their repulsive force. A decrease in the counterweight produces a decrease in the current. The arc on the counterweight lever is made adjustable because the repulsion exerted by a given current in the coils is not the same for all positions of the coils, being greatest when the primaries and secondaries are close together. By means of the adjustable arc, the effective radius of the balancing weight is made to change as the coils move through their working range. When the current value is reduced below the normal the mutually repelling forces diminish and the primary and secondary coils approach each other, thus restoring normal current. The opposite action occurs when the secondary current exceeds normal. The transformers, therefore, maintain the constant current for which they are set, regardless of the external resistance to which the coils are connected. The efficiency of these transformers at full load with arc lamps at 60 cycles varies from 96 per cent. for the 100-lamp type to 94.6 per cent. for the 25-lamp type. The two primary coils of the 75- and 100-lamp apparatus are connected in series for 2,200 volts and in parallel for 1,100 volts. An exception to this connection is found in the 100-lamp transformer in which the coils are arranged with a number of taps for use under partial load and full power factor. These transformers are built for 1,100 or 2,200 volts, and the connections of the primary windings must not be changed. With the large transformer with two secondary coils, each of these coils is connected to its own circuit, while with the small type all lamps are connected in series on the same circuit.

The connections of a transformer having two primary and two secondary windings (75 or 100 lamps) are shown in Fig. 159,

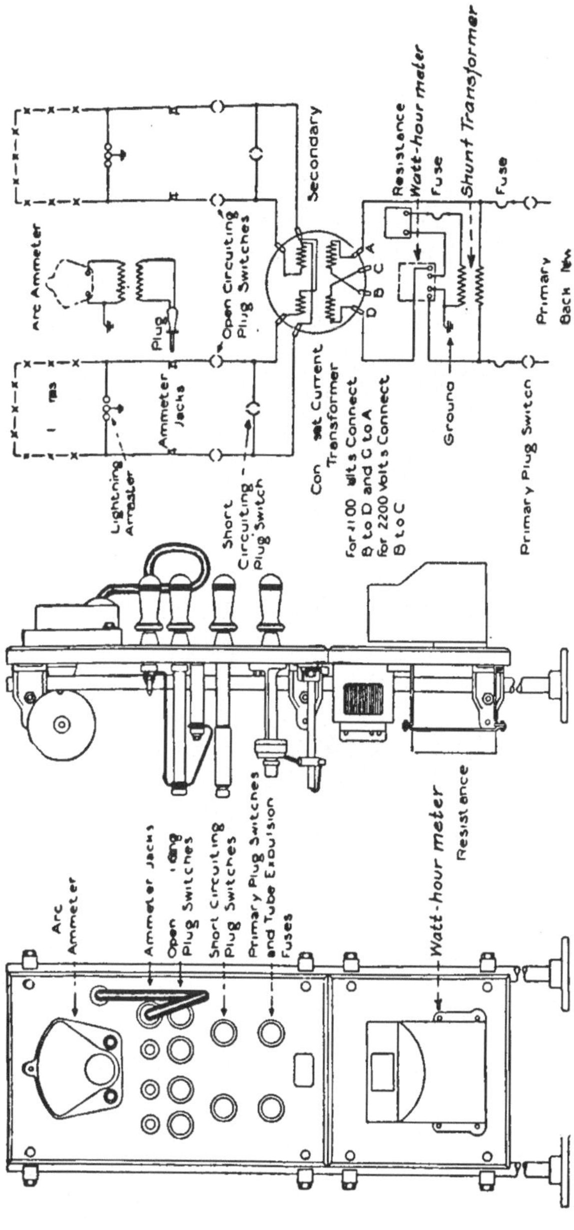

FIG. 159.—Connections and panel equipment for a constant-current transformer for 50, 75 and 100 amps.

together with the circuit and switchboard. The primary side is connected to the outer circuit by primary plug switches. In a similar manner the individual lighting circuits can be connected with the transformer secondary or can be short-circuited by means of a series of plug switches. The series transformer for the ammeter is connected into the desired circuit through an ammeter jack. The front and side elevations of the board show the manner of mounting the plug switches and instruments together with their transformers and resistances.

FIG. 160.—100-lamp oil-cooled, constant-current transformer, interior.

Each transformer calls for the following equipment:

1 ammeter.

1 series transformer (which may be omitted on boards for less than 35 lamps).

1 ammeter jack plug with necessary leads (for boards controlling two lighting circuits).

4 ammeter jacks (for boards controlling two lighting circuits).

2 sets open-circuiting plugs and receptacles (for each lighting circuit).

1 set short-circuiting plugs and receptacles (for each lighting circuit).

2 primary plug switches with receptacles.

2 plug racks for the reception of idle plugs.

2 primary fuses.

The plug switches on the primary side are often replaced by oil switches, on account of the high voltage.

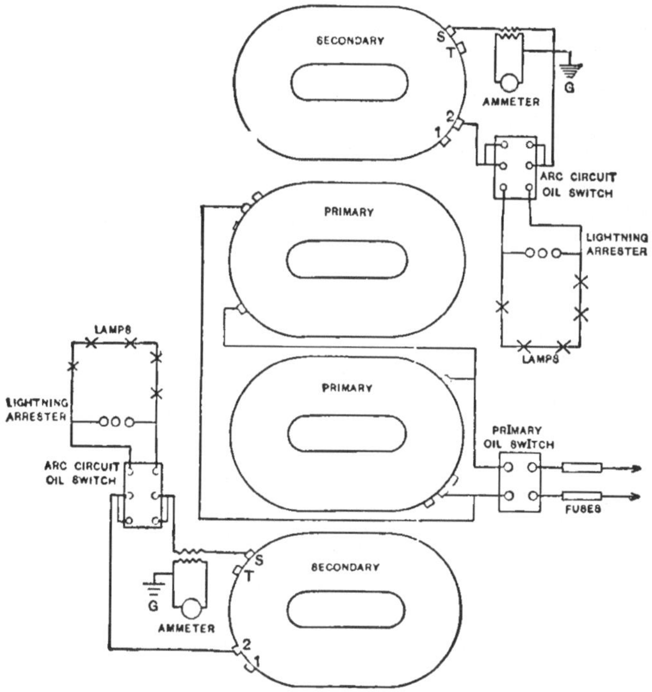

FIG. 161.—Diagram of connections for 100-lamp air-cooled, constant-current transformer.

The size of the fuses is such that they will protect the transformer from short-circuit. The above equipment is mounted on a panel of blue Vermont marble 28 inches high, from 16 to 20 inches broad, and set up 36 inches above the floor. An extra watt-hour meter is often inserted in the primary circuit to indicate the total amount of energy delivered to the transformers. This instrument with its resistance is mounted on a base of the same width as the panel and 12 to 16 inches in height.

The interior connections and appearance of a Westinghouse 100-lamp oil-cooled, constant-current transformer are shown in Fig. 160. It has one stationary primary winding and two movable secondaries, each of which feeds its own lighting circuit. Both the primary and the secondaries are provided with taps for connection to different voltages (within certain limits) and to

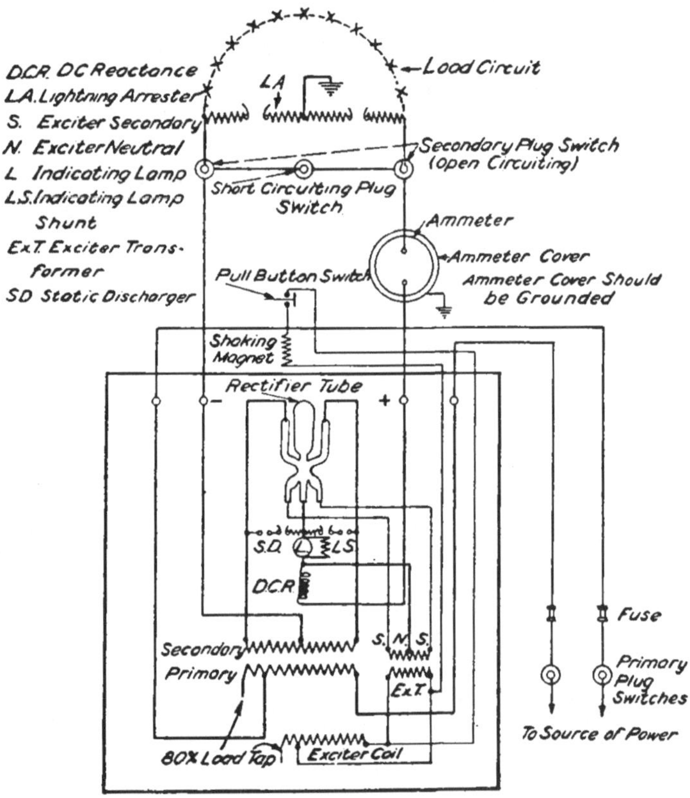

FIG. 162.—Wiring diagram of air-cooled, constant-current transformer shown in Fig. 157.

partial load at full power factor. To set the apparatus for a given amperage small weights are added to the main counterweight. In constant-current transformers for 25 to 75 lamps with only one movable coil, an addition to the counterweight produces a falling off of the current, while in the type for 100 to 200 lamps with two movable coils this causes the current to be decreased in one coil and boosted in the other.

Figure 161 shows the connections of a 100-lamp air-cooled regulating transformer on a single-phase circuit. The connections of the primary side with the outer line, and of the arc lamp circuit with the secondary, are made through oil switches. The oil switches for the lamp circuits are a combination of a double- and a single-pole oil switch with separate handles, in which the double-pole part serves to cut the lighting circuit in and out, and the single pole to short-circuit the secondary transformer winding when starting and stopping.

To permit the operation of direct-current series arc-lighting systems from a single-phase alternating-current circuit the use of a combination of a mercury-rectifier outfit with a constant-current transformer is recommended. This combination may either consist in one boiler-iron tank containing all the parts of the transformer, rectifier bulbs, and necessary connections immersed in oil (Westinghouse Company), or the rectifier tubes are submerged in a separate steel tank independent from the transformer case. The tube tank is provided with a cooling coil through which water can be circulated, while the transformer outfit is of air-cooled type. (G. E. Co., see Fig. 157.)

These combination units are made for capacities of 25 to 100 arc lamps, for primary voltages up to 13,000 and for any frequencies.

The rectifier tube possesses two anodes in the upper part connected to the secondary winding of the transformer. The cathode of the tube at the bottom is connected to the direct-current circuit through a reactance. The starting cathodes are energized by a separate exiting transformer. Figure 162 shows the wiring diagram of the combined unit represented by Fig. 157.

CHAPTER XIX

STARTING COMPENSATORS

Induction or synchronous motors require a starting current several times as large as their full-load current. With motors rated at more than 5 hp. such a starting current produces a considerable voltage drop and load variation in the circuit to which the motors are connected and causes a disturbance in the general service. To prevent a rush of current during the starting period, starting compensators are connected in between the line and the motors during this interval.

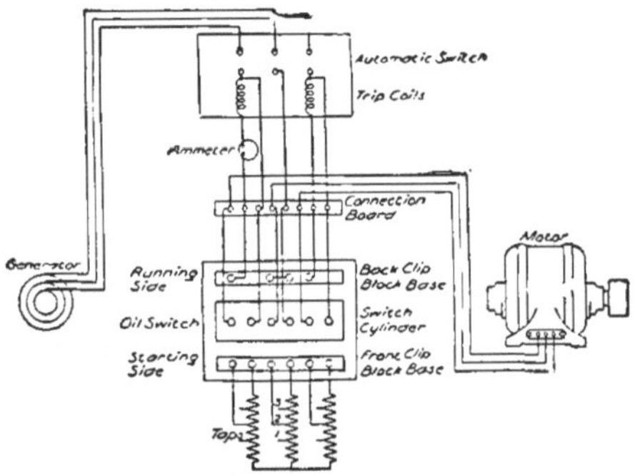

FIG. 163.—Connection of a starting compensator with a three-phase induction motor.

Compensators for starting a.c. motors consist of inductive windings, one coil for each phase, which are provided with several taps and which supply a large current at reduced potential. Their effect is equivalent to that of a step-down transformer, and the product of e.m.f. and current on the line circuit is approximately equal to the corresponding product on the motor circuit. Each coil is placed on a separate leg of a laminated iron core and is provided with several taps to obtain a number of sub-voltages for permanent connection to the starting switch of the motors according to their requirements. The three coils of the three-phase winding are connected in Y, and the line is joined to the three free ends of the coil by a controller switch or, for large machines, by an oil switch. During the starting period the motor is connected by means of the controller switch or a double-throw oil switch to the taps of the coil, and directly to the line during service. If the control switch connects the motor to the line, fuses are used to guard against short-circuit or overload, while if the oil switch is used, the switch, itself provided with an automatic trip coil, affords the necessary protection.

Figure 163 shows the induction motor and compensator connections for a three-phase machine. The lever of the controller has three positions, namely, "off," "starting," and "running."

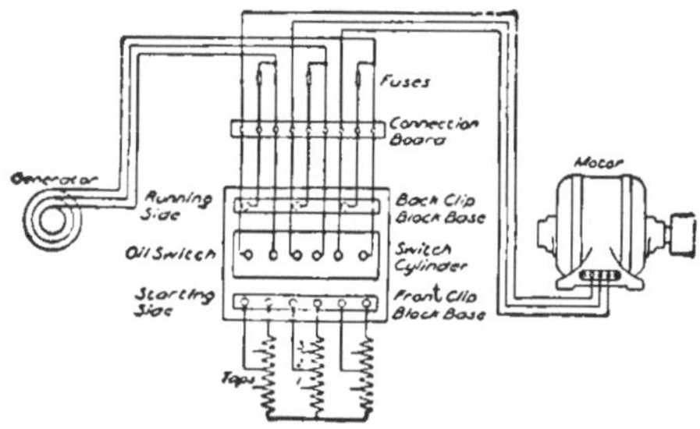

Fig. 164.—Connections of a starting compensator with a separate automatic switch to three-phase induction motor.

In the "off" position the compensator and motor windings are disconnected from the line, in the "starting" position the switch connects the line to the free ends and the motor to the taps of the compensator windings, and in the "running" position the compensator winding is cut out, and the motor is connected to the line through the fuses or circuit breaker, which are mounted directly above the compensator.

Figure 164 shows the connections of a three-phase induction motor with a starting compensator and automatic oil switch on the line side. In this diagram the line is connected to the ends of each coil, and the starting connection of the motor to one of these ends and the tap.

In Fig. 165 there are shown the connections for starting a number of three-phase induction motors from one starting compensator, which is also joined to the line through an oil switch called a magnetizing or primary switch. The motors can only be started separately.

Extracts from an information label included with a General Electric starting compensator is given below:

"The following directions apply whether the primary switch of the compensator is placed in the compensator box or on the switchboard. With either arrangement a double-throw switch is placed on the switchboard to connect the motor to either the compensator or the line." This double-throw switch can also be included in the compensator box.

To start the motor the following procedure should be observed: "Close the primary switch and then throw the motor switch into the starting position. The motor should reach practically full

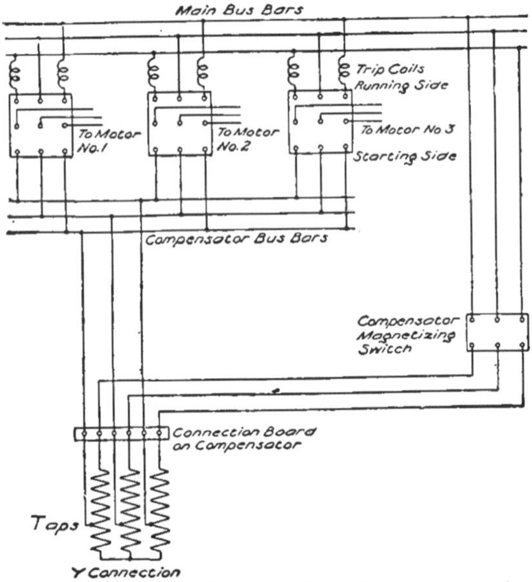

FIG. 165.—Connections for starting a number of three-phase induction motors from one starting compensator.

speed in one minute with the switch in this position. Just before it reaches full speed, throw the motor switch into the running position and open the primary switch of the compensator. This primary switch should never be opened with the motor switch in the starting position. If the motor does not start immediately, open both switches and adjust the compensator as follows: Connect the motor line in the compensator to the next higher connection on the coils, being careful to have all of the cables lead to corresponding taps on all the coils. If the motor fails to start again, connect the motor lines to the next higher taps and proceed as before. If it does not start with the highest tap connection, the load is either too great for the motor or the line voltage is low." Where separate oil switches are used, the same mode of procedure should be followed.

CHAPTER XX

LIGHTNING ARRESTERS

Lightning arresters serve the purpose of protecting transmission lines, machines, and apparatus against the destructive influence of abnormal phnenomena in voltage and frequency. All phenomena of this sort are defined collectively as lightning. Such disturbances may be caused in three different ways.

1. They may be the result of exterior occurrences, such as electrical discharges between clouds or between clouds and earth. A special case of this is when the discharge strikes the transmission linc. Thc influence of electrostatic induction of charged clouds or atmospheric strata and the collection of static charges from wind, rain, snow, or mist is noticeable in the more or less dangerous phenomena observed in the rise of voltage and frequeney. They may also be produced by sudden changes in temperature at sunrise and sunset.

2. Interior processes within the circuits, machines, and apparatus will cause the same phenomena in changes of pressure and frequency as the outside disturbances, so that they are properly included under the term lightning. Such processes may be caused by load variations, opening and closing of circuits, throwing machines into or out of circuit, or discharges by faulty insulation, short-circuit, or grounding.

3. Interior or exterior lightning phenomena may be serious or quite harmless in their consequences, but their appearance in a system carrying considerable energy, though unimportant in itself, may cause an abnormal surge liable to prove detrimental to the installation.

The phenomena resulting from the causes just described may be classified thus:

1. A steady stress or gradual electric charge.
2. A strong impulse or traveling wave.
3. A stationary wave or oscillation and surge.

Under the first conditions a series of discharges takes place in the lightning arrester, or if this apparatus does not operate properly, which is equivalent to its total absence, the insulation of the conduits is punctured at its weakest point. The harmful action, therefore, lies in the destruction of the insulation of the system and the serious consequences attendant upon short-circuiting or discharging thus made possible. It therefore becomes a function of the lightning arrester to take care of any excessive pressure in such a way that no disturbance shall be created in the system.

A strong impulse or traveling wave is generally caused by lightning striking a line or by discharge through arcing of accumulated static electricity. It may also result from spark discharge, sudden variation of load, or switching of apparatus.

Switching produces, in general, an abrupt change in the value of the electromotive force and of the current, or of either of these quantities, and therefore it produces a sudden change in the amount of energy stored in the circuit. The severity of oscillations thus produced depends on the difference between the running conditions before and after the switching is done. For instance, if a dead line of zero potential is suddenly connected to an a.c. generator, the oscillation produced will be most severe when the switching makes the connection at the instant at which the electromotive force has its maximum value. Likewise, if a short-circuited line is disconnected from an a.c. generator, the oscillation is most severe when the circuit is broken at the instant at which the current passes through its maximum value.

The oscillations are two kinds: They may occur between inductance massed at one point at the circuit and capacity massed at another point of the circuit, as, for instance, when a dead line which acts practically as a condenser is connected to a generator which acts practically as an inductance, or when a line (capacity) connected at one end to a step-down transformer (inductance) is disconnected from the generating system.

They may also occur between the distributed inductance and capacity of a part of the circuit, as, for instance, when a line is disconnected from a generator and the energy stored in the line gradually dies out, changing from electromagnetic energy stored in the distributed inductance of the line to electrostatic energy stored in the distributed capacity of the line, and *vice versa;* or when an unloaded step-down transformer is disconnected from the line, its windings acting similarly to a line.

The oscillations do not, in general, constitute a great danger to the circuit if the contact in the switch is made once, positively, for every switching operation.

However, if each switching operation is accompanied by arcs, so that the circuit is repeatedly closed and opened, then the pressure rises may become dangerous, and destructive cumulative oscillations may be set up in the apparatus. Furthermore, at the instant at which contact is made traveling waves are sent out from the point of switching along the line in both directions. At the instant of their appearance such waves may possess considerable potential, and their influence may extend over a longer or shorter portion of the line, depending upon the inductance and condenser capacity of the line concerned. These traveling waves with steep front follow the rules of reflection, and whenever they strike a localized oscillatory circuit they excite in it oscillations at its natural period.

For instance, if a dead line is suddenly connected to a live line having the same constants, a wave of charge, equal in value to one-half of the potential of the live line, starts along the dead line from the switching point. If the dead line is open at the far end, the wave is reflected back at double potential, etc. At the same time a wave of discharge starts along the live line from the switching point with a value equal to one-half of the original voltage of this line. If the live line is connected at its origin to transformers, this wave of discharge is reflected back with double voltage and may excite local oscillations in the transformer windings or in the circuit of the power house, etc.

A break in the propagation of a wave or series of waves at the entrance to a station is made evident by electrostatic discharges, sparks, and arcing at the cables, switches, and switchboard. Moreover, oscillations or fluctuations of varying frequency will occur in the transmission lines, which result from a tendency to reëstablish equilibrium of the energy flow after a disturbance has taken place. The danger from these fluctuations increases with the amount of energy carried by the system, the frequency depending upon the resistance, inductance, and capacity of the line.

The phenomena may occur singly or in combinations, one often causing the other.

Dr. C. P. Steinmetz defines the purposes of lightning arresters:

1. As guards against the entrance or origin of disturbances.

2. As prevention against spreading of disturbances.

3. And as means of taking up an existing disturbance and rendering it harmless, without in any way affecting the system, i.e., in regard to raising of voltage, etc.

The arresters must therefore prevent excessive although transitory rises of potential between lines as well as between lines and ground; they must restrain the flow of the dynamic current from line to ground following the discharge path and extinguish the arc when normal potential is restored. The lightning arrester ought to discharge high potentials covering a wide range of frequency.

The phenomena with which we have to deal under abnormal pressure and frequency are of such complicated nature, both in respect to the size, length, and form of wave, and to the time and relative order of their occurrence, that it has hitherto been impossible to design a protective device which will meet all the requirements imposed upon it for given conditions of voltage and energy flow of the system. The various devices in use are of value only under the conditions for which they are designed. The controlling factors in their construction are the voltage and energy of the system, the kind of load, and the overhead and underground installation of the transmission lines.

Of the devices tabulated below we will discuss only those occurring most frequently in practice, as their constant use has brought out some noteworthy improvements.

The following are the most common forms of protective devices:

1. Multigap lightning arrester, without resistance, with series or shunt resistance, or with both.
2. Horn arrester, with or without series resistance, and with or without fuses.
3. Magnetic blow-out arrester.
4. Electrolytic arrester:
   (a) Aluminum arrester, with or without gap.
   (b) Liquid electrode arrester.
5. Condensers.
6. Reactive coils.
7. Overhead grounded wire.
8. Overload switches.
9. Water jet from line to ground.
10. Coherer type.

Multigap Lightning Arrester.—This device consists of a number of small brass cylinders mounted in a row on an insulated base with small gaps between them. One end of the row is connected to the line and the other end is grounded. *D.* B. Rushmore and D. Dubois[1] explain the action of the cylinders as that of small condensers, which are charged with varying amounts of potential according to their distance from the line connection, being maximum at that point and zero at the grounded end. When the potential difference between the first pair of cylinders increases beyond a certain limit, a discharge takes place between them. The potential of the second cylinder, being connected to the first by an arc, rises and may rise to such a point that it breaks down to the third cylinder, and the third to the fourth, and so on, till the arc has passed entirely across the arrester. This action shows the successive discharge of the arrester at a voltage in excess of the normal line pressure. As soon as all the gaps are bridged over by arcs, the line current starts and the distribution of potential is changed, following a straight line from a maximum value at line potential to zero. This indicates that the potential difference between all the cylinders is the same and is much less than before the breakdown. It is well known that to maintain an arc of alternating current across a gap the voltage must be at least high enough to break down the dielectric of the air gap each time the arc goes out at the end of a half cycle. The dielectric is greatly weakened by the heat produced by the passage of the arc. Therefore the voltage to maintain the arc need not be so high as that which originally broke across. The weakening of the dielectric depends upon the heat of the arc. and this in turn upon the boiling temperature of the metal of the arc cathode. A metal or alloy having a low boiling temperature is therefore chosen which at the same time will preserve its cylindrical form under the arcing action. Metal meeting these requirements is called non-arcing metal.

The size, material, and number of the cylinders and the spacing between them are so chosen that at the end of a half cycle the potential difference between the cylinders is insufficient to break down the dielectric resistance offered by the air gap. The arc therefore is extinguished after the first half cycle.

Another explanation for the non-arcing property of multigap arresters is given by Mr. Thomas in the discussion of the above-mentioned paper: "It is a known fact that the arc starts by ionization through potential strain of the gases between the electrodes, by which ions are liberated. They are forced to move extremely

[1]"Protection Against Lightning and Multigap Lightning Arrester," *Proc. A. I. E. E.*, March 29, 1907.

rapidly by the high potential and produce other ions until finally there is such an increase of temperature and such an increased ratio of ionization that the quantity produced is sufficient to carry the normal arc current at a low voltage. At the instant the discharge and the normal current cease at the end of an alternation, in view of the close neighborhood of the conducting cylinders, the ions which would otherwise hold over until the return of the voltage are freely absorbed by the metal. Such ionization is very closely related to temperature, the high temperature very much increasing the ionization. Metals of low boiling point are used to keep down the temperature."

Arcing between any two successive cylinders consumes a certain amount of e.m.f. The successive losses in voltage thus incurred can reduce the discharge voltage between a certain pair of cylinders to such a value that no further discharge is possible between the remaining cylinders. The discharge therefore is only partial over a certain number of gaps. A complete discharge takes place when the initial voltage is so great that the successive discharges do not bring it to a value less than that of the spark voltage, or when the drop across the gaps is so small that the sum is not sufficient to affect the initial discharge voltage.

The drop in voltage between any two cylinders depends upon the value of the current. Messrs. Rushmore and Dubois state that "as this current is that due to the capacities which have been considered, it will be greater at high frequencies, and the fall of potential between the first and second cylinders will therefore be less. As the arrester gaps break down successively the fall of potential from one cylinder to another is less, and therefore such an arrester will discharge at a lower voltage for a higher frequency than for a lower." Multigap arresters without resistance discharge more readily at high frequencies than at low. In the same way that high frequency lowers the breakdown of multigaps by increasing the current of the spark, high resistance, by absorbing e.m.f. when this current exists, decreases breakdown. Resistance is most effective at high frequency in increasing breakdown voltage, as at high frequency the charging current is greater, and therefore there is more voltage drop in the resistance. A resistance in series or shunt with the arrester limits the current and decreases the number of gaps. Series resistance limits the current under all conditions, but although in this way protecting the arrester, it is dangerous in case of a surge. Its action is questionable on account of the inductivity of the resistance, as in case of a high-frequency stroke series resistance will prevent free discharge, so that no effective protection is afforded to the line and connected apparatus. By shunting a different number of gaps through different resistances the protection offered by multigap arresters for both high and low frequencies is assured, and the current between cylinders, as well as the number of cylinders themselves, is reduced.

An arrangement of this sort is shown diagrammatically in Fig. 166. As stated above, a discharge across the gaps is facilitated by high frequency. This kind of discharge therefore follows the direct path across the gaps and not through the shunt resistances. With low frequency the discharge passes through a low resistance and the remaining gaps in series. If the frequency is still lower, it chooses a path over a greater resistance with a less number of gaps in series, etc. This is explained by the fact that a discharge of low frequency is less opposed by a resistance than one of high frequency, the voltage being the same, and that it is still less opposed, the lower the frequency. When the discharge of low frequency breaks down, say over the high resistance, the entire voltage minus that lost in the discharge gaps acts upon the next gap division with its resistance, which is in turn broken down. A drop in voltage again follows, and the remaining pressure acts upon the next division with its still lower resistance, where another discharge takes place, etc. In this way the discharge is accomplished in rapidly following parts from one end to the other of the gap row, and ceases either when the voltage is no longer sufficient to break down the next division, on account of the current in the resistance, or after the entire lightning arrester has been broken down. Breaking down of the arrester is, of course, not to be interpreted as destruction of the instrument, but simply as a discharge over the apparatus.

This arrangement therefore acts as efficiently for low frequency as for high.

Figure 167 shows the arrangement of a multigap arrester with shunt resistances for a delta or star connection without grounded neutral. Each of the three main legs consists of form V-gap units and is connected to one of the lines through a spark gap shunted by a fuse. The other ends of the legs are joined to each other by a common or multiplex connection, from which, for ungrounded systems, a fourth leg connects to the ground. In the arrester for grounded star systems no fourth leg is added, the multiplex connection being connected to ground. The reason for introducing the fourth leg lies in the design of the arrester, which calls for two legs between line and line. If one line became accidentally grounded, the full-line potential would

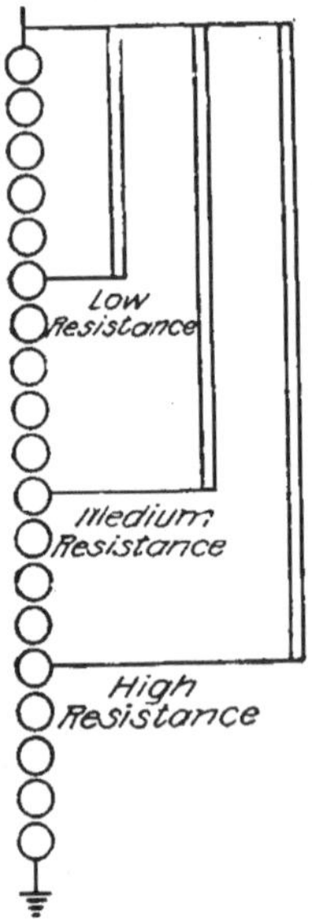

FIG. 166.—Multigap lightning arrester.

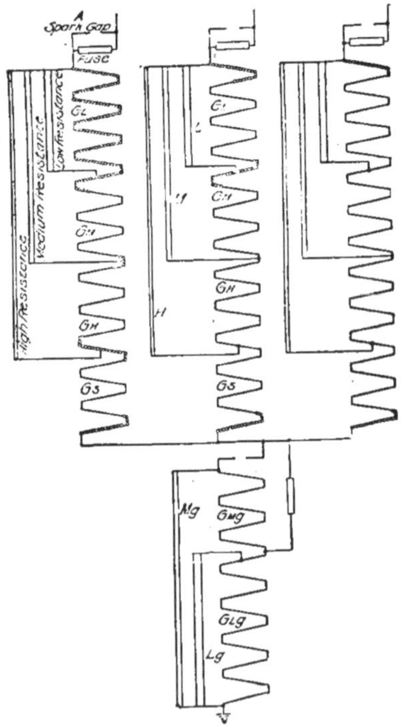

FIG. 167.—Arrangement of multigap lightning arrester for 35,000 volts, delta or ungrounded star transformer connection.

be thrown across one leg if the fourth or ground leg were not present. On a Y-system with a grounded neutral the accidentally grounded phase causes a short-circuit of the phase and the arrester is relieved of the strain by the tripping of the oil switch. The fourth or ground leg of the arrester is usually employed when, for any reason, the system could be operated, even for a short time, with one phase grounded. The above diagram shows the three shunt resistances with their respective connections to the legs. When for some reason or other a high voltage in the line is not reduced by the discharges through the arrester, the device becomes exposed to destruction on account of the prolongation of the current. To prevent this the fuse is inserted, which by blowing out throws its spark gap into series with the arrester. This allows the arrester to be adjusted to discharge at but a small percentage above the line voltage and thus to afford real protection. The blowing out of the fuse does not eliminate the protection afforded by the arrester, because the spark gap still preserves the connection between the arrester and the line. It merely adds a factor of safety against the destruction of the arrester. It may be repeated that the shunt resistance forces discharges of different frequencies to seek different paths to ground, so that under all conditions about the same discharge voltage exists in the apparatus.

With a low-frequency surge, when the voltage rises, and before it reaches a dangerous value, the gaps G_S arc over. It will take care therefore of accumulated static and also of all disruptive discharges of small ampere capacity. With these gaps broken down the current of the arc across them is limited by the resistance to about one-sixteenth ampere, which gives about 80 volts drop per gap. The remainder of the voltage is consumed in drop across the resistance rods, and is thus applied across the gaps in G_H. Although this voltage is less than that which broke across G_S, the series resistance is less, and approximately the same number of gaps will therefore break across this lower voltage. With G_S and G_H broken down the increased current gives a smaller drop in the gaps, but twice the number of arcing gaps are now in series. Therefore, the number of gaps in G_M is made the same as in G_S and G_H. These three sections should all arc over in succession at very nearly the same voltage. It has been found that when gaps are shunted by resistance and the resistance is low enough, dynamic current will not follow a high frequency across those gaps, but will shunt at once to the resistance, i.e., over L, and not over G_L. This shows that the resistance L is practically a series resistance as far as the safety of the arrester is concerned. Should the arc pass across all the gaps following a static discharge, the number of gaps is sufficient to extinguish the arc, and this holds with each resistance and with the gaps connected with it. That is, any current which may start across any of the resistances and the corresponding gaps would immediately be extinguished by that combination alone, without the shunting effect of other resistances, thus rendering the extinguishing of the arc doubly secure. The object of the multiple connection of the single legs of the arrester is to admit of discharge between the lines and between the lines and ground.

An equivalent needle gap is a gap which when connected in parallel with a lightning arrester just fails to discharge, forcing the

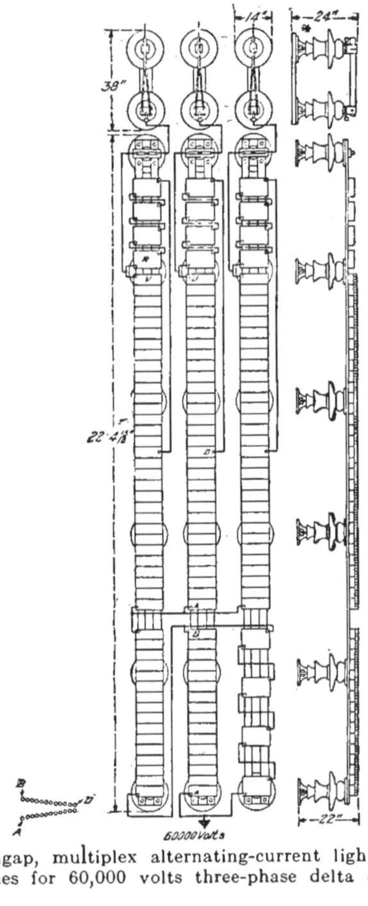

FIG. 168.—Multigap, multiplex alternating-current lightning arrester with single-blade switches for 60,000 volts three-phase delta or ungrounded star-connected circuits.

discharge to pass through the arrester. It is also defined by Prof. E. E. F. Creighton as a gap which when in parallel with the arrester possesses such a value as to cause at least 90 per cent. of the spark discharges to pass through the device and not more than 10 per cent. across the gap. This equivalent needle gap is an indicating device for the efficiency of a lightning arrester under discharges of varying frequency. The smaller the value of the gap, the more efficient is the protective device. If the laying out, material, and mounting of the cylinders are correct, the equivalent needle gap must be less than the sum of all the cylinder gaps.

Figure 168 shows the dimensions of type GE form V shunt-resistance, multiplex a.c. lightning arresters with single-blade switches for station use for 50,000 and 60,000 volts. The arrangement is for a three-phase delta- or star-connected circuit without grounded neutral. On account of its great length, the fourth leg between the multiplex point and ground is divided into three parts at the lower end of the three legs. A disconnecting switch is necessary to facilitate disconnection from the line for inspection or repairing. The gap units *V* and resistances *R* are mounted on

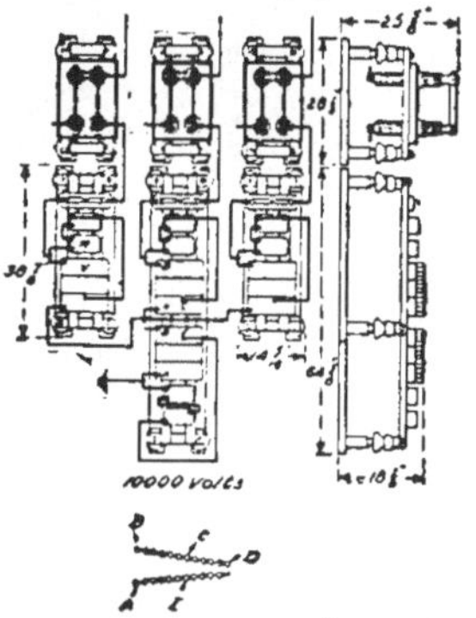

FIG. 169.—Multigap, multiplex alternating-current lightning arrester, with single-blade switches for 10,000 volts, three-phase delta or ungrounded star-connected circuits.

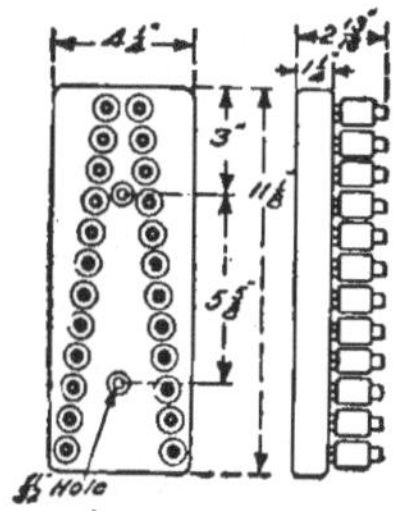

FIG. 170.—Form V gap unit.

porcelain bases supported by means of wooden strips on line insulators. The number of gap units depends upon the length of the line, upon the nature of the country through which it passes (this referring to elevation, neighborhood of trees, passage through cities or open country, etc), upon the insulation of the line, and also upon the voltage and load of the system. Each transmission line therefore calls for special consideration. More precise adjustment of the arrester is made by means of the spark gap between line and arrester.

Figure 169 gives the dimensions of a similar device with double-blade switches, built for 10,000 volts. Of the two disconnecting switches, which are mounted on a common base, one is for the lightning arrester and the other for the line itself.

One of the gap units of form V is shown in Fig. 170 mounted on a porcelain base.

Figure 171 shows the adjustable spark gap which is used with each leg of a multigap, multiplex arrester.

A double-pole multigap device for 3,000 volts is shown in Fig. 172. In order to save space the shunt resistances are placed in slips directly over the gaps. Between poles the porcelain base is shaped to form a barrier. This device can be made suitable for

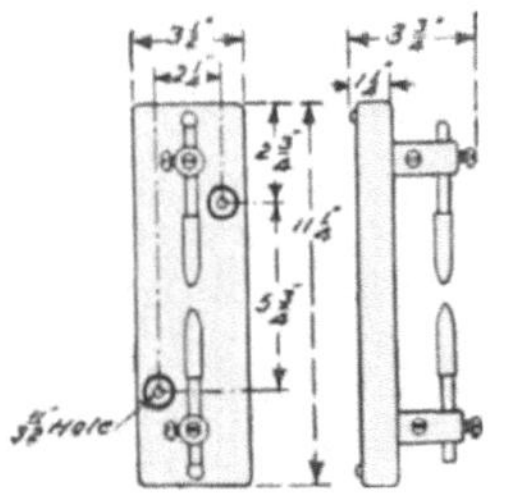

FIG. 171.—Spark gap.

FIG. 172.—1,000- and 2,000-volt multiplex, multigap lightning arrester.

use with 1,000 or 2,000 volts by short-circuiting one or more series gaps on each side of the ground connection, as shown in the cut.

Cables installed underground are generally lead-covered so as to protect the insulation against rough handling or chemical injury. Although such cables are not exposed directly to atmospheric disturbances, they are nevertheless subject to oscillations and surges caused by switching, load variation, etc., which may puncture the insulation at weak points. This is liable to produce sparking between the cables and their grounded lead covering, which, on account of the energy back of them, may result in burning up of the insulation of these and adjacent cables. Puncturing of the insulation and sparking may also take place due to the gradual accumulation of static charges on the lead covering, which are discharged from time to time without detection until finally the entire insulation becomes damaged. It is therefore desirable to ground the lead covering as often as possible. Sparking between cable and covering or ground brings about a series of successive impulses which must be eliminated from the cables by static dischargers. In a combination of overhead and underground transmission lines of high capacity and inductance oscillations of high frequency may occur without detection by the ordinary switchboard instruments which may seriously endanger the insulation of the whole system. Protection against such contingencies is afforded by the static dischargers. Multigap arresters also possess the capacity of protection against static discharges, a function which is accomplished by the high resistance with a few gaps in series, the remaining gaps and resistances not being necessary when only static dischargers are required.

Figure 173 illustrates a static discharger for 15,000 volts built on the lines of the multigap arresters.

In regard to the installation of arresters a number of points must be kept in mind. For 5,000 volts or more as much space as possible should be provided on the wall and in front of the arresters for their inspection and the safe operation of the disconnecting switches. Very often specially constructed high compartments or separate towers are provided for safe and efficient mounting of the high-tension apparatus.

The following table gives the proper spacing between lightning arresters as recommended by the General Electric Company:

SPACING BETWEEN LIGHTNING ARRESTERS
General Electric Company

Volts	Distance in inches between live parts of adjacent phases	Minimum distance between centers (see note), inches
7,600	8	28
12,250	8	28
13,500	8	33
17,000	10	35
22,000	12	37
27,000	18	48
32,000	22	52
37,000	26	56
40,000	28	62
45,000	32	67
50,000	36	72
60,000	40	78

NOTE.—If barriers are used, the width of barriers should be added to distances given.

The place where the arresters are mounted should be dry and warm, and before mounting all wooden parts and insulators should be thoroughly dried. It is advisable to place brick, asbestos, or soapstone barriers between the legs on the line side of the multiplex connection. The latter connection and shunt leads for the resistances must be kept away from the barriers, and the arresters must be separated from the barriers by a space corresponding to the normal line voltage, as the barriers are to be regarded simply as fire protectors rather than absolute insulators. There should be no doors in the front of barrier compartments. For single-phase only two legs, for three-phase three legs, and for two-phase, four-wire circuits four legs are necessary, the leg between the multiplex connection and ground remaining the same in each case. Good ground connections are essential to proper operation of lightning arresters. These connections and the arresters themselves must be inspected from time to time to make certain that they are in proper condition.

The Westinghouse Company produces a lightning arrester similar to those above described, under the name of low-equivalent a.c. lightning arrester. This device consists of three parts, namely:

1. Series gap, a number of gap units in series.
2. Shunted gaps, and shunt resistance in multiple.
3. Series resistances.

The connections between the three parts are as in Fig. 174. The gaps are formed by cylinders of non-arcing metal mounted between two porcelain holders, and all gap units and resistances are mounted on a marble base. (See Fig. 175.) The action of this arrester is as follows: When a discharge takes place in which all the series gaps are broken down, it meets opposition in the shunt resistance and passes over the shunted gap to earth through the series resistance. The arc which momentarily follows the discharge is then withdrawn from the shunted gap by the shunt resistance, and aided by both resistances is suppressed by the series gaps The potential at which a discharge takes place is determined by the number of series gaps, a sufficient number of which are used to withstand the normal voltage and yet give a proper factor of safety for the severest service. The use of shunted gaps is to provide a by-pass for the lightning discharge, which otherwise would meet opposition in the shunt resistance. The use of the shunt resistance is twofold, first, to withdraw the

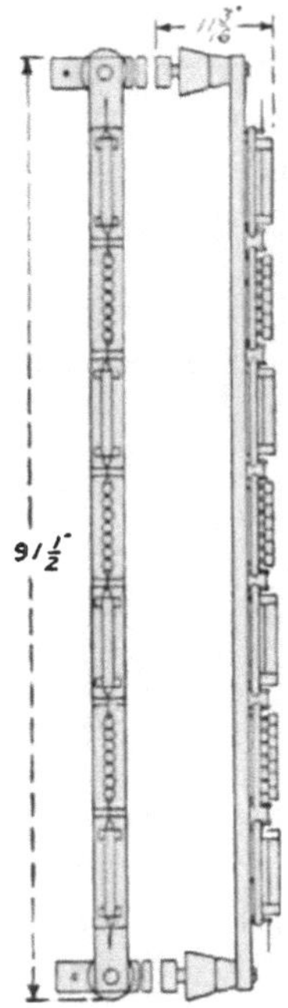

FIG. 173.—Alternating-current static discharger.

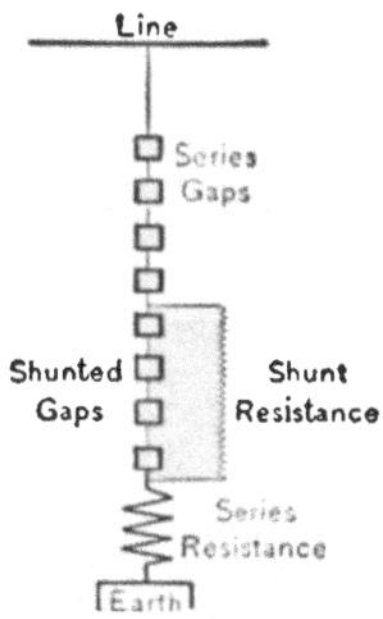

FIG. 174.—Diagram of low-equivalent alternating-current lightning arrester.

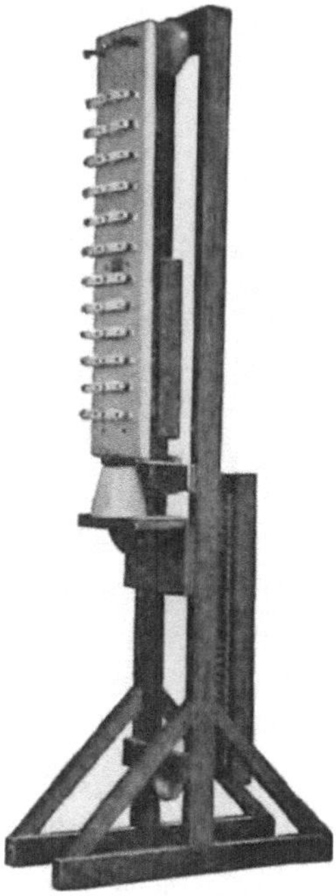

FIG. 175.—Low-equivalent lightning arrester.

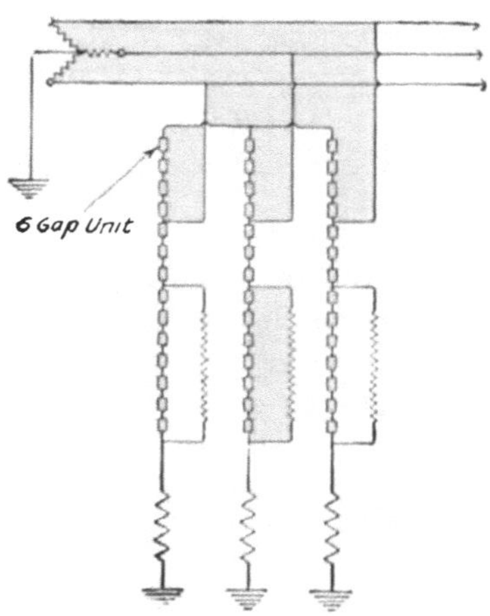

FIG. 176.—Connections of low-equivalent lightning arresters to an 18,000-volt, three-phase, star-connected circuit with grounded neutral.

arc from the shunted gaps after the passage of the discharge, and secondly to reduce the volume of the arc so that the series gaps, too few in number to act successfully unaided, can, with this assistance, suppress the arc. The small series resistance limits the initial current that follows the discharge and thus prevents burning of the cylinders. An auxiliary spark gap used in connection with the arrester permits adjustment within certain limits. Figure 176 is a wiring diagram for the Westinghouse low-equivalent lightning arrester in connection with a three-phase line with Y connection and grounded neutral. Provision is made for easy discharge between the lines themselves. The space between active parts of adjacent arresters connected to different sides of the circuit should not be less than the distance designated in the following table:

Spacing Between Lightning Arresters

Voltage		Distance between active parts, in inches
Exceeding	Not exceeding	
5,700	8,500	6
8,500	12,500	7
12,500	18,000	9
18,000	25,000	12
25,000	29,000	15
29,000	37,000	20

For low tension, and particularly for distributors for lighting and traction, the Westinghouse Company uses type C arrester. It is a double-pole multigap device, with non-arcing metal cylinders, and is built for tensions of 500 to 2,500 volts. It consists of seven independent cylinders of non-arcing metal carried on porcelain supports, forming a unit which is mounted in a weatherproof cast-iron case. This type is used on circuits of a capacity not exceeding 200 kw. within a radius of two miles of the source of power.

Another type of multigap arrester is shown in Fig. 177 as produced by the Stanley Electric Company. It consists of a nest of concentric cylinders of brass or other high melting-point metal with flaring upper ends. The line terminal is at the center of this group, and the ground connection at the outside. When line current follows a static discharge, it takes the narrowest gap space of the arrester. At the same time a current of air is established through the many small holes in the bottom and top supporting porcelains. This draft pushes the arc upward when, by reason of the attenuation of the arc and the greater cooling surface of these gaps, the short-circuit is broken.

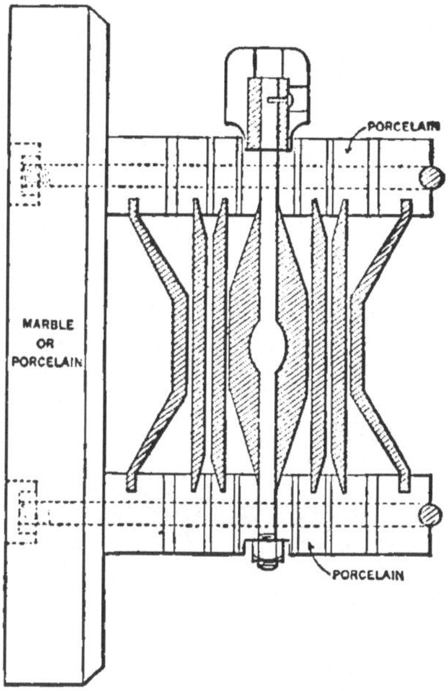

FIG. 177.—Metal multigap type of lightning arrester with diverging sides (Stanley Electric Company).

In practice it has been shown that the multigap arrester has been giving good service, particularly on moderate voltages, but it has natural limitations. For instance, while it is sensitive and effective in taking care of momentary disturbances, it cannot be expected to discharge continuously for long periods as in the case of accidental ground on a normally ungrounded circuit.

Horn Arrester.—This type of protective device has not proved of any great value in practice. It serves as an emergency device rather than as a normal protective apparatus. If the horn is connected to the line without resistance the arc short-circuits the apparatus for a time until it is ruptured by being driven to the upper ends of the horns through the magnetic and heat effects. If it is connected to a resistance of sufficient size to prevent it from causing considerable voltage drop by diminishing the current, its protective value becomes smaller. When a horn device discharges to ground without resistance the machines are thrown out of synchronism and must be restarted. If a fuse is joined with the horn, the combination can offer protection for maximum voltage only as long as the fuse is not blown out, which results

FIG. 178.—Horn-gap arrester with a disconnecting switch.

in lack of protection in storms when the fuses cannot be replaced. However, several fuses may be joined to the horn, so that when one is blown out another can be inserted by means of a switch. But this arrangement makes the fuses, and not the horn, the actual protective device. The arc in the horn gap can cause more serious damage than the original disturbances. Under certain conditions it is apt to cause very high strains, so that the horn becomes the cause and not the preventive of a disturbance. The only proper application of horn lightning arresters is along tramsmission lines, where they serve to protect the insulators, and not in the station. To protect an insulator, a gap of such value is chosen as will cause the sparking to occur across the horn rather than around the insulator. At all events, if trouble occurs around an insulator, the system is disturbed whether the gap is there or not, the only difference being that with the horn the trouble is of short duration, while without it the line may be tied up for a longer period of time until the broken insulator is found and replaced.

If we endeavor to deal with static charges by means of horn arresters, we must choose a very high resistance in series with the horn. In consequence of the spark gap, the apparatus will not commence to operate until the static charges have risen to a value which is already dangerous for the plant. In general, all arresters

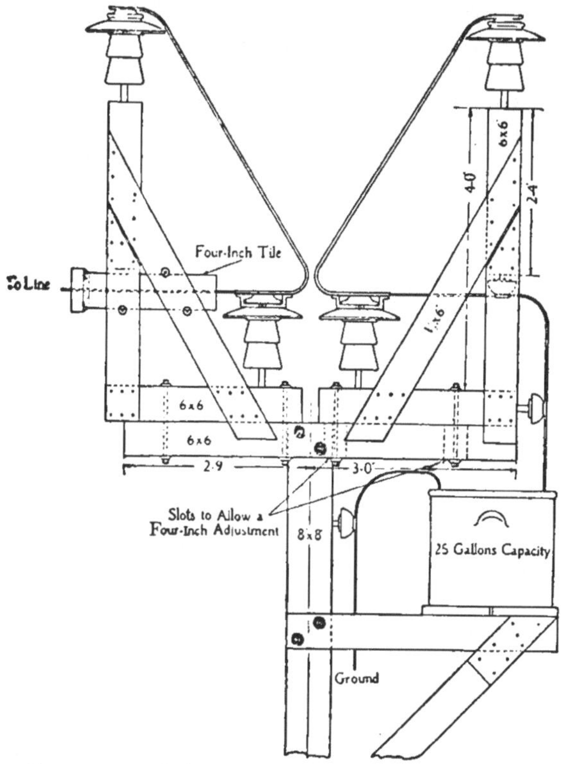

FIG. 179.—Construction of the horn-gap arrester used by the American River Electric Company.

including a spark gap do not operate properly for high-frequency discharges the voltage of which is below that for which it has been adjusted. Hence high-frequency discharges which may have only a pressure of 1,000 volts may already be extremely dangerous. This explains why some users of horn arresters have noticed that their transformers or machines have burned out without any discharge having taken place across the horn arrester, as the potential difference at the high frequency is applied between two wires situated in the same tunnel of the rotor or in the same coil of the transformer.

In constructing horns and proportioning their air gap care must be taken that the arc is driven upward and not downward, as not only the heat of the arc, but also its magnetic effect, tends to drive it outward.

The most important factors which affect the horn-gap settings are the density of the surrounding atmosphere, the inherent characteristic of the circuit, the wave form of the impressed voltage, and the antenna effect of grounded structures near the horns. It has been demonstrated by actual practice and by

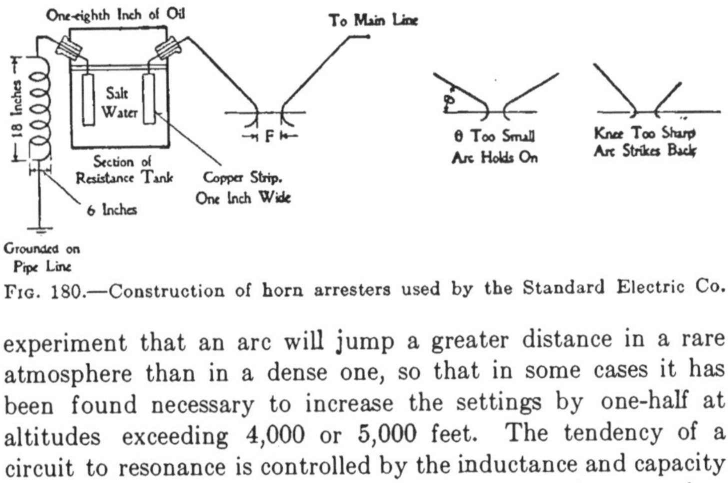

FIG. 180.—Construction of horn arresters used by the Standard Electric Co.

experiment that an arc will jump a greater distance in a rare atmosphere than in a dense one, so that in some cases it has been found necessary to increase the settings by one-half at altitudes exceeding 4,000 or 5,000 feet. The tendency of a circuit to resonance is controlled by the inductance and capacity of the circuit. This affects the horn-gap settings, because when there is a tendency to resonance or surges the gaps are more readily jumped. In such cases the cause should be removed; if this is not possible, the gaps must be set farther apart. If the wave form is peaked, it is necessary to use a greater horn-gap setting than if the wave form more nearly approaches the sine wave, since it is the peak voltage and not the effective voltage which determines the distance which an arc will jump.

The presence of grounded objects near the horn gaps affects the horn-gap settings because they change the potential gradient across them. This is sometimes termed the antenna effect. It may either increase or decrease the tendency to spark over, depending upon the exact conditions.

In a Siemens horn the arc is slow in rising on the horns owing to insufficient magnetic field. By inserting an iron sheet it is claimed that the magnetic field is increased and that an arc formed will rise on the horns and disappear much more quickly.

Another disadvantage of this type of arrester is the fact that at low voltages the gap must be made so small that it becomes difficult to maintain it constant, for dirt, dust, raindrops, and insects quite readily collect in it and change its width, which materially reduces the value of the apparatus. In fact, it is impossible in practice to use a smaller gap than ¼ inch. To produce an arc ¼ inch in length a potential difference of 17,000 volts is necessary. Thus arrester protection is illusory when used for a line voltage far below 17,000 volts.

By using the so-called radioactive substances the sparking potential is reduced so that a spark gap is produced which possesses considerable length, even at small potential. A body of radioactive material is supported in a small receptacle in proximity to the spark gap in such position that radiation therefrom will pass into the space between the spacing electrodes. Radiation from this source acts upon the spark gap or sparking members by ionization in such manner that a potential of given value can pass through a considerably longer air gap than otherwise could be the case. As the effectiveness of the radioactive substance does not change or weaken within a reasonable time, this source of radiation requires no attention and may be regarded as practically constant.

Figure 178 shows a horn arrangement used with an aluminum lightning arrester. Note that one side is built out in the form of a movable disconnecting switch lever.

The construction of the horn type of arrester as used by the American River Electric Company is shown in Fig. 179. The horns are of galvanized-iron gas pipe and separated 2.25 inches for 40,000 volts. A jar of water covered with oil is used as a resistance in the ground wire. The oil, of course, is to reduce the evaporation. The arresters are mounted on wooden poles outside of the station.

Another diagram is shown in Fig. 180. This is the type used by the Standard Electric Company, of California, for 40,000-volt transmission lines. It consists of a gap formed between two horns and a tank of salt water, with an induction coil resistance in the ground connection. The horns are of No. 0000 copper wire mounted on ordinary line insulators. Copper strips are immersed in the salt-water resistance. The reactor is a coil 6 inches in diameter and has 18 turns, and the air gap for the 40,000-volt device is made 3 to 3.25 inches wide. The curve of the knee is of great importance in the operation of the device. Figure 180 illustrates two horns which are bent improperly, the angle of the first being too small, so that the arc remains in the gap, and the second having too sharp a crook in the knee, in which case the arc either remains stationary or jumps back again across the gap after having been driven upward.

Figure 181 shows a single-loop series horn which consists of a sharply bent coil connected in series with the line. All the working current flowing in the line flows around the horn which

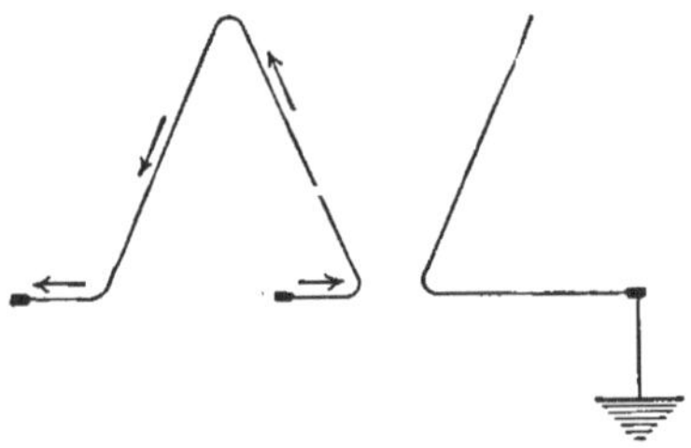

FIG. 181.—Single-loop series horn.

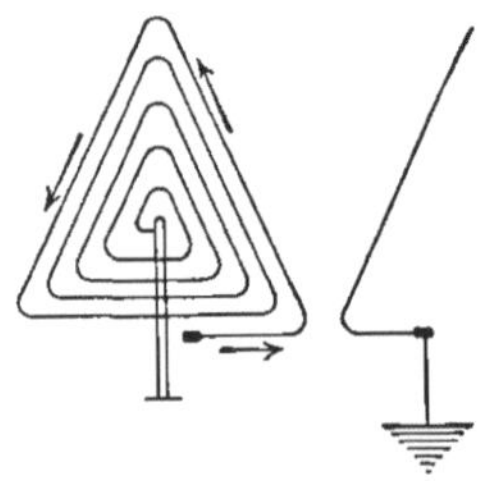

FIG. 182.—Horn with five coil turns.

also acts as an effective reactive coil. The other horn arm is grounded as usual. The device is so constructed that a sharp bend is made in the line at the arcing point, and it is evident that current flowing in the line will meet its first obstruction at this point and consequently have a maximum tendency to jump over the gap. It is claimed that with this construction the accurate setting of the air gap is no longer of prime importance, as an arrester of this construction can have an air gap 50 per cent. greater and still be more sensitive than an arrester of the old type, thus allowing a longer air gap to be used and enabling this type of arrester to operate successfully on a voltage as low as 2,200. It is further claimed that this arrester is very effective in relieving the line from static. The magnetic field will be increased by the amount of working current which is flowing around the horn. On low-voltage lines carrying 25 amp. and above this additional increase in the electrodynamic field is sufficient to allow the arrester to operate without any resistance in the ground lead, the arc rising and lengthening on the horns so rapidly that sufficient resistance is almost instantly interposed and the current limited in this manner.

To increase the choke-coil effect instead of one coil turn there are used several (five) turns, as shown in Fig. 182. The electrodynamic field is thus increased to such an extent that this arrester will operate successfully with very little current flowing in the line up to 45,000 volts, without any limiting resistance connected between it and the ground. A further development of the arrester shows the grounded arm of the horn built up also as a spiral reactance. The ohmic resistance of this reactance being exceedingly low, there will be little opposition to the discharge of static, but when an alternating or line current attempts to follow, the reactance element becomes effective and limits the amount of current which can flow, and also greatly increases the electrodynamic field, causing the arc to rise on the horns very rapidly.

They are several further developments of the above horn arrester. In one the arc is broken into two parts by combining the series reactive coil with two horn gaps in series. Another development has a third gap which is shunted around the grounded reactance and acts as a relief gap. If a very heavy high-frequency discharge breaks down the gaps and the reactance offers too much opposition to the discharge, the third gap will be broken down, giving a direct flow to ground until the current is reduced, when the third gap will break, leaving the discharge to pass through the two gaps and the reactance.

Magnetic Blowout Protective Devices.—One of the first lightning arresters to come into practical use was that invented in 1884 by Prof. Elihu Thomson. This apparatus was so arranged that if a current followed a static discharge to earth, it was made to pass through the winding of an electromagnet, which then excited a strong magnetic field about the gap, with the result that the arc was immediately blown out.

With direct-current this type of device is now used almost exclusively. The gap is in series with a resistance of low inductance. By providing a direct path for the discharge the possibility of short-circuit in the box in which the device is enclosed is reduced to a minimum when the discharges are especially severe.

The connections to d.c. converters and feeders are shown in Figs. 1, 5, 12, 13, and 34. When the device is installed on the line it is usually enclosed in a wooden box supported on the poles carrying the transmission line. This type is used for voltages up to 6,000. A spark-gap on top of the porcelain casing is used to adjust for voltage.

Electrolytic Lightning Arresters.—(*a*) Aluminum devices.

The construction of the aluminum type of arrester is based on the following principles: If an aluminum plate and another plate of some other metal be immersed in a suitable electrolyte, the resulting cell will have the property of passing current in practically only one direction. Only a very small fraction of the current is passed in the opposite dircetion until the applied voltage reaches a certain value. After this limit has been exceeded, however, the current rises much more rapidly with respect to the e.m.f. than Ohm's law would indicate. This action is explained by the presence of a thin dielectric hydroxide film on the surface of the aluminum plate. If both plates were of aluminum, the action of the cell would be analogous to that of a steam safety valve. The device prevents the current as long as the pressure lies below a certain limiting value, but as soon as this limit is exceeded a very large current is established, which continues until the pressure again falls below its critical point. With a suitable electrolyte the above-mentioned dielectric film will be capable of resisting from 335 to 360 volts tension for alternating current.

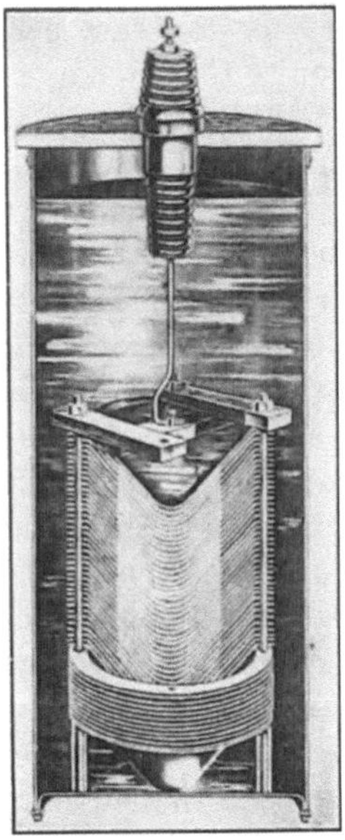

FIG. 183.—Electrolytic lightning arrester.

However, this voltage will vary considerably with the nature of the electrolyte. When a cell is connected permanently to the circuit two conditions are involved, one of which may be distinguished as the temporary critical voltage, and the other as the permanent critical voltage. If the cell has volts applied to it constantly below the permanent critical voltage and then this voltage suddenly increases, there will be a considerable rush of current until the film thickness has been increased to withstand the increased voltage. This action will occur at any potential up to about the permanent critical voltage. Above this value the film is broken down at innumerable points and a flow of current will thus be established.

If the voltage is again reduced below the permanent critical value, say to 300 volts, the excess thickness of film will be gradually dissolved.

The permanent critical voltage of an aluminum tray having a fixed value, it is possible by connecting the trays in series to provide collective resistance to any desired degree.

The film of aluminum hydroxide on the surface of the trays constitutes also a dielectric of a condenser. The flow of current due to capacity varies inversely as the frequency of the current. At commercial frequencies the flow of current in the normal arrester due to the condenser action is small. Lightning discharges are, however, of very high frequencies, and with a discharge at a frequency of 10,000 cycles the commercial arrester would permit a flow of something like 100 amp.

In the commercial arrester the aluminum plates referred to take the form of trays such as shown in Fig. 183. These trays serve as receptacles to hold the electrolyte, the sloping sides permitting the escape of such gases as may form during a discharge of the arrester. The film on their surface is formed by immersing the trays in a tank of electrolyte and passing current from one tray to the other. At first there is heavy flow of current, but as the film builds up the current decreases. The film gives the trays a dull gray appearance; if it is not well formed, they present a slightly shiny appearance. The trays are placed one above the other, each one being separated from the next by a porcelain spacer placed upon the periphery. The number of trays which are placed in a stack depends upon the voltage of the line upon which the arrester is to operate. In practice there is usually one tray for each 290 volts alternating current, or each 350 volts direct current, impressed across the cell. The assembled stacks are placed in steel tanks. The oil in which they are immersed prevents evaporation of the electrolyte, increases the insulation from the trays to the tank, disposes of heat by convection, and also absorbs considerable quantities of heat. This property of oil is very essential because considerable quantities of heat may be developed during a discharge. The natural circulation is aided by the micarta barrier, which also increases the insulation from the tray to the tank.

When an aluminum cell is disconnected from the circuit a part of the film is dissolved by electrolysis. Thus, when such a cell is again reconnected to the circuit, there is a momentary rush of current which reforms that part of the film which had dissolved. This current rush will have increasing values as the intervals of rest of the cell are made greater. Suitable means have to be provided with the arresters for throwing them directly on the line and thus keeping the film always in good condition. On the other hand, when a voltage less than critical voltage is impressed across an electrolytic arrester a small charging current flows, which in the normal commercial arrester is less than 0.8 amp. Its components are at 90° from each other. In an arrester in good condition the direct "leakage" is less than 0.02 amp., while the remainder is capacity current. Thus it will be seen the actual power used in the charging operation is very small.

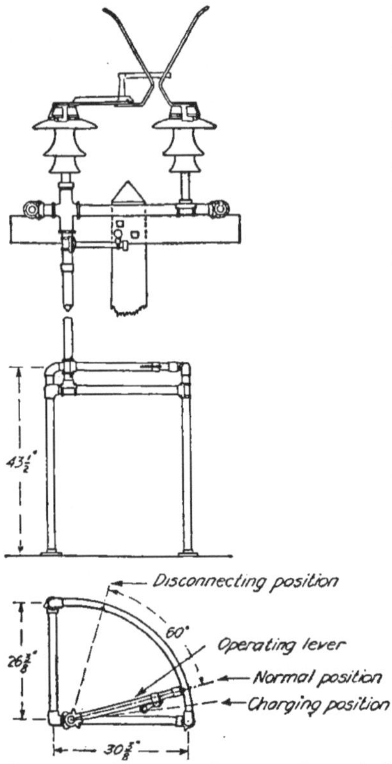

FIG. 184.—Horn-gap disconnecting switch for aluminum arresters.

This flow of current would in time heat the cell of an electrolytic arrester, and to prevent this horn gaps are inserted in the line between the electrolytic element of the arrester and the line to be protected. These horn gaps have to serve therefore a triple function:

First.—As fixed gaps in series with the cells, they prevent the arrester from being subjected continually to the normal line voltage, which would result ultimately in overheating the cells.

Second.—They may act as disconnecting switches to disconnect the arrester from the line for repairs, inspection, etc.

Third.—They can be used as a connecting switch for daily testing and charging to rebuild the dissolved part of the film.

The construction of the horn gap to perform the foregoing functions varies to suit the conditions of line voltage. (See Figs. 178 and 184.)

Some horn gaps possess a common rotating shaft with three metallic projectors which, when the shaft is rotated, bridge the three gaps of the three horns simultaneously so that the three phase legs are charged at the same time. For higher voltages, say over 7,250 volts, each pair of horns is clamped firmly to petticoat insulators, one insulator being fixed rigidly to the pipe frame, while the insulator carrying the other side of the horn can be turned about on its own axis. All three of the movable insulators are joined by a connecting rod so that they move simultaneously. To prevent a discharge on a leg of the arrester from continuing indefinitely, fuses must be inserted to protect the arrester. These detachable fuses are also used as a disconnecting feature of the arrester.

The various manufacturers give tables of setting for the horn.

The number of units or cells containing the aluminum trays used differs with the voltage of the system and depends on the grounding or ungrounding of the neutral.

In an arrester for a grounded neutral circuit each stack of trays normally receives the neutral potential when the arrester discharges, but if a phase becomes grounded accidentally, the line voltage is thrown across each of the other stacks of trays until the oil switch opens the circuit. The line voltage is 173 per cent. of the neutral or normal operating voltage of the cells and therefore about 150 per cent. of the permanent critical voltage of each cell. This means that when a grounded phase occurs this 50 per cents excess dynamic potential is short-circuited through the cell. until the oil switch opens. The amount of energy to be dissipated in the arrester depends upon the kilowatt capacity of the generator, the internal resistance of the cell, and the time required to operate the oil switch. The greater the amount of resistance in the neutral, the longer will be the time required for the oil switch to trip.

Therefore, when the grounding resistance in the neutral is great enough to prevent the oil switch from opening instantaneously, an arrester for a non-grounded neutral system should be installed. No mistake will be made in adopting the four-tank arrester even on grounded Y-circuits.

Arresters for circuits with thoroughly grounded neutrals have three stacks of trays. The bases of the stack of trays are connected to the tanks and grounded. Up to voltages of 7,250 (the Westinghouse Company uses up to 13,200 volts) all the trays are mounted in a single tank, but for higher voltages each stack is mounted in a separate tank. For non-grounded circuits the arresters have four stacks, the bases of which are connected in multiple. The fourth stack is between the multiplex connection and the ground, the object being to give the same protection between line and line as between line and ground. The fourth stack usually carries a transfer device which permits the interchange of the ground stack with one of the line stacks during the operation of charging.

The location of an electrolytic lightning arrester depends, to a considerable extent, upon the station lay-out. All electrolytic arresters for voltages less than 10,000 should be mounted indoors, because for these voltages the manufacturers provide the tanks with indoor entrance bushings and because the horn-gap setting at these low voltages becomes so small that the gaps would be bridged over by rain or snow. For voltages between 10,000 and 39,000 the arrester may be placed either indoors or outdoors as desired, or according to available space, both locations possessing their respective merits. Above 39,000 volts it is advisable to place the horn gaps outdoors so that the arcs can do no damage. In a great many installations the horn-gap structures are placed on the roof and the tanks inside the station.

Aluminum arresters should not be installed where they are exposed to excessive temperature, as the film on the surface of the trays deteriorates more rapidly at high temperature. The disintegration of the film with heat makes charging necessary more frequently in warm than in cold weather. Water-cooled lightning arresters have been successfully operated under temperatures which were said to be as high as 136° F. The longer the trays remain in contact with electrolyte and the higher the temperature, the greater is the rush of current when the arrester is charged. These are the reasons why transformer and generator rooms which have arresters installed near the top where the heated gases are liable to accumulate should be thoroughly ventilated.

Aluminum arresters have been operated very successfully

outdoors even in climates as severe as Canada, Norway, and Sweden. The electrolyte freezes between 20° and 25° F., but there is a considerable amount of heat stored in the jacket of oil surrounding the trays and electrolyte, and it takes a long exposure to cold weather to penetrate this jacket of oil. The electrolyte when frozen increases its resistance considerably, becomes mushy, and temporarily decreases the discharge rate. Hence the efficiency of the arrester. Fortunately, there is very little or no lightning at the time of year when the electrolyte is frozen, and its conductivity, even when frozen, is sufficient to take care of internal disturbances on a system.

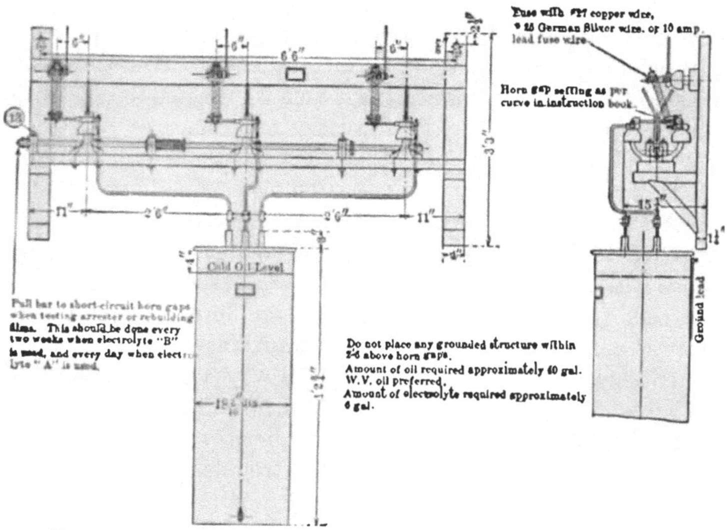

FIG. 185.—Aluminum arrester for 7,000 volts and grounded neutral.

Cleanliness is very essential in the handling of treated aluminum trays and electrolyte, and on the installation of arresters the tanks should be cleaned and the dirt and excelsior thoroughly removed from the trays. When a tray filled with electrolyte is being lowered into the tank, care should be used not to spill the electrolyte, as the failure of any tray to be in contact with the adjacent one through the medium of electrolyte will render that column inoperative. When filling the tank with oil it should be poured against the side of the tank, or siphoned or pumped in, to avoid splashing, and should come within a few inches of the top of the tank to cover the top of the trays.

Electrolyte arresters should be inspected at frequent intervals to see if they are in good condition, especially after a year's service. If at any time they are found not to be in good shape, they should be overhauled, the trays should be cleaned, by washing with gasoline or good soap, rinsed in clear water, damaged trays replaced, and all trays filled with new electrolyte.

Figure 185 shows an aluminum arrester for 7,000 volts and grounded neutral. Figure 186 represents a 47,000-volt aluminum arrester for a three-phase ungrounded neutral system.

Equipment installed in a 60,000 volt plant has these characteristics. Each branch of the three-phase circuit is provided with three horns. The first is grounded by a 6-foot fuse, which is positive and effective in operation. The second is in series with a resistance consisting of concrete blocks. The third is in series with an aluminum arrester which is expected to take up all static discharges and impulses in times of storm.

The recent addition of a charging resistance will insure a material improvement in the operation of the aluminum arrester, as well as an increased life of trays and electrolyte. The horn gap with charging resistance has an auxiliary horn mounted above and insulated from the regular horn in such a manner as to intercept the arc if it rises on the regular horn. Enough resistance is connected in series with this auxiliary horn so that the current flow and arc across this gap are always limited to a moderate value. Since the mechanism is so arranged that the charging is always done through the auxiliary horn, the current rush is limited during the charging, and thus troubles from carelessness or ignorance are avoided. It also gives a more uniform charging current. Light discharges will pass the auxiliary gap through the series resistance to the cells. If the discharge is heavy, the resistance offers impedance to cause the spark to pass the main horn. If the cells through either negligence or some untoward condition are in poor form, the dynamic current may follow the discharge across the main gap and the arc will rise to the safety horn and be extinguished through a resistance.

(*b*) Liquid electrode arrester.

The above name has been adopted for this type of apparatus to distinguish it from aluminum arresters, which, though they also contain a liquid electrolyte, operate in an entirely different manner. As was noted above, the discharges in the aluminum type are regulated by a dielectric film. In the liquid electrode type, however, according to the inventor, *Prof.* E. E. F. Creighton,[1] the electrolyte itself plays the main part. The following

FIG. 186.—Electrolytic lightning arrester for 47,000-volts, three-phase; ungrounded neutral with horn gaps open.

discussion must necessarily be restricted to a somewhat general

[1] "New Principles in the Design of Lightning Arresters," E. E. F. Creighton, *Proc. A. I. E. E.*, March 2, 1907.

description, as the apparatus has hardly emerged from the experimental stage, so that no results of practical importance have as yet been obtainable. The arrester discharges only at a critical limiting voltage, and at the same time reduces or entirely suppresses the machine current, at normal voltage, without a series resistance. A very high pressure of about 1,500 volts is required to establish a current through the electrolyte from one

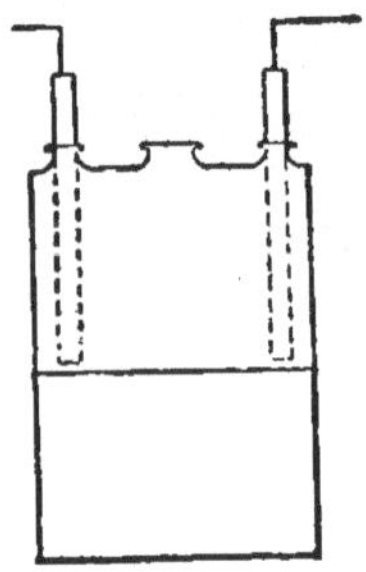

FIG. 187.—Liquid electrode cell.

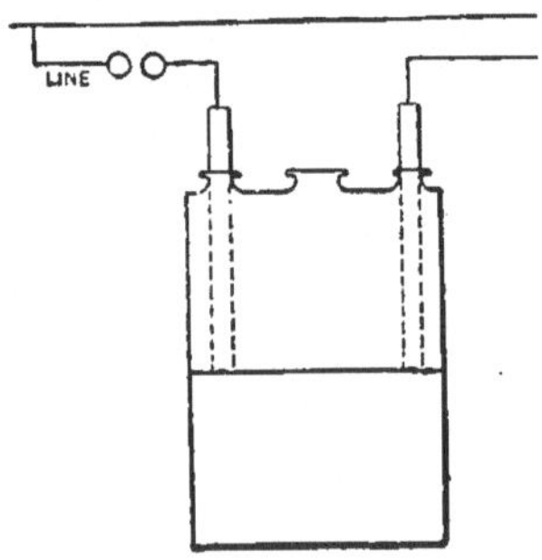

FIG. 188.—Liquid electrode arrester.

electrode to the other, and the current is limited by the counter-e.m.f. of the arc. One cell of a liquid electrode arrester is shown in Fig. 187, in which the electrodes form small gaps with the electrolyte. Experiments have shown that several hundred static discharges of 1,000 amp. starting current will pass between the electrodes before any considerable machine current follows the discharges. The combined counter-e.m.f of all the cells is greater than the line voltage. Therefore sparking will not cause arcing. The cell has a critical limiting voltage below which no discharge is possible and no current can exist. By adjusting the spark gaps the critical voltage of the cells may be regulated. If the electrodes dip into the electrolyte, an outside spark gap must be connected in series with the cell in order to keep back the normal pressure, which is set for the given voltage limit. (See Fig. 188.) When a high pressure breaks down the spark gap and reaches the electrodes, arcs are formed at their ends which drive the electrolyte away, thus automatically increasing the length of the arcs. Since the arc voltage is greater than the impressed voltage, the current quickly dies out. The series gap keeps out the normal pressure, whereupon the electrolyte recovers its original horizontal position touching the electrodes.

The arc voltage depends upon the length of the arc, and this in turn upon the value of the current causing the depressions in the surface of the electrolyte.

Condensers.—High-frequency discharges confined only to overhead lines do not constitute such danger from the point of view of any high potential which they may produce, but only because their effect on the alternators or transformers is necessarily concentrated upon the first few turns of the windings. Even pressures which are lower than the working pressure greatly endanger the windings if they are of high-frequency nature. It is therefore necessary to produce a zero potential for high frequency at the entrance of the lines into the station and a suitable self-induction (see "Reactive Coils") which reflects back the wave on the line without permitting it to enter the station. These reflected waves diminish in amplitude and are finally absorbed by the ohmic resistance of the line.

A suitable condenser has to be inserted to produce the required zero potential. This condenser is placed on the ground, and in consequence of the short ground wire between the protective apparatus and the ground plate the same is at zero potential. The condenser is known as the only apparatus having a resistance which is inversely proportional to the frequency of the current, and therefore it produces a zero point of potential for every high-frequency disturbance at the point where the line enters the station; that is to say, it constitutes a dead ground for all kinds of surge tensions, while simultaneously offering an enormous resistance to the passage of the line current to ground.

The Société Générale des Condensateurs in Fribourg, Switzerland, produces the Moscicki condenser, which is especially adapted for the foregoing purpose.

The older condensers, which are made up of some dielectric with coatings composed of metal foil, with contact between the metallic coating and the dielectric, are not entirely satisfactory, because the enclosed air particles and the gases produced by the decomposition of the sticking medium are set in vibration and eventually they separate the coating from the dielectric and produce a puncture of the latter. In all condensers which have the edge of the coating exposed to air, brush discharges are produced which pass from the outer coating to the inner coating over the surface of the dielectric. In a little while these discharges bring about the destruction of the condenser. These troubles with the older condensers checked the commercial use of such apparatus for protective purposes until the Moscicki condenser came on the market. This condenser consists of a glass tube (see Fig. 189) thickened at the neck, where the edges of the coating are situated, because it has been found that at this spot the static tension is much greater than elsewhere. The coating is produced by the chemical deposition of silver which is afterward plated with copper, and it does not enclose any air space.

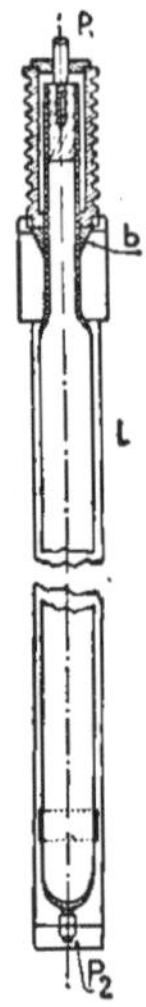

FIG. 189.—The Moscicki condenser.

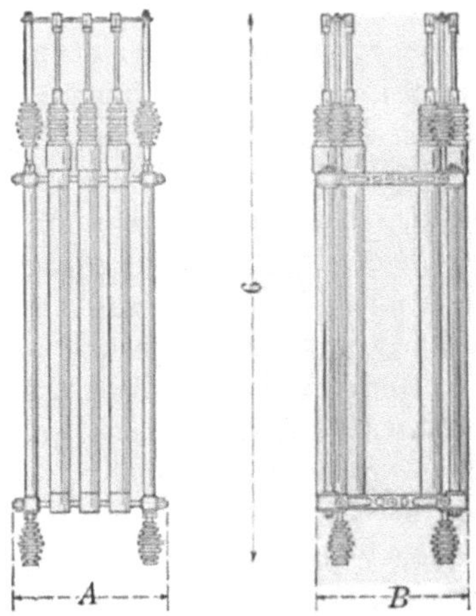

FIG. 190.—Moscicki condenser battery.

The neck of the condenser is fixed inside a high-tension porcelain insulator, and the space between the neck and insulator is filled with an insulating compound. The edge of the coating is completely buried in this mass in order to avoid the surface discharge.

Each condenser is mounted in a brass or copper tube, the space between the condenser and tube being filled with a non-freezing mixture of water and glycerine, so that any local heating which might bring about the perforation of the dielectric is rendered impossible, as the liquid distributes the heating over the whole surface.

These tube condensers can be conveniently grouped according to the protection requirements of different-sized batteries.

(See Fig. 190.) Each condenser is provided with its own fuse in such a way that should one tube puncture, the fuse will blow and the other condenser tubes will still remain in the circuit.

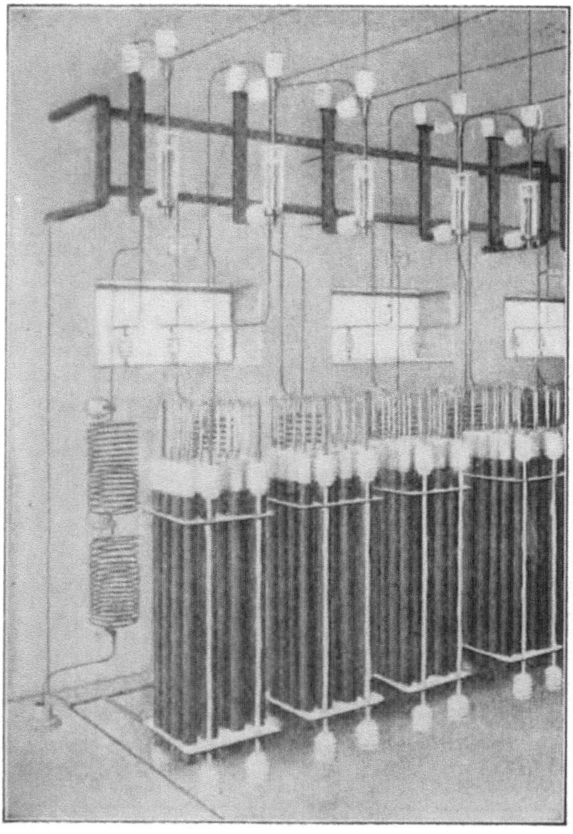

FIG. 191.—Condenser batteries on outgoing lines of a 13,500-volt generating station.

The upper paralleling bars of the batteries are connected to the line and the frame of the apparatus to ground. Figure 191 shows the installation of condenser batteries on outgoing lines of a 13,500-volt generating station. The generating station belongs to a system comprising 434 miles of overhead lines and more than 250 transformer sub-stations, all entirely protected by condenser batteries. The halftone shows also the use of choke coils and liquid disconnecting switches. These switches make it impossible that at the closing or breaking of the connection between line and condenser battery there should be capacity in series with a

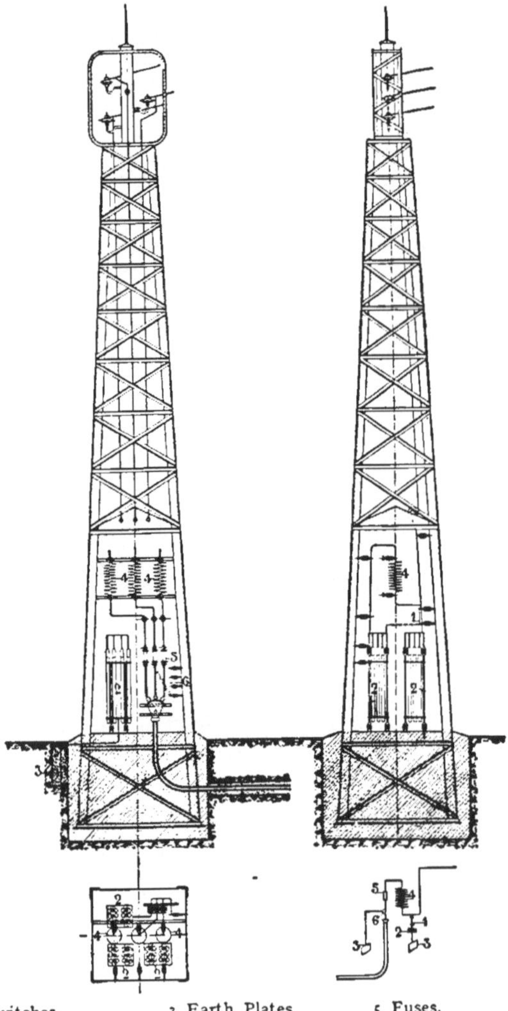

1 Switches. 2 Condensers. 3 Earth Plates. 4 Choking Coils. 5 Fuses. 6 Earthing Switches.

FIG. 192.—Tower construction for junction between cable and overhead line with condenser batteries.

spark gap, an arrangement which would temporarily reproduce dangerous surges.

Figure 192 shows a tower construction for junction between cable and overhead lines with condenser-battery protection.

Reactive Coils.—Reactance in alternating-current circuits may be divided into three general classes by the function they are expected to perform, as follows:

1. For protection against high-frequency disturbances.
2. For voltage regulation and for providing synchronizing power.
3. For limiting the current in case of breakdown of lines or apparatus.

Class 1.—Refers to reactive coils such as are used in connection with lightning arresters or other protective devices for keeping out high-frequency disturbances.

High reactance in the line will break down any surge or wave trying to enter a station. Part of the wave is reflected, and part is allowed to pass. The latter portion must not exceed a value determined by the insulation of the station apparatus. Another way of expressing this function is: To lower the frequency of the oscillation so that whatever charge gets through the choke coil will be of a frequency too low to cause a serious drop of potential around the first turns of the end coil in either generator or transformer. A high reactance will also hold up a wave in point of time long enough to give the lightning arrester an opportunity to discharge.

A reactor is a device of high reactance designed to give protection against disturbances occurring when the line is carried overhead. Such coils are of no value for disturbances in underground cables or in the station. Each feeder in the station, however, should be provided with one of these coils. They are always used with lightning arresters, one arrester sometimes being placed on each side of the coil to hold up both inside and outside disturbances.

The coil should be so arranged that if continual surges are set up in the circuit, a resonant voltage due to the presence of the choke coil cannot build up at the transformer or generator terminals. This factor is often a menace to the insulation of machines. The coil has to be installed in such a way that surges which originate in a transformer will be prevented from passing to the lightning arrester.

Some engineers recommend the use of reactors of few turns, that is, of medium reactance. The turns are insulated from each other by air space which prevents permanent short-circuit, and they must therefore not be spaced too closely. In constructing these coils great attention must be paid to providing sufficient radiating or cooling surface, in order that the heat generated may not weaken the insulation and cause a breakdown at high pressures.

In such coils iron is not permissible because eddy currents set up by high frequencies would probably be sufficient practically to eliminate the effect of the reactance.

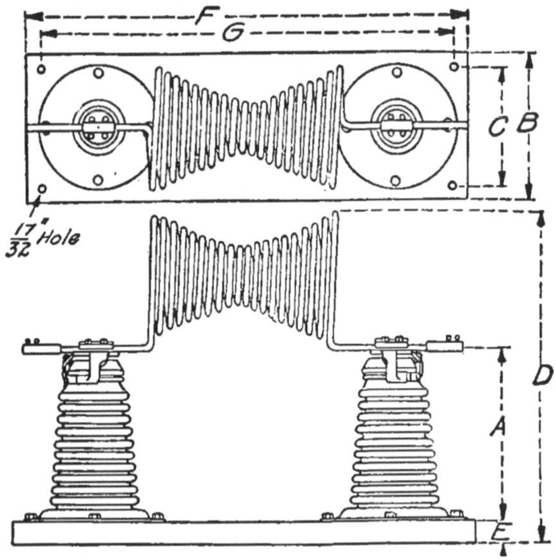

Fig. 193.—Choke coil up to 60,000 volts.

Figure 193 illustrates a type of choke coil used up to 60,000 volts. Those on circuits not exceeding 6,600 volts are made of several turns of insulated wire, while for voltages above 6,600 volts the hour-glass type with air-insulated turns is used and the coil is mounted on a steel, slate, or marble base.

The Westinghouse Company produces a similar choke coil made of aluminum rod wound into a helix about 15 inches in diameter and having twenty turns. The coil is thoroughly insulated, supported on two insulators which are mounted on a wooden base. Such coils are used for potentials of from 2,200 to 110,000 volts on transformers and station wiring, but they are not recommended, on account of their lesser effectiveness, for the protection of generators.

In Figs. 31 and 32 there are shown a number of copper coils which serve the same purpose as reactors and which are used for d.c. feeders.

Figure 194 shows the Westinghouse form 7 reactance coil, built for 2,500 to 25,000 volts and 260 amp. capacity. They are air-cooled and have a large number of turns, and therefore high inductance, since for high-tension apparatus a greater "reactive

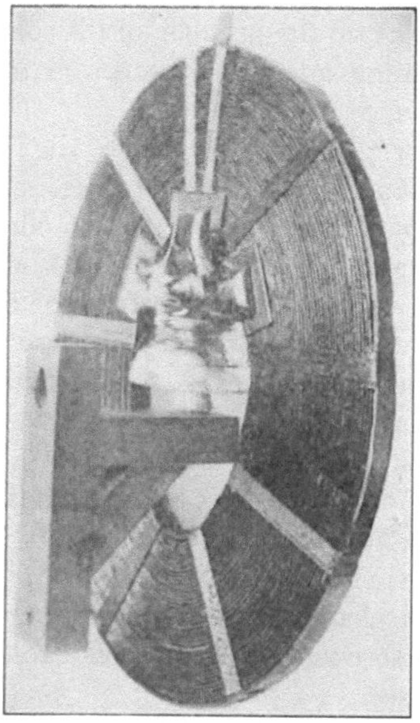

Fig. 194.—Type of choke coil with support.

effect" is required. For voltages over 25,000 the Westinghouse coil is oil-cooled. It contains a much larger number of turns than the air-cooled device and hence has a higher inductance. Since there is a tendency with high inductance for the discharge to jump across from turn to turn, a high insulation of the winding is essential.

To sum up: A reactor will protect apparatus in the station against high-frequency fluctuations, but it offers absolutely no protection against static discharges or low frequencies. This is due to the fact that the reactance of the coil is limited as it must be less than that allowable for the normal line voltage.

Class 2.—Refers to reactances used in connection with compound-wound rotaries for voltage regulation, or inserted in lines in connection with rotative apparatus, in order to give stability to synchronous motor operation and to aid in tying generator stations together over lines of high resistance. Such reactances, which also include reactance coils in arc lamps and series arc regulators, are built with iron cores.

Class 3.—The rapid increase in the size of electric plants and systems, as well as the tendency to use larger units, and cousequently the concentration of large amounts of power on the busbars, has brought up the problem of the constantly increasing difficulty of obtaining adequate protection from short-circuits and similar accidental faults.

A short-circuit or a ground on a feeder, the failure of a current transformer, the hanging of an arc at a fuse, or the throwing on the busbars of a generator slightly out of phase, have resulted in tearing out generator and transformer windings, blowing oil switches to pieces, and even in the destruction of masonry walls.

The effects of short-circuits noticed on the apparatus are:[1]

1. Severe mechanical stresses along the path of short-circuit, becoming particularly damaging in the distortion of generator armature windings.

2. The inability of the existing normal types of oil switches to open short-circuits on large plants successfully.

The generator windings were not the only path of the short-circuit in which mechanical stresses were observed. Bending of busbars and the throwing of cables off cable racks were also experienced at times.

One of the elements that govern the amount of mechanical stress set up in armature windings is the normal capacity per pole of the generator. With the advent of turbo-generators came much higher speeds, therefore fewer poles, as well as a marked increase in generator capacities. The problem of winding stresses, therefore, has reached a maximum with the coming of the large-capacity, high-speed modern turbo-generators.

Analysis as well as experiment show that during the first few cycles of a short-circuit the amount of current that may be taken from a generator or other synchronous machine is vastly in excess of the value to which it settles after a few seconds. It is the existence of this enormous first rush of current that gives rise to the mentioned difficulties. The natural remedy that

[1] P. M. Lincoln: "Protection Against Short-circuits," *The Electric Journal*, December, 1913.

suggested itself was to employ reactive coils in generator leads as a means of cutting down this first rush of current.

In large electric plants power is distributed in general through underground circuits. Such a plant will have, normally, many more feeders than there are generators. Assuming a short-circuit on a feeder on such a system at a point close to the busbars, an enormous current will then flow through this short-circuit, being at maximum during the first cycles. During the first few cycles practically the only impediment offered to the passage of current is the inherent reactance of the generator windings themselves. The voltage that will be taken up in the arc at the point of the short-circuit may possibly have a value as high as a few hundred volts at the utmost. This will be but a very small proportion of the total voltage that existed just before the short-circuit occurred. The remainder of the busbar voltage must be accounted for in other ways. Some of it appears in the resistance and reactance of the busbars themselves and of the generator leads up to the point of short-circuit. The amount of resistance and reactance between the generator terminals and the point of short-circuit is, however, very small compared to the inherent resistance and reactance of the generator windings themselves. By far the larger part of the voltage, therefore, will be taken up in overcoming this generator impedance. Since the generator reactance is many times as great as its resistance, most of the voltage, therefore, will be taken up in overcoming inherent generator reactance. The condition of short-circuit will give rise to excessive currents through the generators, the value being limited only by the design of the generator as regards inherent reactance.

The current through the feeder switch where the short-circuit occurred will be still more excessive, since the current from all of the generators will pass through this switch. Not only do the generators feed current into such a short-circuit, but every synchronons machine on the entire system does the same. The effect of the distant synchronous machines will be limited, owing to the fact that they will probably have a certain amount of feeder cable between them and the point of short-circuit, but their effect is such as to increase the strain on the switch over what it would be if the load did not consist of synchronous apparatus. The shock on the feeder switch will, therefore, be tremendously severe. The shock on the generator switches, if these are automatic, will also be severe since each generator switch will momentarily deliver all of the current that it is possible to obtain from the generator, and, if the generator switch is ever to trip, it will trip under such conditions. It is obvious, however, that the shock on the generator switch cannot be as severe as that upon the feeder switch, since the latter must pass the current from all synchronous apparatus, whereas the generator switch is limited to the maximum possible output of a single machine.

There is an effect that from the standpoint of protection to service is still more important, and that is the dropping out of step of synchronous apparatus during the interval that the short-circuit persists. Since practically all of the voltage of the busbars is taken up in circulating current through the generator windings, the proportion that remains on the busbars during the persistence of the short-circuit is very small. This means that no current can circulate between synchronous machines during the short-circuit interval, and they are, therefore, free to drift apart during this interval as conditions thereon may dictate.

A synchronous motor will begin to slow down rapidly, owing not only to the drag of its load, but also to the energy which the motor itself begins to pump into the short-circuit. The generators themselves may also begin to slow down, owing to the fact that they are pumping energy into the short-circuit. The rate of slowing down of the various synchronous machines will not be the same, since it will depend upon the amount of drag, as well as upon the momentum of the revolving parts. Therefore, if the short-circuit persists more than a very brief length of time, the voltage when it reappears will find another short-circuit, owing to the fact that the synchronous apparatus of the system is in improper phase relation. If the amount of synchronous apparatus on the system is relatively small, it may be dragged back into step within a few seconds. However, in order to drag synchronous apparatus into step with a generator, the amount of current taken is so excessive that it is impossible for the generators to supply sufficient current for this purpose, if all of the load on the generators consists of synchronous apparatus.

The result obtained by using reactive coils may be regarded from two viewpoints, namely, as a protection to apparatus and as a protection to service. To the operating man protection to service is paramount; if need be, he is willing to sacrifice protection to apparatus to a considerable extent to obtain this end.

The reactive coils may be applied as follows:

(*a*) In the generator or transformer leads.

(*b*) In feeder circuits.

(*c*) In the station busbars.

(*a*) The total reactance for high-speed generators will vary with the size of the unit, since it is desirable to limit the total energy which may flow and which can safely be handled by an oil switch. The following values are considered conservative, the reactance in each case being total reactance, part of which may be in the machine itself and part in the external coils:

	Per Cent.
For generators up to 8,000 kw. capacity	6
For generators of 8,000 to 15,000 kw. capacity	8
For generators of 15,000 to 20,000 kw. capacity	10
For generators over 20,000 kw. capacity	12

The above-estimated percentages apply to individual units and may vary materially when several generators are operated in multiple.

For transformers located in main stations it is recommended that the same amounts of reactance as for generators be applied.

The effects that the presence of reactive coils in the generator leads will have in the case of a short-circuit on one of the feeders at a point close to the busbars are the following:

It is evident that the rush of current through the generator will be reduced, owing to the fact that the current is now limited by external as well as internal reactance of the generator windings. Bearing in mind that the stresses in the windings vary as the square of the current therein it is evident that the problem of winding distortion is well met by this remedy.

The amount of current through the generator and feeder switch has been decreased. Whether or not the easing up of the oil switches is sufficient to allow them to act properly is entirely a question of degree and depends upon the conditions surrounding the case. In any event any feeder switch is in position to receive the maximum possible stress that may be caused by this system, and all of the feeder switches should be capable of withstanding the maximum shock that the system is able to give.

The disappearance of voltage on the busbars during the persistence of a short-circuit is more complete than if there were no reactive coils in the generator leads. The tendency, therefore, for synchronous apparatus to stay in step is materially reduced. It is the prevention of this dropping out of step of synchronous apparatus that is of the utmost importance from the viewpoint of continuity of service. Protection to generator windings and to switches constitutes more of a protection to apparatus than a protection to service, whereas the maintenance of voltage on the busbars during the persistence of a short-circuit constitutes a protection to service rather than to apparatus. This effect of the more certain drop out of step of synchronous apparatus constitutes a very decided disadvantage of the placing of reactive coils in generator leads.

(*b*) When the reactive coils are placed in the feeder leads instead of in the generator leads they are usually made from 3 to 6 per cent. of reactive value or possibly less. Therefore, while the number of the coils is considerably increased, the aggregate kva. capacity of the reactive coils is reduced compared with the corresponding value when installed in the generator leads only.

When a short-circuit occurs in one of the feeder circuits near the busbars, the strain of the generator windings will be reduced to a much lower value than in case *a*. This follows from the consideration that the capacity of a single generator is much greater than the capacity of a single feeder, and consequently the value of a 3 per cent. reactance in a feeder circuit of say 3,000 kva. capacity will be as high as that of even a 10 per cent. reactance on a generator capacity of say 10,000 kva. Moreover, all of the current which passes through the feeder reactance must come from the generators, and this total current is divided into as many parts as there are generators operating.

The stresses on the feeder switches are very much reduced by this arrangement. The switches will have therefore a much ampler margin of safety, or smaller and cheaper switches may be used and the same margin of safety over the arrangement described in (*a*) still be retained.

The voltage on the busbar will stay at nearly normal, and there is little tendency for the synchronous apparatus of the system to drop out of step.

This arrangement provides reactive protection only for those short-circuits which occur on the feeders beyond the point where the reactive coils are inserted, and it does not give protection to short-circuits which occur on the busbars or upon the generator leads themselves. The objection is, however, largely offset by the fact that the probability of a short-circuit occurring is vastly greater upon feeder circuits than upon the conductors back of these feeders. In the ordinary system the amount of feeder circuit is somewhere around 100 to 1,000 times the aggregate length of the circuit back of the feeder reactive coils.

It is possible also to design a construction and installation of busbars and generator leads such as to reduce the possibility of a short-circuit thereon almost to the vanishing point. Busbars, for instance, are normally run in brick or concrete compartments, so as to make it almost impossible to obtain a short-circuit on them. Generator leads may be run with single-conductor cables in such a manner that a short-circuit between them must be preceded by a ground to the lead sheathing, and a proper installation of grounding resistance may be made such as to eliminate such a ground before it can develop into a short-circuit.

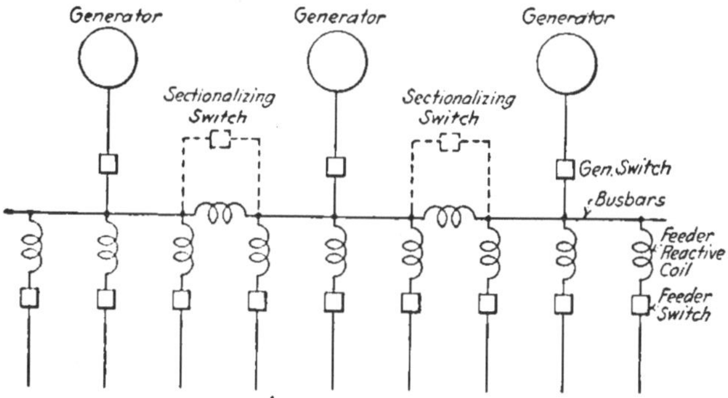

FIG. 195.—Reactive coils used to separate busbars.

Many engineers, therefore, will be content with the protection which is offered by the use of feeder reactance and will not deem it necessary to install protection against the very remote contingency of a short-circuit upon the busbars or the generator circuits.

(c) Figure 195 represents an arrangement where the busbars are separated from each other by interposing reactive coils. The feeders possess also reactive coils described as in (b). The number of the bus sections corresponds to the number of generators. The feeders are likewise divided into the same number of groups of as nearly equal capacity as it is possible to arrange. The bus section reactive coils can be made of such a value that even with a short-circuit on one section of the busbars the voltage will be sufficient to insure that synchronous apparatus on the adjacent sections will still be held in step.

An objection to this arrangement may be raised that there will be also sufficient voltage drop through these coils from the normal current that may circulate between bus sections to cause an inadmissible difference of voltage between adjacent sections. This can be met by the use of quick-acting oil switches and by an arrangement of bus section switches to short-circuit the reactances.

It is quite possible to make the speed of such oil switches high enough so that the current limiting function of the reactance coil will come in action in less than 0.15 second after a short-circuit has occurred, and in general this is not a sufficient length of time to cause synchronous apparatus to fall out of step. This scheme cannot prevent service from being interrupted upon the group of feeders connected to the short-circuited section of the busbars, but it should be capable of preventing the interruption of service on all other groups. The sectionalizing switches should be set for such a current that they will not trip at a value smaller than the maximum that can be passed by a single feeder reactance. Thus the sectionalizing switches will not come into action for any condition except a short-circuit either upon a busbar or upon generator leads.

Another method of shunting bus reactance coils when they are not needed is by the use of a ring and transfer bus. The main bus is a ring bus provided with reactances between adjacent generator sections, while the transfer bus may be either a straight or a ring bus and is without reactances.

With this system of connections, the middle one of the running generators is connected to both main and transfer buses, while all other running generators are connected to the main bus only. A simple system of interlocks can be provided between generator oil switches to prevent more than one generator at a time from being connected to the transfer bus. On every bus section the oil switches connecting the generators to the main bus should be interlocked with the oil switches connecting the feeders to the transfer bus so that it will not be possible for the operator to shunt the reactances when they should be in service.

An objection found to the scheme shown in Fig. 195 is that the generator feeding the short-circuited section of the busbar has no reactive-coil buffer to ease the shock on the armature winding caused by the short-circuiting current, and further, that when the sectionalized reactive coils are short-circuited by switches, as shown by the dotted lines, a similar shock will be experienced by every generator on the system, since it is impossible for these sectionalizing switches to act soon enough to prevent such a shock upon the windings. In such case, when the generators are not self-protective, it is wise, by all means to use reactive coils in the generator circuits. However, modern generators are being built so that they can and do withstand the shock of short-circuits without the assistance of reactive coils in their circuits.

Generator reactive coils do not accomplish the same ends as do feeder and bus sectionalizing reactive coils, since with generator reactive coils only it is impossible to prevent drop in voltage upon the busbars, and the consequent pulling out of step of all synchronous apparatus, in case of short-circuit on the main busbars. In brief the function of the generator reactive coils is to give protection to apparatus while the function of the feeder and the bus sectionalizing reactive coils is to afford protection to service.

The external reactance put in service in addition to the internal reactance of the transformers, generators, etc., must be made as nearly fireproof as possible, and very strong, since the greater part of the shock caused by short-circuit has to be absorbed in the reactance. A form of reactance with a hexagonal concrete core on a porcelain block has been developed. Radial arms around which bare cables are wound are embedded in the concrete. The cable is insulated by mica sleeves at the point of contact with the radial arms, and usually two or three layers are wound on the cores. (See Fig. 196.) One of the largest installations of this kind is in the Quarry Street and Fisk Street stations of the Commonwealth Edison Company, Chicago, where each machine is protected by its own reactance, thereby limiting the flow of current in case of short-circuit to about twelve times normal. The busbar is divided in the Fisk Street station, and between the two sections a busbar reactance is installed. The tie feeders of the Quarry Street station are each protected with reactance, and the entire system is thus operated in parallel.

The Mississippi River Power Company at Keokuk, Ia., is using the reactive coils in the generating station in the following way: The main busbars of 11,000-volt tension are in duplicate, one extending solidly through the station while the second busbar is broken up into sections each of which groups together three or four generators and step-up transformers (each 9,000 kva.-, 11,000/110,000 volts). These sections are tied to one another by automatic oil switches and reactive coils which limit the flow of current from one section to another in case of a fault. The reactive coils are, therefore, located in the tie links between bus sections of four generators. They are constructed of 1-inch copper cable wound on wooden forms and are adjustable for reactances of 4, 6 and 8 per cent. at 9,000 kva.

Fig. 196.—Reactance with hexagonal concrete core.

Owing to the slow-speed design of the main generating units (57.7 r.p.m.) it has been possible to include considerable reactance in the windings of the large machines themselves, so that on short-circuit these units will deliver only from three to five times their normal full-load output. The transformers are also wound for 5.7 per cent. reactance.

While it is evident that the individual units are thus well protected against the occurrence of heavy discharges in case of short-circuit, the sheer size of the whole plant (at present 135,000 kw., ultimately 270,000 kw.) is such that the total short-circuit output of a number of these machines might cause tremendous damage.

Through the installation of the bus-sectionalizing reactive coils the total current input into a dead short-circuit on the low-tension busbars is limited to a maximum of from thirty to forty times the full-load current of one generator unit, depending upon the per cent. reactance used on the above reactive coils. The total current is independent of whether the station consists of the present fifteen or the ultimate thirty units; that is, only half of the units in the ultimate station will be affected by a dead short-circuit in the low-tension busbar.

In case of a short-circuit on one of the high-tension lines close to the power station not more than two sections on each side of the one short-circuited will be affected. This result is attributable to the intervention of transformer reactance in series with the reactance of the busbar reactive coils. The total current input into a short-circuit, whether high-tension or low-tension, due to the high generator reactance is practically the same whether taken at the instant of short-circuit or after a considerable time delay.

The bus-sectionalizing switches are opened by inverse time-limit relays operated by the current in the transformers connected to an outgoing line, so that in case of line trouble the section affected will be automatically isolated. The bus-sectionalizing switch is in series with the bus reactive coil and does not serve to short-circuit the same.

The settings of the bus-sectionalizing switches are high in order to defer the opening of the circuit until no possibility remains of clearing up the trouble easily.

The reactances of the different reactive coils are so proportioned that the busbar voltage on a section adjacent to the one short-circuited will always be higher than 75 per cent. of normal voltage. By parallel operation of all generators, therefore, there is obtained without spreading disturbances to adjacent sections a stability in operation and a flexibility which reduce to a minimum the duplication of equipment required to maintain reliable service.

An additional automatic feature, intended to secure the quickest possible clearing of arcing grounds or crosses, is a special contact in the Tirrill regulator so arranged that the opening of the oil switch will introduce reactance into the generator excitation, at once depressing the voltage in the group and line so that the arc will go out from lack of potential to sustain it. By this means it is hoped to clear a certain class of troubles without even interrupting the operation of synchronous machines whose rotating inertia will carry them over until full voltage is, restored.

Another method of limiting the trouble in case of short-circuit has been to ground the neutral of the system, in some cases with resistance and in some without. As a rule, grounding the neutral of the generators on an underground system will protect the system from a complete short-circuit across the phases, since most of the trouble on underground cables starts first as a puncture from one conductor to the lead sheets. With a ground on the neutral of the system, the oil switch of the particular feeder would be caused to open. Selective action on a number of outgoing underground feeders is thus secured. Without the ground connection, the oil switch of the feeder would not open until the puncture had developed sufficiently to cause a short-circuit across the phase conductor. This usually takes a certain definite time, and much burning may, therefore, take place before the short-circuit develops.

Resistance is inserted in this neutral ground circuit to limit the flow of current in case of breakdown of any of the cables, although too much resistance would mean that the switches would not open, and consequently the fault would probably develop into a short-circuit across the phase. It is, therefore, advisable to limit the amount of resistance installed, and in all cases it should be so designed that in case of short-circuits a sufficient amount of current can flow to trip the switch with the lowest overload setting. Even with this amount of resistance in circuit, the feeder will not be disconnected from the system as promptly as it would be without the resistance; since, of course, in case of puncture from one conductor to the lead sheets there is a certain amount of resistance in circuit. With the installation of reactance in circuit there is less need for resistance in the neutral circuit because the current flow is limited by this reactance.

The resistance to ground should be non-inductive to

eliminate the danger of high-frequency oscillations between line and ground.

The Grounded Wire.—Although this treatise is properly confined to central and sub-station installation arrangements, it may not be out of place to devote some attention to a protective device which serves mostly to shield the transmission lines themselves, but which affects the stations indirectly. This device is known as the grounded wire for overhead lines. In a body possessing perfect conductivity no disturbances of outside origin can be produced. If, therefore, a transmission line were incased in a grounded electrical conducting shell, electrostatic and other atmospheric influences would be eliminated. Precisely this is the condition attained in cables buried underground. But as the cables cannot always be laid in this position, the ground can be raised to the wire by an auxiliary grounded wire or series of wires. The theory of the shielding action of ground wires, according to Ralph D. Mershon,[1] is as follows:

The electrostatic induction of the clouds induces a bound static charge on the transmission line opposed in sign to that in the clouds, and at the same time liberates a free charge of the same sign. The free charge has a tendency to pass to earth. It will pass by gradual leakage over and through the insulation of the system provided the approach of the cloud is slow enough to give time for such leakage. If not, it might puncture the insulation and thus pass to earth. The intensity of the charge will depend upon the potential of the line wires due to the charge of the cloud. Suppose that there be near the transmission wires other wires parallel to them and grounded at frequent intervals. They also will be subject to the inductive action, and the charge set free upon them will pass to earth as fast as liberated, the bound charge of the opposite sign to that of the cloud remaining and depending for its magnitude on the potential due to the cloud and the electrostatic capacity of the grounded wires. Under these conditions the intensity of the charge on the transmission wires will no longer depend only upon the potential due to the cloud, but upon the combined action of the charge of the cloud and the bound charge of the grounded wires. The potential of the line wires will be equal to the difference of the potentials due respectively to the cloud and the grounded wires and

[1] Ralph D. Mershon: "The Grounded Wire as a Protection Against Lightning," *Proc. A. I. E. E.*, 1903.

will in general be less than that due to the cloud. Suppose now that the cloud be discharged by a lightning flash to earth. The bound charge on the line will suddenly be set free, and if the ground wire were not present, this suddenly released charge might puncture the insulation in order to reach ground. But as auxiliary ground wires are provided there will be less tendency toward the puncture of the insulation of the system, because of the fact that the impressed potential of the line wires is less with the grounded wires than without them. The liberated charge in the ground wire passes to earth. This charge is obstructed more or less by the inductance of the discharge path, the effectiveness of this inductive obstruction depending upon the suddenness with which the cloud discharges. The worst condition would be that under which the charge on the ground wires could not pass to ground at all, in which case the sum of the two charges of the line wires will be just equal to that which would have existed if there were no ground wires.

The grounded wire affords even more effective protection against electrostatic accumulations from wind and rain and against the potential difference between the atmospheric strata due to the different altitudes through which the line passes. The use of grounded wires, therefore, greatly relieves the stress which can be thrown on the station apparatus and reduces the duty required of the station lightning arresters. It also protects the poles from direct stroke by leading the discharge to ground without damage to the line insulators. These wires are, to a certain extent, a protection against electromagnetic as well as electrostatic effects of lightning discharges in the neighborhood of the transmission line, as they are interlocked inductively with the main line, so that a part of the energy of the wave train is absorbed. The result is a rapid fall of the wave.

The grounded wires are usually of galvanized iron of sufficient size to offer a slight resistance to the discharge and to be self-supporting. The latter property is important, for a break of the wire might cause considerable trouble in the transmission lines below it. The conductivity of the ground wire has material influence since upon it depends the protection against the inductive effects of oscillations and sudden discharges. A 3/8-inch standard steel wire grounded at least every 500 feet, and strung above the highest transmission wire within an angle of 45° to 60° to the outside wires, as recommended by *D*r. C. P. Steinmetz, is the best form of this kind of protection. Two auxiliary wires are sometimes added on each side of the lines to increase the shielding action. All ground connections must be carefully made. Barbed wire is sometimes employed instead of the smooth kind.

The transmission system of the Southern Power Company has a normal line voltage of 100,000. It uses overhead ground wires for line protection, namely, one from peak to peak of the twin-circuit steel towers upon which the two three-phase circuits are arranged in vertical planes, and one on each side of each tower. As the towers are not set in concrete, the wire is clamped at each tower and thus grounded, because each of the four legs is set about 8 feet in the earth. The ground wire is 3/8-inch Simens-Martin steel strand.

Water Jets.—Protection with water jets is similar to that afforded by large resistances permanently connected into the line, with the additional advantage of self-maintenance in case of a breakdown after a discharge. Water jets prevent to a certain extent the slowly accumulating static charges on the line. Their application is, however, restricted more or less to European use. The principal drawback is to be found in the fact that they pass as much generator current as static current and consequently waste a fairly considerable amount of energy all the time.

Coherer Type of Arrester.—This class includes the M. P. arrester, also called multipath arrester, as made by the Westinghouse Company for alternating and direct current. The static discharge passes in a large number of small streams over a carborundum block, the voltage across each gap being very small. The arrester is used for voltages up to 1,000. It discharges static accumulations and comparatively high tensions, but opposes the current at normal voltage.

A few words in regard to the installation of lightning arresters are in place at this point. Lightning arresters are just as essential for protection against interruptions in service as any preventive of fire or accident in the station. In large systems a certain percentage of the cost of installation and of the yearly income is set aside for the purchase and maintenance of the best lightning arresters on the market.

In order to obtain maximum protection at minimum cost it is necessary:

1. To determine the location of the stations, transmission

lines, and apparatus, and to become familiarized with the character of the surroundings with respect to geographic position, physical characteristics, frequency of thunderstorms, strength of prevailing winds, etc.

2. To know the nature of the system as to voltage, load and distribution, as, for instance, if low-tension a.c. or d.c. or high-tension distribution with sub-stations, or if grounded or ungrounded neutral is to be employed.

The Westinghouse Electric Company recommends that to obtain absolute protection arresters be placed at all points where apparatus is located. In circuits not exceeding 2,500 volts it will usually be sufficient to place arresters at various intervals where good ground connections are available. These arresters should be so placed as to leave no considerable length of circuit unprotected, and should be more numerous in neighborhoods where circuits are exposed, as is the case in outlying districts where the lines are not protected by buildings and trees. Under average conditions satisfactory protection will be secured if no point of the circuit be more than 1,000 feet from the arrester. For voltages exceeding 2,500 arresters should be placed as nearly as possible at or near apparatus on exposed lines.

In all cases of circuits with ungrounded neutrals arresters rated at the voltages between line wires should be chosen, that is, for the maximum working voltage, and not for the voltage between line and ground. If the circuit has a grounded neutral, arresters should be chosen for a voltage 20 per cent. greater than the maximum voltage between line and ground. For example, for a circuit with grounded neutral having 16,500 volts between line and ground (approximately 28,000 volts between line wires) arresters for 20,000 volts should be chosen. If, however, the transformers are connected in star in both high-tension and low-tension windings, arresters should be chosen as when the neutral is not grounded. They should always be placed on the line side of all apparatus.

Protection and maintenance of protective apparatus is of great importance for effective protection of the system. Broken or damaged parts should be replaced after each storm, including arresters in the station as well as insulators, poles, horns, and other protective appliances located out of doors. All disturbances are recorded by placing paper sheets behind the arresters, which also indicate if proper protection is obtained and if any adjustment of the spark gap and resistance is necessary. Proper ground connections, which must be inspected from time to time, are essential. Without these all protective measures are of no avail. One method of making good ground connections is as follows: The ground wire, or what is even better, a copper strip, should lead directly to ground with as small a number of bends as possible, and the ground connection should be next to the arrester. In case the ground under the device is not suitable for a good connection, a proper ground is made at some other point, which is then connected with the ground under the arrester. Copper sheets are recommended for the ground, thick enough to prevent wasting away, approximately 1/16 inch thick, and having at least 4 square feet of surface. The ground wire must be carefully soldered and riveted to this plate and then buried in powdered coke or charcoal in soil which is always damp. Dry, sandy soil should be kept wet by artificial means if this is the only soil available for the ground connection. Where plates are placed in streams of running or dead water, they should be buried in mud along the bank. Where there are metal flumes, pipes, or rails, it is advisable to rivet or solder the ground wires to them in addition to the connection to the copper plates, and when rails are utilized they should be thoroughly grounded. In view of the fact that it is advisable occasionally to examine the underground connections, it is desirable when the ground plates are installed to lay out exact plans of their location and that of the ground wires and joints.

CHAPTER XXI

HIGH-TENSION SWITCHBOARDS AND WIRING DIAGRAMS

This chapter is essentially a continuation of the chapter on low-tension a.c. switchboards where 250- to 600-volt circuits were treated. The intermediate chapters are inserted for the sake of discussing the devices and provisions for high and extra-high tensions before a description of their mutual interrelation and connection with station arrangement is undertaken.

Generator.—The switching arrangement for 1,150- to 2,300-volt circuits corresponds in many ways to the 600-volt arrangements. In most cases the oil switches are mounted on the board proper, and the measuring instruments are directly connected

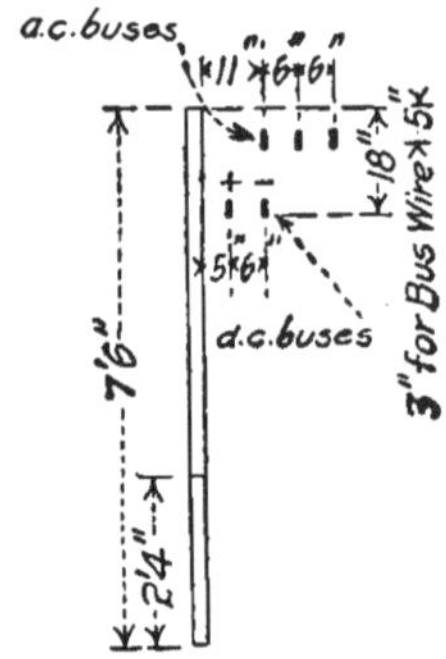

FIG. 197.—Location of busbars or bus wires for voltages from 600 to 6,600.

to the main lines, but with higher tension it is advisable to supply them through series or shunt transformers. Generator oil switches are non-automatic, while the feeder switches are provided with automatic trip coils. Three-phase generators or feeders require one or three ammeters according to whether the load is balanced or unbalanced. Potential regulators are often used with single-phase feeders, in which case a compensating-voltmeter mounted on the board is required, in order to indicate the regulated voltage. For busbars either copper wire or copper bars are used according to the amperage of the plant, and these are mounted on insulators at the back of the switchboard near the upper edge. See Fig. 197, which shows the location of the a.c. and d.c. buses relative to the board. The illustration shows

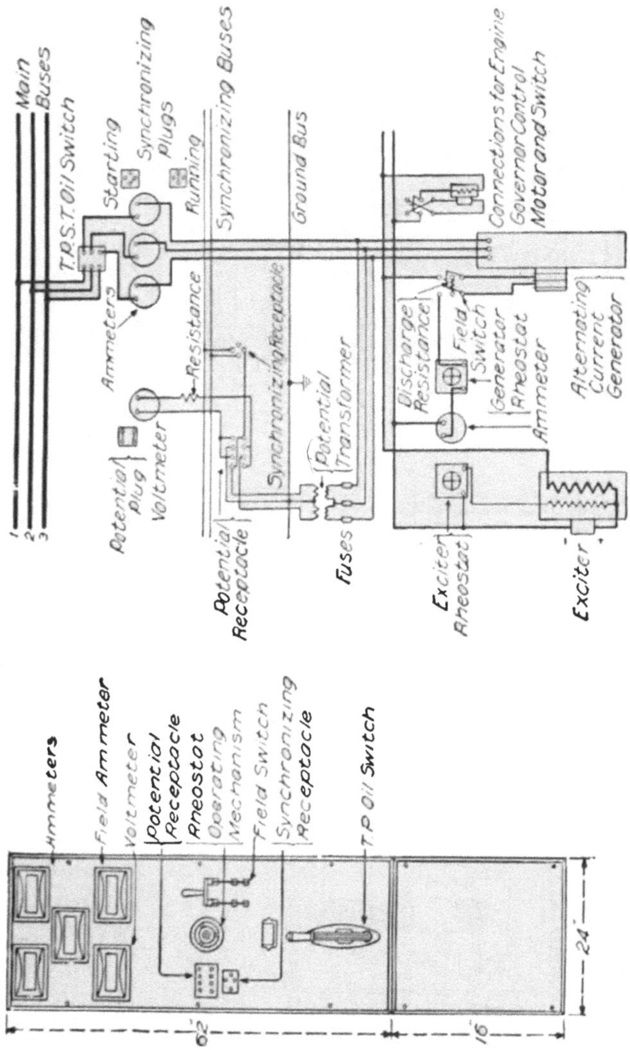

Fig. 198.—1,150- and 2,300-volt, three-phase generator panel.

an example of the case where several generators are excited from d.c. busbars fed by one or more exciters.

A generator feeding one set of busbars is shown in Fig. 198.

The voltage between any two phases is indicated by a voltmeter connected through a plug switch. The generator rheostat is

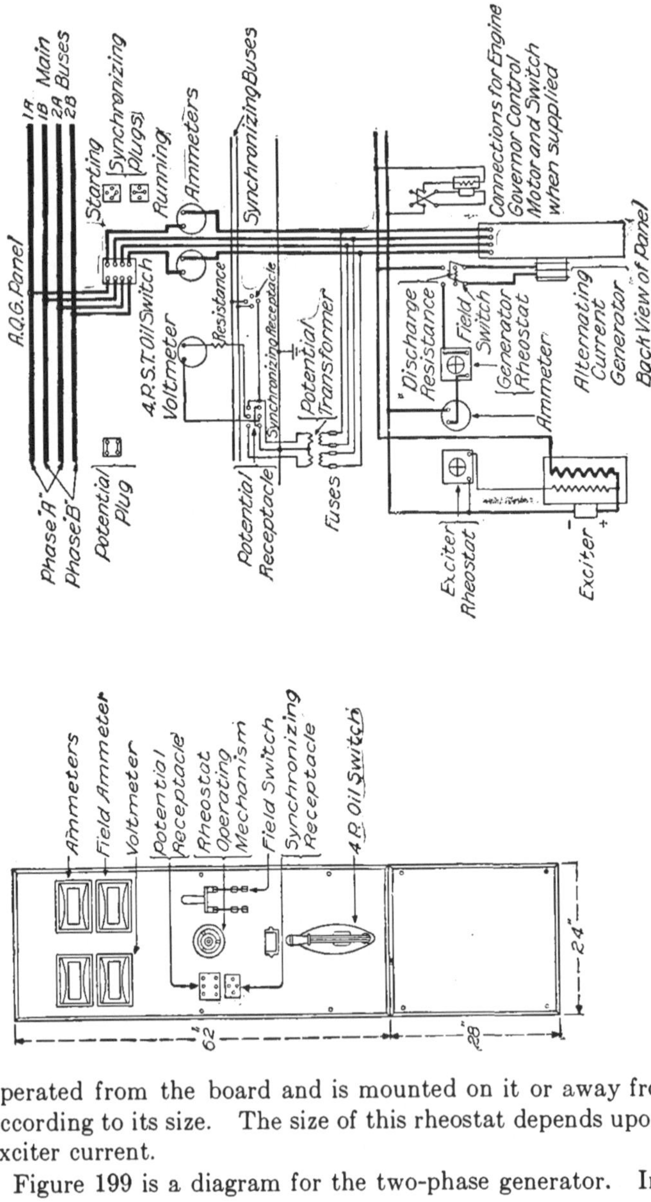

FIG. 199.—1,150- and 2,300-volt two-phase generator panel.

operated from the board and is mounted on it or away from it according to its size. The size of this rheostat depends upon the exciter current.

Figure 199 is a diagram for the two-phase generator. In the

two figures note the difference in the position of the synchronizing plug switches for starting and running conditions of the machine. One synchronism indicator with these plug switches suffices for all the generators.

A single-phase feeder connected to the busbars through an oil switch is shown in Fig. 200. A potential regulator is provided,

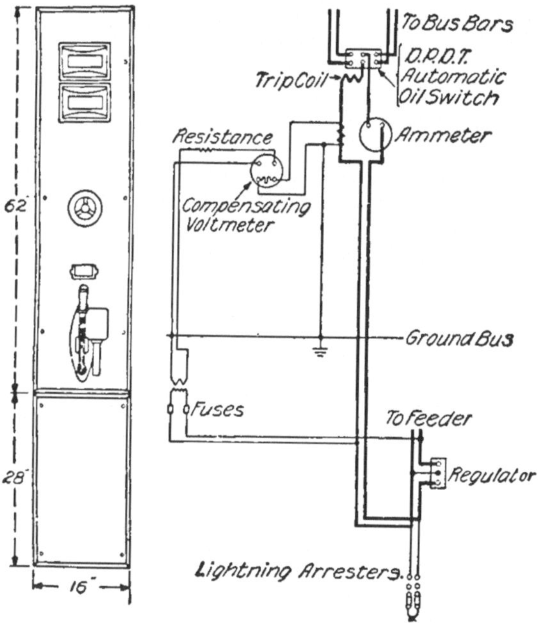

FIG. 200.—1,150- and 2,300-volt, single-phase feeder panel with feeder regulator

which is operated from the board through chains. Note the connections of the shunt transformers to the compensating voltmeter.

Figure 201 is an arrangement similar to the one described above, with the exceptions that the oil switch is replaced by plug switches and that protection is afforded by using fuses instead of trip coils. The type of fuse employed is that shown in Fig. 83. Lightning arresters are provided.

Two systems of connection for a high-tension generator are given in Fig. 202. In the one the oil switch is operated by a motor through a small double-throw switch, while in the other a toggle mechanism is used. A solenoid-operated oil switch might also be used to replace the motor-operated H_3 switch.

In both cases the switchboard has the same equipment. One each of the following instruments and apparatus is used: Ammeter, voltmeter, wattmeter, watt-hour meter, small double-pole, double-throw, switch for the wattmeter, four-point receptacles for synchronizing connections, handwheel and chain-operat-

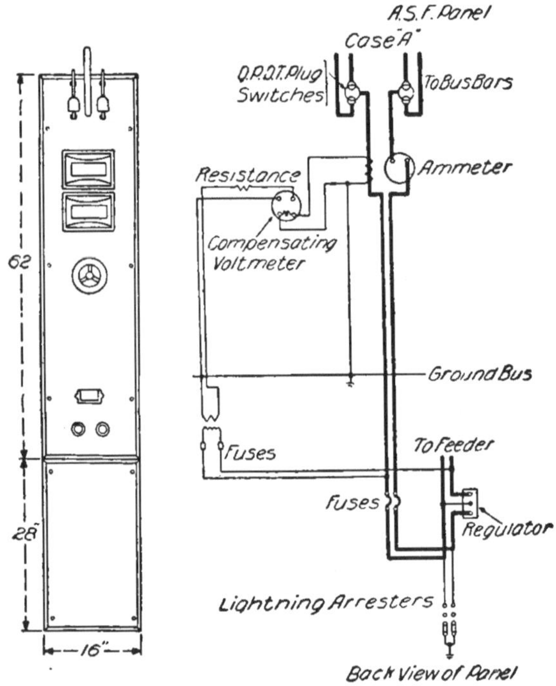

FIG. 201.—1,150- and 2,300-volt single-phase feeder panel with plug switches and expulsion fuses.

ing mechanism for field rheostat, single-pole, single-throw carbon-break field switch with discharge clips, double-pole, double-throw engine-governor controlling switch, single-throw, triple-pole non-automatic oil switch, and series transformer; also two shunt transformers. For a manually operated oil switch, the motor control switch is replaced on the board by the handle of the toggle mechanism. The series and shunt transformers are mounted away from the board because of the high voltages to which they are connected, and all high-tension wires are also kept well away from the board. The exciter, synchronizing, operating, and ground buses run across all the panels carrying instruments connected to them.

Figure 203 is similar to Fig. 202 except that step-up transformers are connected to the generator. . This arrangement is often used for long high-tension transmission lines in order to save copper in the lines. Series and shunt transformers are joined

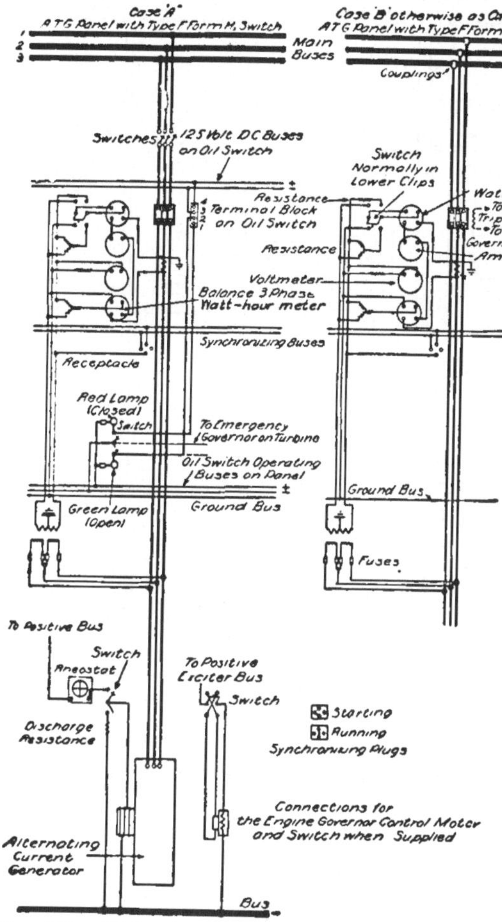

FIG. 202.—Wiring diagram of alternating-current generator panels.

to the primary sides of the main transformers. Oil switches and busbar compartments should be mounted near to each other and in fireproof cells and compartments. (See Chap. XXII.)

If an alternator is to be thrown into parallel with a circuit, four conditions must be fulfilled:

1. The machine must run with precisely the same frequency as those in the circuit.

2. This frequency must remain constant.

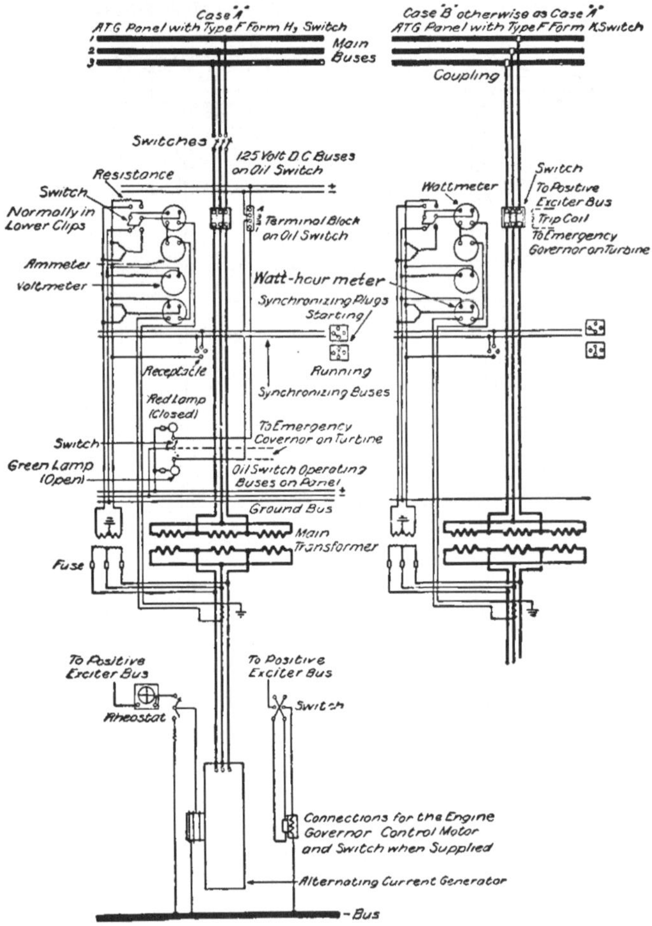

FIG. 203.—Wiring diagram of alternating-current generator panels with step-up transformer.

3. The machine voltage must be in phase with that of the system.

4. The terminal e.m.fs. of the alternators must be exactly equal.

Successful parallel running of alternators results in proper load division of the machines, which should run without hunting.

The generators should preferably be paralleled on a common low-tension bus, the generator switches being non-automatic, or if automatic protection is desirable, the switches should be provided with definite time-limit relays, set very high. Reverse power relays are also occasionally installed, but are generally arranged to ring an alarm gong in case of reversal of the power, and will not trip out the switch.

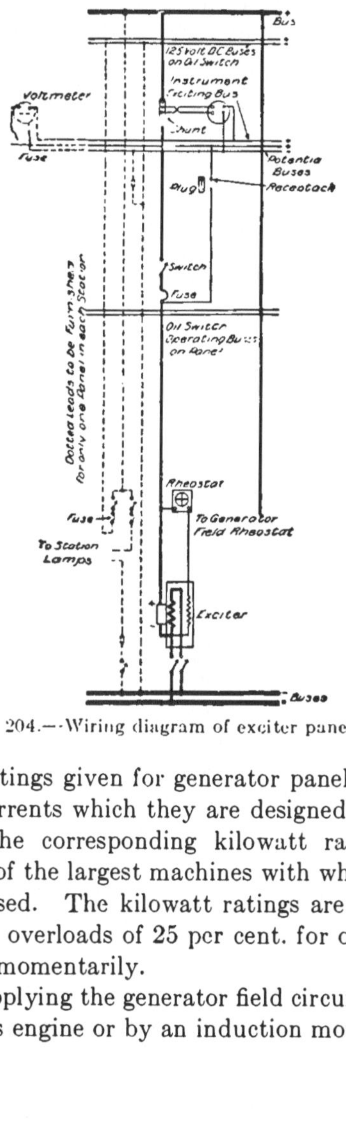

FIG. 204.—Wiring diagram of exciter panel

The ampere ratings given for generator panel equipments are the maximum currents which they are designed to carry for one or two hours, the corresponding kilowatt ratings being the normal capacity of the largest machines with which these equipments may be used. The kilowatt ratings are based on unity power factor and overloads of 25 per cent. for one or two hours and 65 per cent. momentarily.

The exciter supplying the generator field circuit may be driven by a steam or gas engine or by an induction motor fed from the main busbars. It is essentially a d.c. generator of low voltage (125 volts), with similar connections. (See Fig. 204.) A fuse is inserted in the positive lead to take the place of the circuit breaker. One voltmeter will be sufficient for all the exciter machines. The positive exciter bus will usually be found on the switchboard, while the negative is placed under the floor in the exciter foundations.

Feeders.—An outgoing feeder with its connections is shown in Fig. 205. For unbalanced load, transformers are inserted in

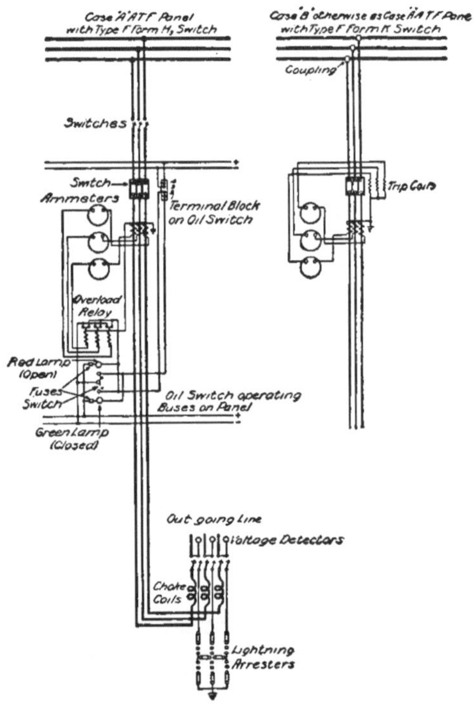

FIG. 205.—Wiring diagram of outgoing line panel.

each leg to be used for the three ammeters. Automatically actuated oil switches, when operated electrically, are tripped by relays, and if manually operated, they can be tripped either by relays or directly from the series transformer.

It is general practice to use an inverse time-limit relay if it is not desired to have the oil switch operate instantly on heavy overloads. Where it is necessary to notify an attendant automatically of the interruption of the supply, say of the motor feeder, a signal lamp or alarm bell can be applied. In case the feeder panel supplies power to drive a motor which is connected to an exciter, the motor circuit should have its overload oil switch set to open only on extreme conditions of loads or short-circuits so as to avoid interrupting the excitation and shutting down the plant.

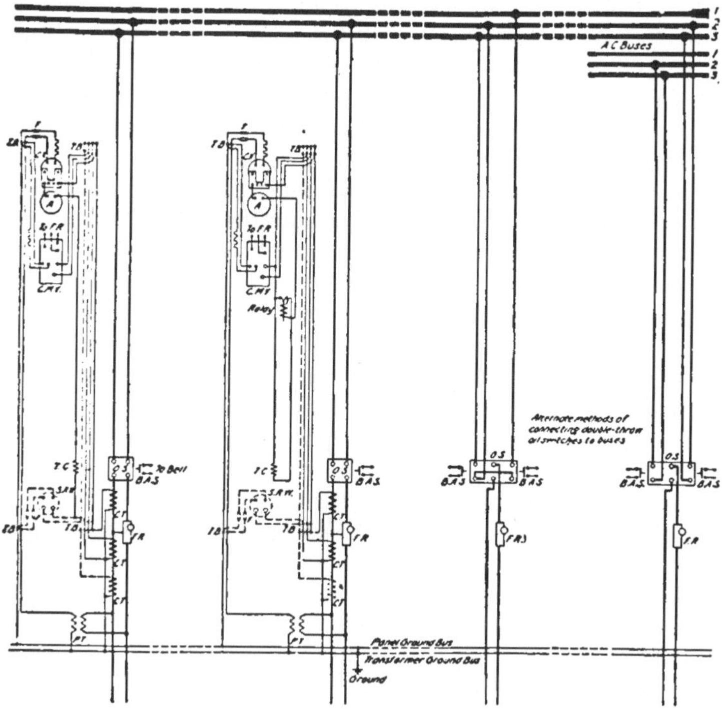

FIG. 206.—Wiring diagram of incoming feeder panels.

For overload lines lightning arresters and reactors are inserted. When ground detectors are to be used, they should be mounted on rigid brackets on top of the switchboard. Additional provision to disconnect the feeders from the line is found in the disconnecting switches.

For incoming feeders only one ammeter with a series transformer is used. In some cases the central-station feeder control is sufficient, so that incoming feeders of sub-stations require no further instruments outside of lightning arresters.

Figure 206 shows a wiring diagram of single-phase feeder panels in connection with automatic feeder regulators.

The ampere ratings given for feeder panels are the maximum current (normal or overload other than momentary) which they are designed to carry. As a rule, panels of larger capacity than necessary should not be chosen in order to provide for a possible future increase of load, since the consequent low-load factor of the current transformers has an injurious effect on the accuracy of the instruments and motors; furthermore, the overload feature of the oil switches is rendered inoperative, except on short-circuits and extremely heavy overloads, owing to the very high ratio of the current transformers as compared with the normal load of the circuit to be controlled.

Synchronous Converters.—The d.c. side of converters has been fully discussed in Chap. III, so that only the a.c. side remains to be treated here. As previously stated, a synchronous converter is built to interconvert direct and alternating currents. The connection to the alternating busbars of the system is made either directly or through a step-down transformer, according to the tension in the system. The latter method is the more frequent because the a.c. side of the converter must usually be joined to a 370-to-430-volt line, and if the converter is in a sub-station, a much higher transmission voltage is led to it. The a.c. generators also deliver current at high pressure. Wiring diagrams for a three-phase converter, with transformer, are shown in Fig. 207.

The starting voltage must always be reduced in order to prevent a sudden rush of current from the transformers, and this is done by means of the auxiliary taps from the transformer secondaries. In starting, therefore, the double-throw switch connects the converter with the auxiliary taps, and as soon as the machine has reached full speed it throws in the entire voltage of the low-tension side of the transformer. Special reactors are inserted between the transformer and the converter in order to prevent too great a starting current. These coils may be separately boxed up, or the transformer itself may possess the necessary reactance, in which case the extra coils are omitted. It is customary to start converters of 300 kw. or less in this way, with the starting voltage at half the normal pressure. The starting voltage for 300- to 1,500-kw. machines is from one-third to two-thirds the normal. If a.c. is used in starting, the armature windings act on the field windings like a transformer primary on its secondary. A large number of field turns in comparison to the number of armature turns may generate a high value of e.m.f. in the former,

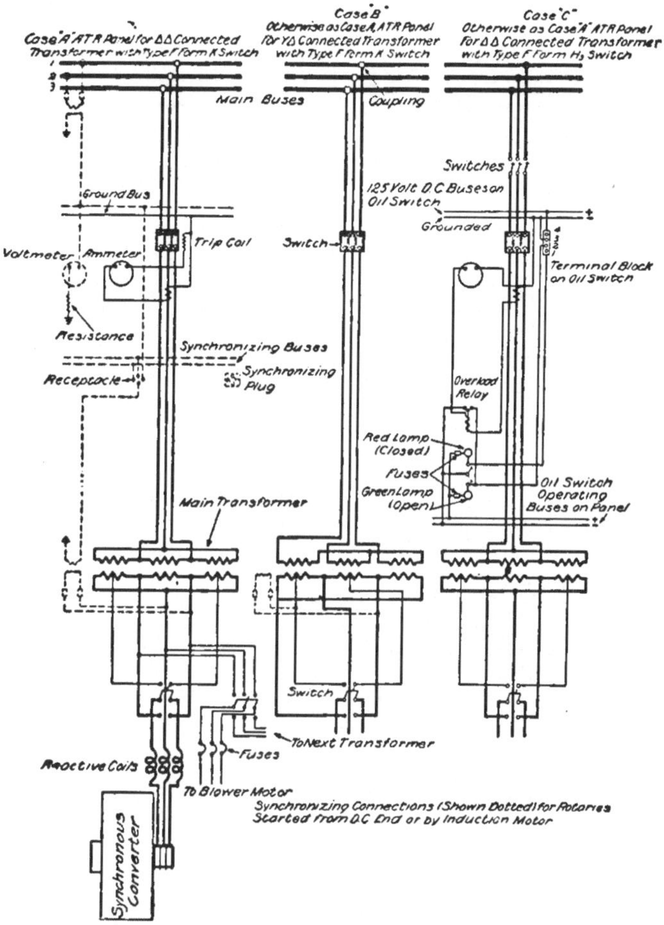

FIG. 207.—Wiring diagrams of three-phase synchronous converter connection, for high-voltage panel, low-voltage starting panel and blower motor panel.

which should be limited. This is accomplished by breaking the field winding at number of points by a multiple switch mounted on the frame of the machine. The connections shown dotted in Fig. 207 are used only when the converter is started on the d.c. side or by an induction motor, in which case synchronous starting is essential.

Figure 208 is a wiring diagram for a six-phase converter with a three-step connection to the low-tension side of the transformer. If the transformers are air-cooled, the connections of the ventilator motors are made as indicated in the figures.

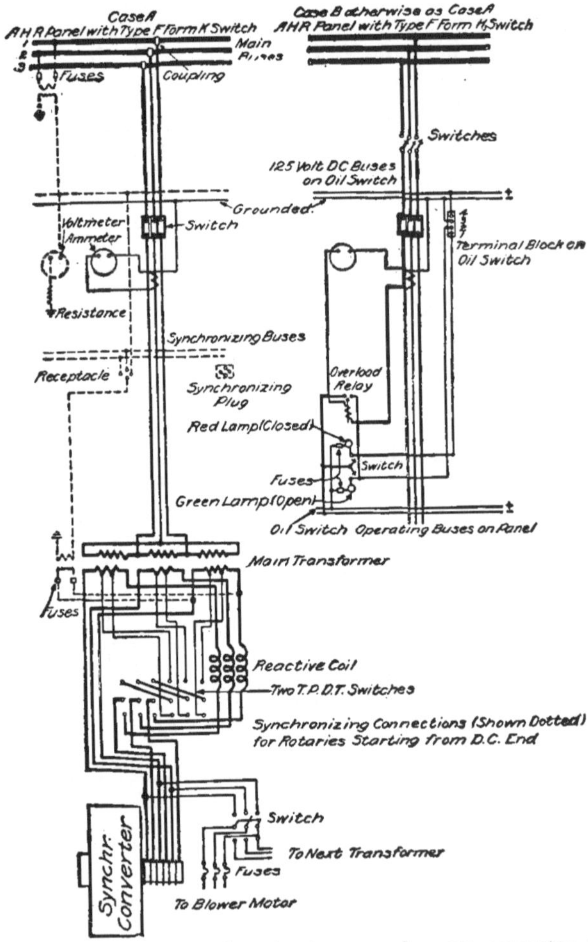

FIG. 208.—Wiring diagram of a six-phase synchronous converter, for high-voltage panel, low-voltage starting panel and blower-motor panel.

The above method of starting is applicable in all cases for converters with low frequency, or when the fluctuations at starting exert no material influence on the voltage regulation of the system.

The instruments and apparatus for the control of the a.c. side of converters are placed on two panels. A main board controls the high-tension side of the transformers and therefore carries an ammeter, a controlling switch for the synchronizing oil switch and plug switches when necessary. The latter are used when starting from the d.c. side or by an induction motor. The second panel is smaller and is mounted independently, near the converter. It carries the multipole switch which controls the low-tension side of the transformer, and hence the machine itself. The d.c. side has a switchboard of its own, as previously mentioned.

If the system is such that it requires precise voltage regulation, it is advisable to start the converter by an induction motor mounted on the same shaft. (See Fig. 209.) This increases the certainty of starting, as each machine has an independent starter. Any accident to the motor, however, will cripple the converter. Before the machine is connected into the circuit it must be synchronized with the line. The synchroscope can be made to indicate whether or not the proper condition of synchronism obtains by properly connecting the shunt transformer and plug switch. A small additional switchboard will be required under these conditions to carry the starting switch for the motor.

The procedure for starting from the a.c. side of a converter is therefore as follows:

1. All switches except the negative switch on the machine must be open.

2. Close the feeder oil switch which connects to the high-tension busbars. (This applies to the sub-stations.)

3. Close the oil switch on the high-tension side of the transformer.

4. Close the starting switch of the low-voltage taps on the low-tension side of the transformer, i.e., the upper side of the double-throw switch.

5. As soon as the converter is in synchronism with the line, close the equalizer switch.

6. Close the series-shunt switch.

NOTE.—If there are other converters on the line, the separate excitation of the series field seeks to establish a correct polarity by means of the equalizer buses. If its polarity is reversed, it can be quickly corrected by throwing over the field-break-up switch, The double-throw switch must be immediately reversed, however. since its lower position is only for reversal of polarity.

7. After the correct polarity has been obtained, throw the field-break-up switch into the upper position.

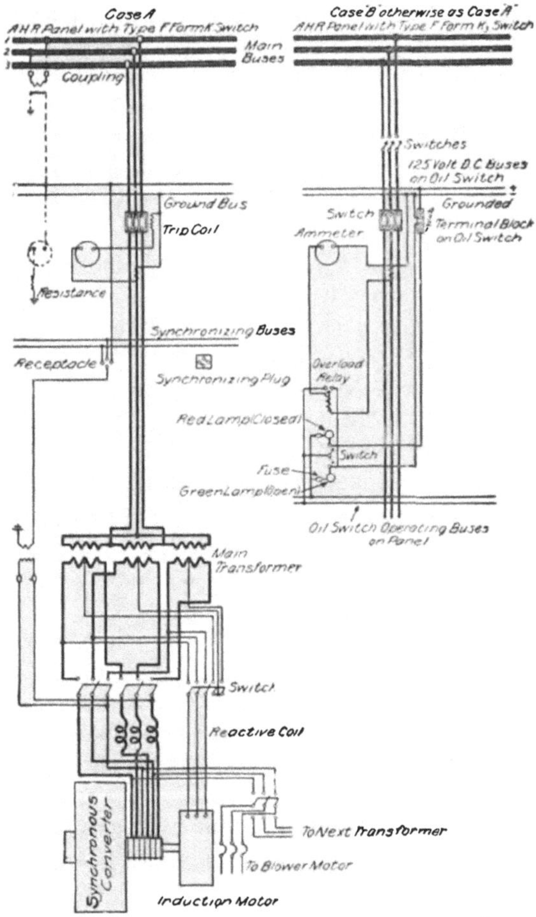

FIG. 209.—Wiring diagram of a six-phase converter with a three-phase induction motor for starting.

8. Throw the double-throw switch on the full voltage of the low-tension side of the transformer, by throwing the starting switch from the upper into the lower position.

9. Push up low-voltage release of circuit breaker and close the d.c. circuit breaker.

10. Regulate the field rheostat.

11. Close the main switch on the switchboard, and again regulate the field rheostat to obtain the proper load factor and voltage.

To shut down a converter open the d.c. circuit breaker; pull out and turn circuit-closing auxiliary switch to stop the ringing of the alarm bell; open the d.c. main switch on the panel; open the high-tension a.c. oil switch; allow machine to run down in speed until volts fall off to about 100 before opening the field-break-up switch or starting switch; open field-break switch, equalizer switch, series-shunt switch, and starting switch.

Note.—The field-break-up switch should not be opened until the machine voltage has run down, otherwise the arc due to opening the inductive field circuit may be blown by the armature blast from point to point, giving a short-circuit, or strain may injure the insulation of the field winding. The starting switch may be opened when the machine voltage has run down to zero. This leaves the transformers without heavy residual magnetism, which is likely to cause an extra rush of current when they are thrown into circuit subsequently. When the converter is connected to the starting taps of the transformer and the field closed, the opening of the oil switch on the high-tension side of the transformer should be avoided, for then the converter tends to build up to full voltage, further tending to run up the transformers to considerably above the normal voltage. The converters can be thrown into parallel on the d.c. side only with closed equalizer switches.

The d.c. circuit breaker is closed always before the main switch on the d.c. side is in closed position, and after a heavy short-circuit the breaker should not be closed without plugging the voltmeter into the converter to make sure that the polarity is not reversed.

Synchronous and Induction Motors.—In the chapter on starting compensators the switching arrangements for this kind of motor and the method used for starting them with the aid of compensators were discussed. If, however, the motor is so small that it consumes only a very small fraction of the station current, it is started and run directly from the busbars. A step-down transformer should be inserted if the voltage of the system so requires.

Two switchboards are required for the induction motor running an exciter. One is similar to the kind used for incoming feeders.

The other is smaller and carries only a double-pole, double-throw starting switch. It is set up near the motor. The connections are shown in Fig. 210. The motor is connected on the low-tension side of a transformer bank whose high-tension side is supplied from the main buses. At starting, the double-throw switch throws the motor in on half the transformer voltage, and when full speed is reached the entire voltage is thrown on.

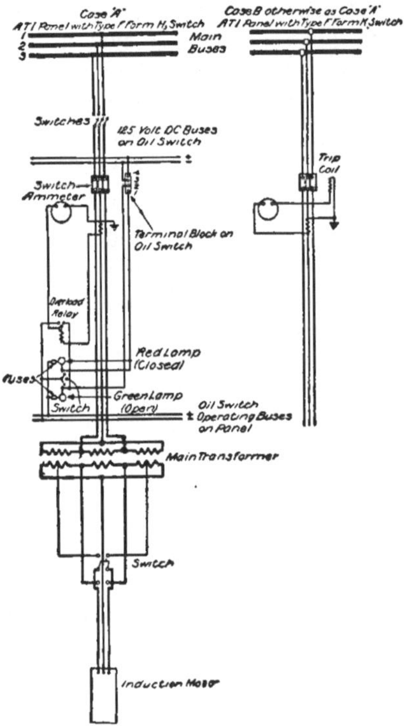

FIG. 210.—Wiring diagram for induction-motor connection, high-voltage panel and low-voltage starting panel.

Figure 209 also shows the connections of the induction motor to the low-tension side of the transformer, and likewise the manner of starting. The normal voltages at which the various induction motors are run are as follows:

CYCLES	VOLTS
60	220–2,080
40	550
25	440

Machines for higher voltages are built upon special order.

The apparent efficiency of an induction motor of given horsepower and voltage varies somewhat with the speed for which the motor is designed, so that panels having a given ampere rating cannot be given a corresponding horsepower rating which will be correct for all speeds. The ampere capacity of a panel equipment required for a given motor may be determined by the following formulas:

For motors with 25 per cent. overload:

$$\frac{\text{Hp.} \times 746 \times 1.25}{1.73 \times \text{volts} \times \text{efficiency} \times \text{power factor}} = \frac{\text{Hp.} \times 539}{\text{Volts} \times \text{efficiency} \times \text{power factor}}$$

For motors with 50 per cent. overload

$$\frac{\text{Hp.} \times 746 \times 1.5}{1.73 \times \text{volts} \times \text{efficiency} \times \text{power factor}} = \frac{\text{Hp.} \times 646.5}{\text{Volts} \times \text{efficiency} \times \text{power factor}}$$

A synchronous motor is usually nothing more than a reversed alternator, whence it follows that the latter is often used. These motors are preferred for motor-generator sets for the conversion of alternating into direct current, or into alternating current of different frequency, potential, and phase.

In an induction motor the power factor is predetermined by its design and construction, and its current is always lagging in phase. With the synchronous motor, on the other hand, the current can be made to lag or lead in phase by varying the field excitation. The synchronous motor, therefore, can be employed to advantage not only for full power factor, i.e., for minimum current, but also to compensate the inductive load on the system, so that for the combined load at the given voltage and power output the system will operate at maximum efficiency. (See Chap. XVII on synchronous condensers.)

This motor is started with the aid of a compensator, or, if frequent starting is necessary, a separate induction motor is used. The latter brings the synchronous motor to full speed, and the field is then excited at the line voltage, whereupon the induction motor is shut off. During the starting period the field and armature windings of the synchronous motor act like those of a transformer. Therefore, if the field winding has a large number of turns, so as to have high inductance, the winding is broken at a number of points by means of a multipole lever switch. (See "Synchronous Converters," Chap. III.)

The apparatus ordinarily required for a synchronous motor for regular power service or for driving a generator consists of the following: One a.c. ammeter, indicating wattmeter, a d.c. ammeter for motor field, a power-factor meter (optional), a field rheostat, a double-pole field switch with discharge resistance, a synchronizing plug switch, an overload oil switch with time-limit relay, and the necessary current and potential transformers.

When the motor is started from the a.c. circuit a double-throw main switch will sometimes be required, one position being for running and provided with the automatic overload release, while the other is for starting and for connection to the starting compensator. If the motor is a part of a motor-generator set and starts from the d.c. side, the switch can be a single-throw switch with automatic overload release.

To reverse the direction of rotation of a three-phase synchronous or induction motor, it is simply necessary to reverse any two of the lead connections. For two-phase machines the connections of both leads of one phase must be reversed.

Instruments.—Instruments will be discussed only in regard to their application in a.c. practice, without going into the respective advantages and disadvantages of the various types.

To determine the output of an a.c. generator, one or more ammeters are necessary. One instrument only is used under balanced load, that is, when the load consists only of rotating apparatus, such as converters or motor-generator sets, etc. Two ammeters are required for two-phase circuits, and three for three-phase under unbalanced load, such as current delivery for lighting. Another way of ascertaining the current in each feeder under these conditions, using only one ammeter, is to join this instrument by means of plug switches and receptacles to a number of series transformers corresponding to the phases of the circuits. With alternating current an ammeter fails to give any information as to the distribution of the load on the various machines. A wattmeter for each machine is therefore required especially for this purpose. A single voltmeter can be made to indicate the voltage of any generator by simply connecting it to the required machine by means of a plug switch. Sometimes a second voltmeter is connected to the busbars. This affords a means of calibrating the scales of the two instruments from time to time and a simultaneous comparison of the busbar and machine voltage when synchronizing. Field ammeters are sometimes used as they give an easy way of ascertaining if anything is wrong in the machine itself. They assist in properly adjusting the field so that the machine will take its load. A power-factor indicator is not absolutely necessary, as a wattmeter can be made to perform the same function by using it with a double-pole, double-throw switch. The wattless component can be read off directly if a polyphase wattmeter is used. One frequency indicator will do for each set of busbars. A frequency meter is always advisable if current is supplied to textile mills or other establishments where careful speed regulation is desirable as such speeds depend upon maintaining the frequency constant. Since the introduction of turbo-generators these indicators have come to replace the tachometers used formerly to indicate the speed. To obtain a record of the generator energy output a watt-hour meter is a very efficient instrument.

The instruments which indicate synchronism for generators and converters are of great importance. By synchronous running of a given machine with respect to other machines with which it is connected in parallel is meant such a condition of running that the frequency and phase relation of the given machine are the same as the corresponding quantities of the other machines. Two machines have the same frequency when the number of alternations of the e.m.f. of one in a given time is equal to that of the other. This occurs when the product of the number of poles by the revolutions per minute is the same for both machines. Two machines have the same phase relation when the relative position of their armatures with respect to their field poles is the same; that is, when corresponding armature turns are opposite corresponding poles at the same time. When two machines have the same frequency, phase, and voltage, there will be no unbalanced e.m.f., and consequently, if they are coupled in parallel, no rush of current will ensue. It is evident, therefore, that before machines can be coupled in multiple they must be synchronized. Instruments must consequently be provided to indicate when synchronous running is obtained, so that the connection may be made at the proper instant. Incandescent lamps and synchroscopes are the devices employed for this purpose. Figures 211 *a, b,* c, and *d* are a number of diagrams for synchronizing machines with voltmeter, lamps, and plug switches. In *a* and *d* the alternator is in multiple with the load. At the instant that the required condition of synchronism is obtained the lamps must be dark because the generator flux is then opposed to that of the system, and both generator and system have the

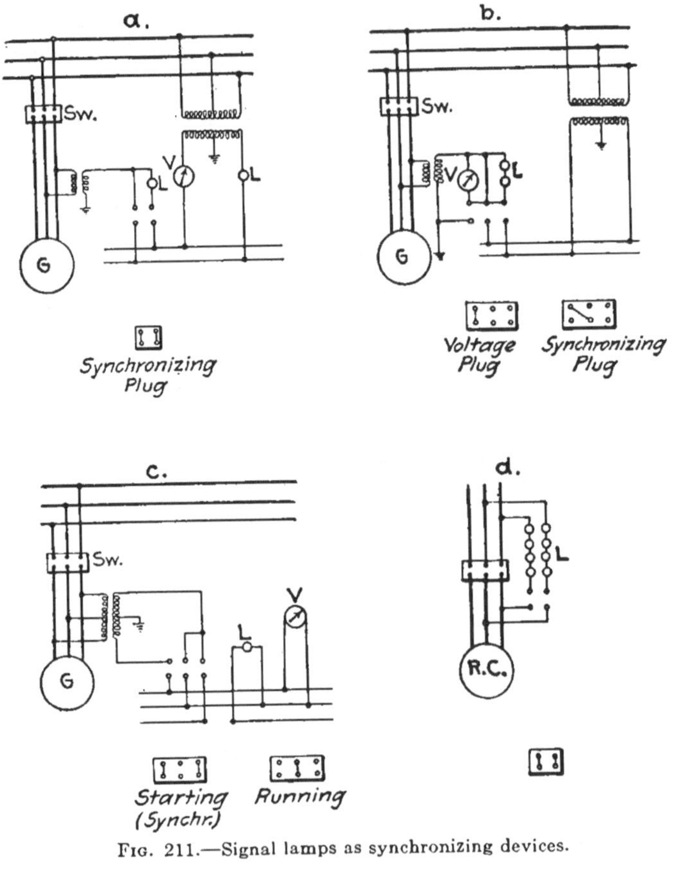

Fig. 211.—Signal lamps as synchronizing devices.

same phase, so that they are balanced in the connection. The same conditions obtain in c, where the alternator is in multiple with another machine. The lamp is darkened at the instant of synchronous running. In case *d* a synchronous converter is to be connected to the system. As synchronism is approached the pulsation through the lamps decreases in number, until the lamps go out at the instant when perfect synchronism is reached.

The use of lamps has the disadvantage that a large difference of phase is required to make them burn, and that they do not indicate if the machine is running too fast or too slow. Moreover, since their use as synchronism indicators depends upon the fact that they become dark at the required moment, it is possible that an undetected flaw in the lamps may give an erroneous indication as to the synchronism of the machine. A specially constructed synchroscope is therefore employed to advantage. The functions of this instrument are to:

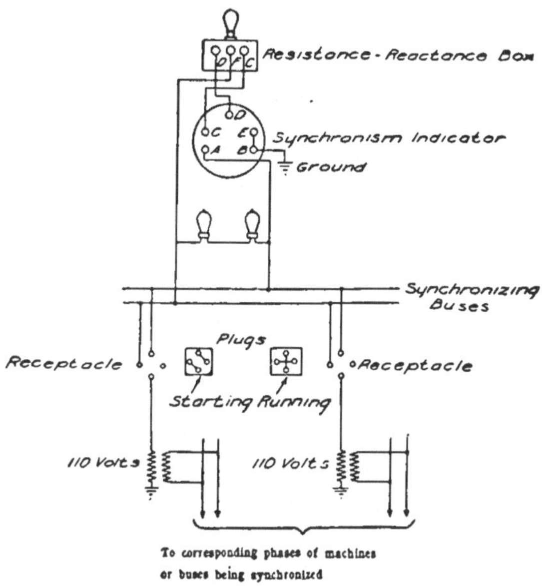

FIG. 212.—Connection for synchronism indicator with grounded secondaries on shunt transformers.

1. Indicate if the machine to be thrown in is running faster or slower than those in circuit.
2. Indicate the difference in the speeds.
3. Indicate the moment when both machines are in synchronism.

These functions cannot be performed adequately by lamps alone. Lamps are nevertheless used in connection with the synchroscope in order to facilitate the reading of synchronism at a distant point. Figure 212 shows a wiring diagram of a General Electric synchronism indicator with a set of lamps in parallel which darken when synchronism is obtained. The plugs are inserted in the receptacles according to whether the machine is used in starting or in running. In this diagram the secondary transformer windings are grounded. Figure 213 shows the transformer secondaries not grounded. The dotted lines are the connections for the lamps when the latter are to be brilliant when synchronism is obtained.

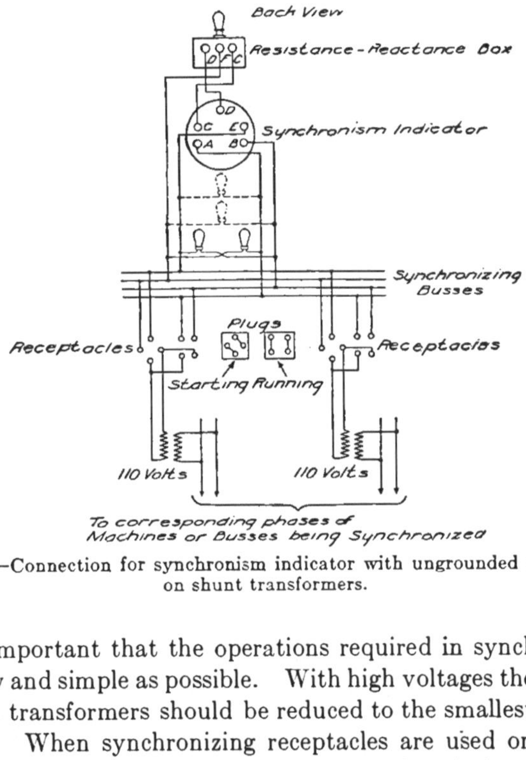

FIG. 213.—Connection for synchronism indicator with ungrounded secondaries on shunt transformers.

It is important that the operations required in synchronizing be as few and simple as possible. With high voltages the number of shunt transformers should be reduced to the smallest number possible. When synchronizing receptacles are used on circuits controlled by electrically operated oil switches, it is customary to have an extra set of contacts on the synchronizing receptacles to close the control circuit of the oil switch when the synchronizing plug is inserted. This feature makes it impossible to close the oil switch without first inserting the plug connecting the circuits to be synchronized, thus minimizing the danger of accidentally connecting generators together without synchronizing. While the control circuit of the oil switch passes through the extra contact of the receptacle, the tripping circuit does not, and hence the circuit may be opened independent of the position of the synchronizing plug.

Lighting and power feeders should be provided with an ammeter, and in case a potential regulator is inserted in the line, a compensating voltmeter must be added in order to indicate the regulator voltage. Recording instruments are sometimes used for plotting the output of any given feeder. Power-factor meters are sometimes used to indicate those parts of the system which require modifications of their load to better the power factor, which in turn reduces unnecessary line losses. All instruments should be so placed that they may not come under the influence of outside magnetic fields. The instruments take their current from series or shunt transformers which are so designed that the current or voltage of the secondaries is the same for all transformers for the required ratio. Five amperes is the normal current for the secondaries of series transformers, and 100 volts the normal voltage for shunt transformer secondaries. The latter are compensated so as to give a correct load ratio, while the former are compensated with regard to the loss with a given current. The load placed upon them should be light and of as small an inductive value as possible. All secondaries of instrument transformers should be grounded in order to eliminate puncturing of the insulation between primaries and secondaries.

For the mounting of the different instruments on the switchboard the following rules are recommended:

1. Indicating instruments should be mounted at or a little above the height of the eyes.

2. Meters (for example ammeters) should be symmetrically grouped on the panel so that phase distinctions are apparent.

3. Meters of a kind should be in alignment with each other on the various panels, for convenience, appearance, and symmetry of wiring.

4. Voltmeter receptacles and similar plugging devices should be located as near the instruments as possible to simplify and decrease amount of small wiring.

5. Rheostats, unless very small, should be sprocket-operated, the face plates and resistance mounted as a unit and located in some convenient place where the contacts may readily be inspected and where the heat from the resistance will do no harm. The handwheels should be located near the center of the panel

at such a height that the operator may readily watch his instruments while adjusting the rheostat and can have one hand free to operate the voltmeter receptacle, field switch, or main switch.

6. The main switches or oil switches, when mounted on the switchboard, should be located at a height most convenient for ease in operation. They must not be so close to the edge of the panel section as to interfere with the framework or to give insufficient distance to the adjacent apparatus, or so near the floor as to prevent free removal of the tanks for inspection.

7. Watt-hour meters or relays may be mounted on the subsections or at the rear of the board, as they require only occasional attention.

8. High-tension busbars, when used on the switchboard, should be mounted near the top of the board to avoid accidental contact, thus obviating the necessity of insulating them to prevent injury to attendants. This arrangement also permits free access to the rear of the board.

9. Voltage and current transformers may be mounted at the rear of the board on suitable supports, but usually at the expense of accessibility to the instrument wiring and auxiliaries mounted at the back of the panel and of the general appearance of the installation. It is recommended that they be placed apart from the board wherever possible—beneath the floor if the leads go down, or on the wall where they go up. The secondary leads are then led to the board together in conduits.

10. Such instruments as graphic recording meters, static ground detectors, and Tirrill regulators should not be placed on swinging brackets or swinging panels.

The Underwriters recommend the grounding of the neutral point of low-tension circuits when the conditions are such that the maximum normal voltage between the point connected and ground will not exceed 250 volts. This means that one side of a 250-volt circuit or the middle point of a 500-volt circuit may be grounded. For potentials above 500 volts, and not exceeding 6,600 volts, a safety gap should be used in the ground connection. This applies to transformers whose ratio of transformation is five to one or greater. The safety gap is, of course, set for the particular voltage to which it is connected.

Switchboard Panels.—In high-tension stations all instruments and control and operating apparatus are placed on a switchboard which may be selected in any one of three forms, namely:

1. Panel;
2. Benchboard or desk, and
3. Control pedestals and instrument posts.

There are certain general conditions which influence a choice of design and which must be recognized before a satisfactory arrangement can be obtained. The most important of these are:

The ultimate capacity of the station.

The number of generator, feeder, bus-tie, and exciter circuits to be controlled.

The relative proportion and importance of feeders to generator circuits.

The scheme of busbar connections and interconnections.

The location and arrangement of switchboard gallery as regards the location and arrangement of the station apparatus.

The first cost and maintenance cost.

The system of operation of various circuits, considered together with the provisional arrangements for emergency use.

The number and kind of instruments and control devices.

The switchboard selected should be such that ease of adding to the board without materially disturbing the existing panels, busbars, and connections may be secured. By arranging the panels in order of their current capacities, with the heaviest-capacity panels at the center of the board, the busbar copper is usually reduced to a minimum. The a.c. busbars do not cross the exciter panels, while the exciter busbars should be of minimum length. The concentration of various panels of a kind at one part of the board assists the operator materially by reducing the number of steps, and consequently the time, required for any switching operation. The automatic current-breaking devices should be suitable for opening the maximum load based on the ultimate kilowatt capacity which may be concentrated upon them, as they will undoubtedly be required at times to open a short-circuit on the station. The character of the plant should also be considered, suitable apparatus selected, and the switchboard so located that noxious gases or fine dust will not cause trouble with the apparatus. If local inspections must be made by fire underwriters or others, their regulations should be considered if they differ in any way from those recommended by the American Institute of Electrical Engineers and the National Board of Fire Underwriters.

The materials which may be used for switchboard panels will depend on the requirements of the installation. These requirements generally involve the following characteristics in the material:

1. Insulating qualities.
2. Tendency to absorb moisture.
3. Adaptability for polishing, cutting, and drilling.
4. Strength of material.
5. Fireproofing qualities.
6. Color and general appearance.
7. Cost.

Of the materials available for switchboard purposes, marble, slate, and soapstone are most frequently used.

Marble is the favorite; it possesses high insulating qualities, and standard practice allows it to be used as an insulating base for conducting parts of apparatus for service up to and including 3,300 volts. It is easily cut and drilled and takes a beautiful polish. It may be obtained with sufficient strength for any switchboard apparatus, if the panels are not made excessively large and are sufficiently thick. Ordinarily the thickest panels used never exceed 3 inches, but 1¼ inches to 2 inches are the usual thicknesses, depending on the size of the panels and the weight of the apparatus.

Marble possesses reasonable fireproofing qualities, and the varieties of color and beautiful markings which may be obtained are infinite. For some special environments, the marble with its natural face highly polished will present the best appearance, but it will not maintain this good appearance for general power-station work unless special pains are constantly exercised to keep it clean and free from oil stains. This is usually very difficult. Marble is usually more expensive than slate or soapstone, although preferable on account of its insulating qualities.

Some marbles have spots of flint or other substances so prevalent that it is difficult to drill holes within very accurate dimensions, because the drills run out of true when such spots are encountered. It is therefore necessary to make an allowance of approximately 1/16 inch for such contingencies, and on this account holes are usually drilled a little larger than absolutely required.

In addition to this, difficulties are encountered in the marble works owing to the state of the art, which prevents the manufacturers from cutting to dimensions without some allowance. This allowance has been determined, the average commercial limit being about 1/32 inch plus or minus from the dimension specified. Most work may be expected to come within these figures, which ordinarily will never be exceeded except for single switchboard panels intended to stand alone, in which case this high degree of accuracy is not required.

On account of the liability of oil stains and dirt disfiguring a marble switchboard, general practice for power-station work has dictated that the marble panels should receive a finish which will not show oil and dirt and which will present a uniformly good appearance after long service. This has been attained by the introduction of the black marine finish. In order to apply it, the marble panel is given a fine honing, after which it is coated with the marine finish, whereby the whole surface is given a beautiful uniform dull velvety-black appearance. This surface may be rendered free from damage by oil spots if it is especially treated, but in such a case it shows a slight luster. This finish is very durable and is easily repaired when damaged.

Another finish which is sometimes used is black enamel baked on. The coating is entirely oil-proof, but the enamel shows scratches easily, and while it is more expensive, it is not so good as the black marine finish, inasmuch as damages are not easily repaired.

It must be borne in mind that polished marble is a natural product, and therefore its colorings and markings are irregular and variegated. It is often very difficult, and in some cases impossible, to obtain at different times panels which will match exactly in color and general appearance. Of all varieties obtainable, the blue Vermont marbles are the most uniform in this regard, but like all marbles they possess many variegations. It should also be noted that when polished marble is exposed to light for a period of time, or put in a dry place, bleaching and drying take place, thus causing the colors to fade. Hence matching up an old panel from records will not insure the duplication of the coloring. Wherever possible the standard black-marine-finished marble should be ordered if a serviceable switchboard is desired and the difficulties involved in matching panels are to be avoided.

In low-potential service slate is used to a large extent for switchboard panels, switch bases, and similar purposes. It has the advantages of being easily cut and manufactured, and it is stronger, in general, than an equivalent thickness of marble. It cannot be recommended for service on pressures above 600 volts because it does not possess the high insulating qualities of marble. One of the disadvantages of slate is its general appearance, which necessitates that it be finished in some way, if good looks are desirable. The finishes most often given it are black marine, black enamel, or some form of oil treatment. When finished, slate presents an appearance which is equal to that of any finished marble, although it lacks the insulating qualities of the latter.

There are several varieties of slate obtainable, those most commonly used being the black slate, gray slate, and purple slate. Some of this slate has spots of various colors which would look bad if the panels were not given some sort of a finish. The slate, as usually obtained, is sand-rubbed and a final finish is put on afterward.

Beside the finishes mentioned, the slate may be given an oil finish which will prevent oil spots from showing. A popular oil-finished slate with some installations is the black Monson slate. This is simply a slate which is given a special oil treatment, the original name being derived from the quarries at Monson, Maine, from which this slate was obtained. A good appearance is often obtained by using black slate well rubbed with oil. This finish can hardly be distinguished from black marine. There is a special finish for slate known as marbleized which consists of applying a pigment in streaks in connection with an enamel, to represent various colorings of marble. This finish is used by manufacturers of face plates for starting rheostats, but it is almost never employed for switchboard purposes.

Soapstone is softer than slate and is therefore easier to cut and manufacture into shapes. On this account it is often turned up into bushings and planed into various forms for structural work in power plants. This material absorbs moisture more readily than marble, although the best varieties are better in this regard than slate; but it is inferior to marble and is rarely used for switchboard panels.

The manufacturers of soapstone specialties have every facility for working this material and shipping it cut to any desired shape, so that it is frequently shipped direct from the quarry to its final destination in carload lots, for use in connection with the bus compartments, etc., of large power installations. The surface of soapstone is so porous and spongy that it does not permit a suitable finish for fine switchboard work. It has a wide field of usefulness as a material for insulating bases, flameproof barriers and shelves in high-tension power installations where a material which can be readily worked with the tools must be used. It is supplied in its natural color and is sand-rubbed for nearly all classes of service.

The installation for high-tension panels is similar to that for low-tension. This type of panel is used when the number of units is small and when the extension of the board does not become too great. They are almost always used for hand- or wire-rope-operated oil switches. Where it is desired to mount the oil switches and accessories on a separate framework, the latter can be either of angle iron or gas pipe. An auxiliary frame of angle iron is sometimes preferred, but it involves a special design for each plant with all ironwork drilled and fitted especially for the apparatus. Gas-pipe frame, however, can be easily put together with standard clamps; it is adjustable and easy replaceable in emergency to suit the local conditions.

The oil-switching devices, meter transformer busbars, and other accessories are located on suitable supports in the rear of the panels or in compartments, the switching devices being operated by means of bell cranks and levers, by ropes, or electrically. The bell cranks and levers can be placed beneath the floor or, if no basement is available, in pockets in the floor with suitable covers, or they may be supported overhead. They may even be arranged to operate on the surface of the floor.

Small stations of 2,500 kva. or thereabouts are hardly justified in employing any other than open arrangement, unless local rules demand that the apparatus be enclosed. Even in this case metal grill work would be much more in keeping and much less expensive than cement structures. Cables to and from the switchboard should be supported in such a manner that their weight will not be suspended from a soldered joint or the terminal of a oil switch. Cables should not be used as connections for switchboard apparatus where bends are necessitated, as they will not keep their form unless carefully supported or enclosed in a stiff covering. Connections are preferably made of flame-proof solid wire, well insulated against accidental contact. For connections of large capacity it is advisable to use bare copper rods, tubes, or straps, which should be well insulated after erection with varnished-cambric tape or its equivalent. All connections, both main and auxiliary, should, when at all possible, be run vertically

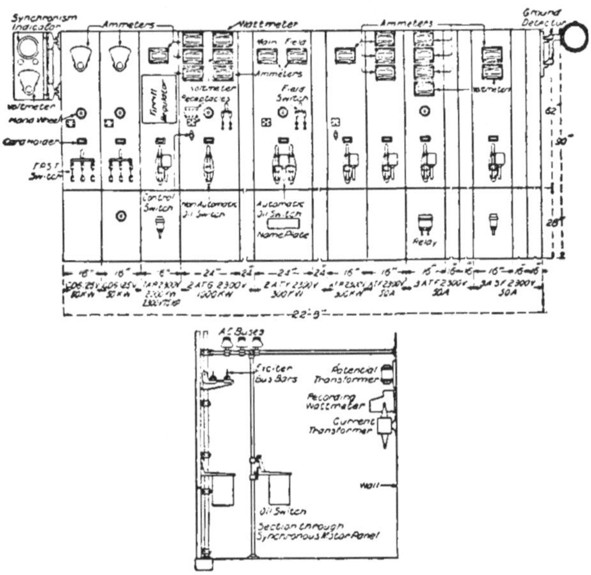

FIG. 214.—Front and end view of a 2,300-volt alternating-current switchboard.

and horizontally with right-angle bends made to a sufficient radius, so that the conductor will not be injured. The same suggestion applies to any straps, brackets, braces, or other members of the switchboard construction.

There is little choice between the horizontal busbar and the vertical busbar from an electrical viewpoint, as perfectly satisfactory designs may be obtained from either. From a mechanical viewpoint the vertical busbar requires greater height and less width, but practically the same amount of supporting framework and the same length of connections. It is obvious that with the mounted vertical busbar the danger of an arc starting at some point on one of the lower busbars and short-circuiting the upper ones is greater than with the open-mounted horizontal busbars. An arc may start at any point where the insulation to ground fails, where current-carrying parts of opposite phase or different voltage are too close together, due to faulty erection or accidental displacement of members; from short-circuits at other parts of the system, or from something falling against or upon the busbars or connections. The entirely enclosed structure is the only real insurance against excessive injury to the equipment in case of arcing grounds.

It is desirable to arrange the panels to correspond with the wiring arrangement, if possible, as this is of considerable assistance to the operator by permitting the use of a dummy or mimic bus, thus forming a single-line diagram of the complete wiring systems. The conduit system for arranging the control and instrument wiring is also greatly simplified.

Figures 214 and 215 show a 2,300-volt a.c. switchboard for the average central station. The switchboard contains panels for two exciters, of which one is driven by an induction motor. Both exciters have their own panels. This arrangement has the advantage that each panel with its exciter can be considered as a unit, and can be disposed of or added as such if any change is made in the equipment. These panels are set up at the extreme end of the switchboard so as to distinguish between the low-tension d.c. and the high-tension a.c. sides of the board. The next panel is for the induction motor which drives one of the exciters. This motor is supplied directly from the busbars at 2,300 volts, which presumes that the other exciter is driven by an engine or from some other independent source. In almost all new plants a Tirrill regulator is installed, which is mounted between generator and exciter, but which may here be placed on the motor panel, as there is sufficient space to receive it. As the regulator is to control two exciters, an equalizer rheostat should be added on one of the exciter panels. The voltmeter for the exciters is mounted on a swinging bracket on the d.c. side of the switchboard.

The next panels in order are the two generator panels, each for 1,000 kw., 2,300 volts, equipped with the following instruments: Three ammeters (unbalanced load); one polyphase wattmeter, from which is read the load division on the two generators; one field ammeter, which, as previously mentioned, facilitates detection of generator troubles; one voltmeter, to read the voltage of the machine between any two phases; and also the potential of the buses by means of an eight-point receptacle; one four-point synchronizing receptacle with the necessary synchronizing plugs; one double-pole, single-throw field switch with discharge clips and resistance; one field rheostat, which, when small, is mounted on the panel and is operated by a handwheel and, when large, is placed away from the boards and is operated electrically or by chain; and one triple-pole, single-throw, non-automatic oil switch.

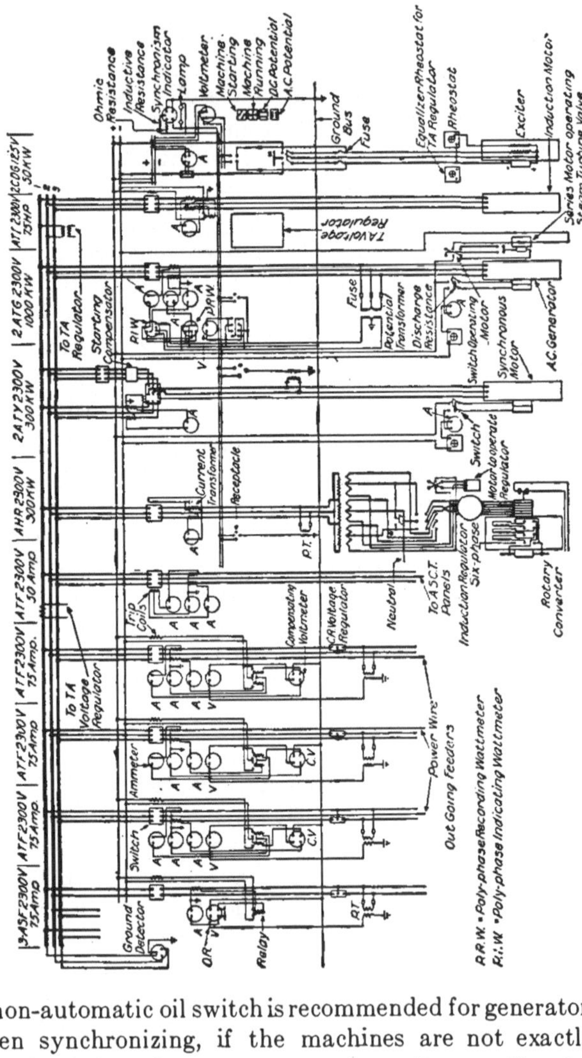

Fig. 215.—Wiring diagram to switchboard (Fig. 214).

A non-automatic oil switch is recommended for generators because when synchronizing, if the machines are not exactly in synchronism when they are connected together, or if a short-circuit or overload occurs on any feeder, an automatic generator switch is liable to open and shut down the plant. Most a.c. generators are so designed that they can carry a momentary short-circuit without injury. When panels are furnished or turbine-driven generators rated at more than 500 kw., a double-pole, double-throw engine governor control switch is furnished for controlling the motor of the governor. Two series and shunt transformers are provided for the instruments, which are generally mounted on pipe supports or on the wall. (See side elevation.) The synchronism indicator with signal lamps is placed on the swinging bracket with the d.c. voltmeter.

When motor-generator sets are used for furnishing either Edison three-wire d.c. service or 500-volt railway or power service, the synchronous motor is supplied directly from the main buses. The equipment for the motor consists of a triple-pole, single-throw magnetizing oil switch, a starting compensator, and a triple-pole, double-throw oil switch. In starting it is supplied with current through the magnetizing switch and compensator, and through the other side of the double-throw oil switch when running. The magnetizing switch and the double-throw switch are interlocked so that the latter is either directly connected to the busbar, in which case the primary switch is open, or to the compensator, in which case the primary switch is closed. Operation is accomplished by two handles so that only one at a time can be in the outer position. In place of a double-throw oil switch two single switches are sometimes used, so that the motor is started through two switches and is maintained through one. In the side elevation there is shown the magnetizing oil switch mounted on a separate pipe support at the back of the board. The equipment for the synchronous motor panel is one main ammeter, one field ammeter, one field switch with discharge clips and resistance, one rheostat with handwheel, one synchronizing receptacle, and one shunt and two series transformers built into the feeders.

The next panel is that for the synchronous converter. When converters are used for feeding Edison three-wire service it is customary to install a regulator on the a.c. side of the converter in order to be able to control the e.m.f. of the d.c. service. This regulator is usually made motor- controlled, and in such cases a double-pole, double-throw switch should be mounted on the panel. Besides this the panel carries one main ammeter, one synchronizing receptacle, one triple-pole, single-throw automatic oil switch, and one shunt and two series transformers.

The diagram shows two kinds of a.c. feeders, one three-phase and one single-phase. The former is used for motor power distribution, and the latter for lighting. If three-phase feeders are used to feed lamps as well as motors, the lighting circuit is connected to one phase of the three-phase feeder, and the regulator is inserted in the same phase.

The three-phase feeder panel for motors carries only three ammeters, one triple-pole, single-throw automatic oil switch and two series transformers. Such panels for both lighting and motor

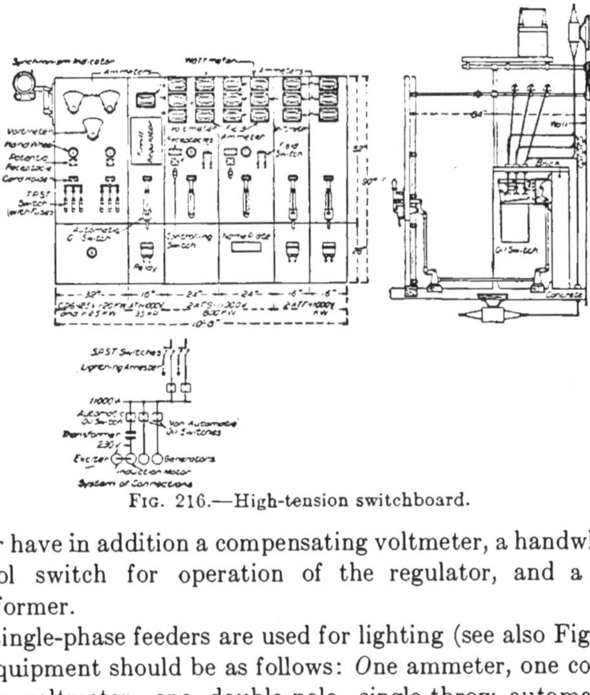

FIG. 216.—High-tension switchboard.

power have in addition a compensating voltmeter, a handwheel or control switch for operation of the regulator, and a shunt transformer.

If single-phase feeders are used for lighting (see also Fig. 206) the equipment should be as follows: One ammeter, one compensating voltmeter, one double-pole, single-throw automatic oil switch, one series and one shunt transformer. All feeders used for lighting have inverse time-limit relays. A ground detector is mounted on a swinging bracket and is connected to the busbars. The latter are supported on the pipe framework which holds the panels. As noted above, the series and shunt transformers are mounted on the same framework or on the wall, according to the arrangement of machine cables and feeders. They must never be mounted on the board, however, since they are connected to high tension wires.

A switchboard for two three-phase, 800-kw., 11,000-volt engine-driven generators, one engine-driven exciter, one induction motor with an exciter, and two high-tension feeders is shown in Fig. 216. The two exciters are controlled from the same panel, which therefore carries a double equipment. The induction-motor circuit and the feeders are each furnished with three single-pole automatic K4 oil switches operated as triple-pole switches. They

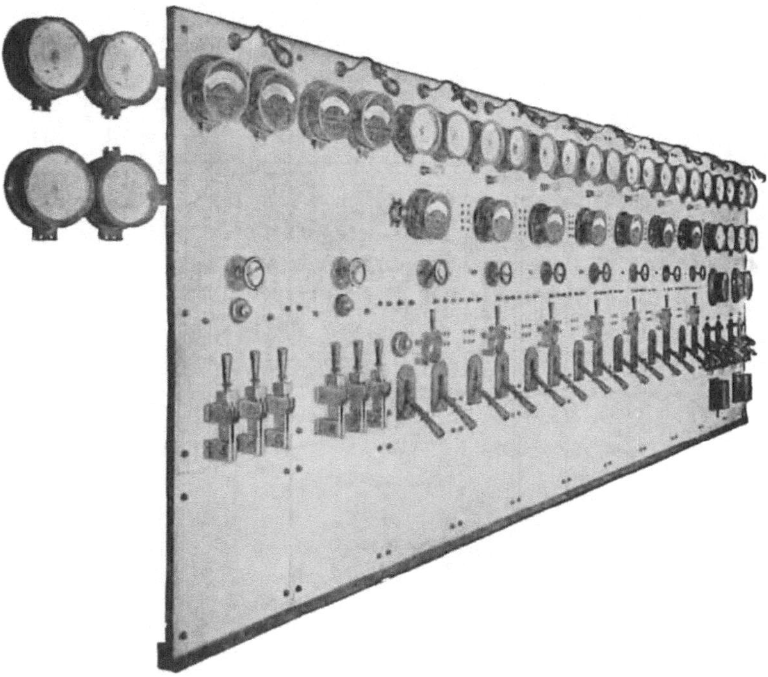

Fig. 217.—1,100- to 2,500-volt alternating-current switchboard.

are mounted in separate cells away from the board. Similar oil switches, but non-automatic, are provided for each generator. All cells should be placed in a row in the same order as their panels. For each overhead feeder there should be a lightning arrester with two disconnecting switches, one to break the line and the other to disconnect the arrester. The diagram shows the connections by simple lines. The side elevation is the same as that for the central station, with the exception that single-pole oil switches are used in place of three-pole. The buses are mounted on insulators above the cells and are quite exposed.

This arrangement is not to be recommended for large installations, for with such high voltages-in this case 11,000 volts—and large kilowatt rating separate fireproof compartments should be provided for the buses. The small wooden gallery over the

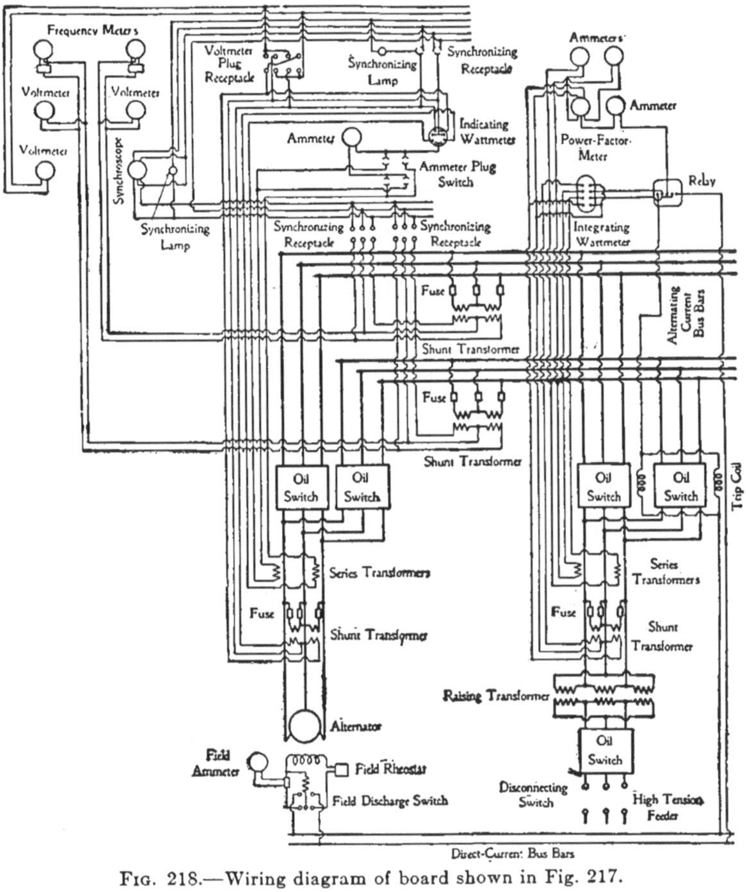

FIG. 218.—Wiring diagram of board shown in Fig. 217.

buses serves for mounting the shunt transformers and for the purpose of operating the disconnecting switches, which for the overhead lines are placed high up out of reach.

Figures 217 and 218 show a Westinghouse switchboard for the control of two 150-kw., 125-volt exciters, seven 1,500-kw., 2,200-volt generators, and two banks each of three 1,500-kw. step-up transformers. The swinging brackets on the left carry three voltmeters and one synchroscope. The next two panels belong to the exciters, and the next seven to the generators, each one controlling two non-automatic type E oil switches to connect the machines to either set of buses. The rest of the equipment is the

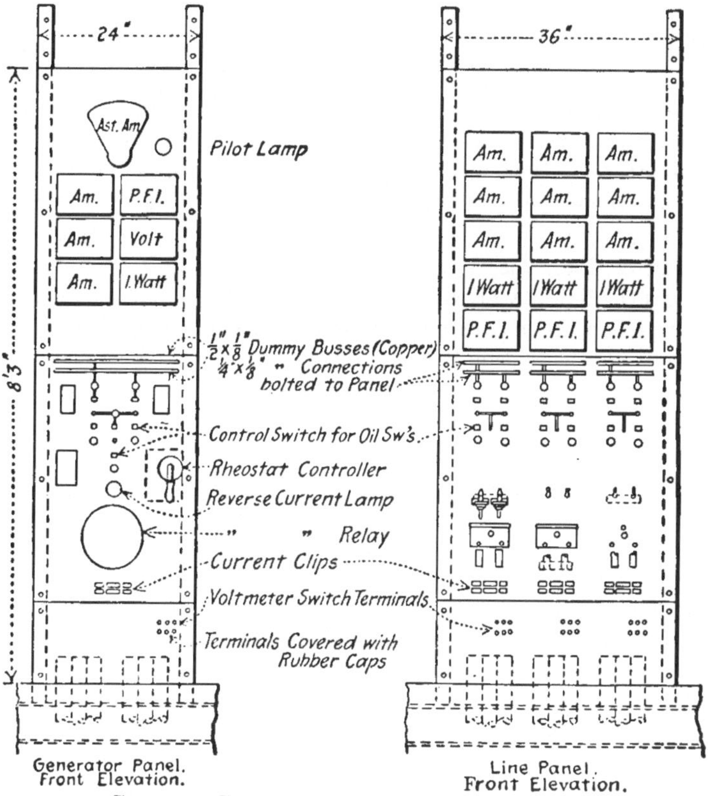

FIG. 219.—Typical generator and line control panel.

same as previously described. The last panels are for the step-up transformer banks, and each carries three ammeters, one power-factor indicator, one polyphase watt-hour meter, and also controls two automatic oil switches with their overload relays. The necessary series and shunt transformers are built into the line. The diagram shows the connections of only one of the generators with one of the transformer banks. All buses, oil switches, and instrument transformers are enclosed in separate fireproof cells and compartments. A voltmeter and frequency indicator are connected to each set of busbars.

A typical arrangement of feeders and generator panels is shown in Fig. 219. The switchboard is set up separated from the oil switches and busbars. In order to indicate to the attendant the functions of the various oil switches controlled and operated from the board, there is mounted on the board next to the control apparatus a set of mimic busbars with connections. The connections to these buses are an exact single-line reproduction of the actual arrangement. Small knife-blade switches or white indicating lamps are used in the connections and mimic buses to represent the disconnecting switches. Whenever a set of disconnecting switches is opened the attendant opens the corresponding knife-blade switch in the mimic arrangement, or an indicating lamp is automatically lighted. This materially reduces the liability of wrong switching. Two signal lamps, one red and one green, are connected to the control apparatus to indicate whether the oil switch is open or closed, and a name plate with the number of the generator, feeder, or transformer bank is placed next to them. The illustrations in Fig. 219 show two parallel copper strips representing two sets of busbars. On the generator panel there are three control switches which operate three oil switches. The generator has one main oil switch and two selector switches, which connect it to either set of busbars. The field rheostat of the generator is motor-operated and is controlled from the board. The feeder panel controls three independent sets of feeders, each of which possesses two electrically operated automatic oil switches. The same is true for the transformer panels.

When it is desirable to reduce the length of the switchboard to a minimum, all control apparatus (for electrically operated oil switches, field switches, field rheostats, governor control, and motor switches, etc.) are mounted on a benchboard. Instruments for the various circuits are arranged so as to enable the attendant easily to pick out which instruments and control apparatus belong together. They may be mounted on an independent panelboard or on the panel which constitutes the back of the bench. A panel on the bench itself or on a frame over the bench is also sometimes used. One or more instrument posts might be employed for the same purpose.

Figure 220 shows a part of a benchboard with instrument board for 100,000 volts. The whole board controls nine 2,500-kw., 6,600-volt generators, two transformer banks of 4,000 kw.

each, 6,600 volts delta to 66,000 volts star, and two 110,000-volt feeders. When installed complete the board will control the entire station, rated at 50,000 kw. On the bench there are mounted the control apparatus for rheostats, field switches, governor control, and the oil switches for the generators and

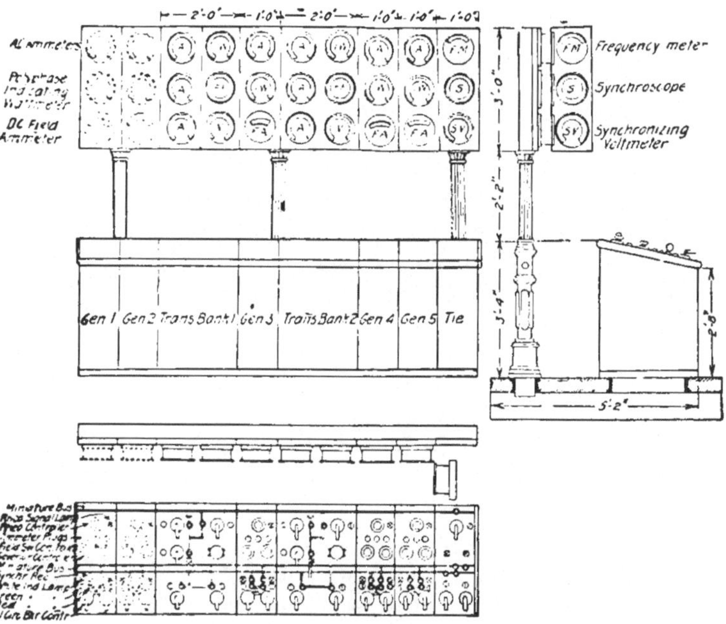

FIG. 220.—Benchboard.

transformer banks. The plug switches for the synchroscope, ammeter, signal lamps, mimic buses, and connections are also carried on the bench. The instrument panel is set up above the bench on a cast-iron frame so that the attendant can see between the bottom of the panel and the top of the bench, and so obtain a full view of the station. The three station instruments on the swinging bracket on the right side can be brought into any desired position with respect to the board, so that the attendant can see them from any position. The supports for the instrument panel also serve as a railing around the gallery. The front of the panel faces the inside of the gallery. The instruments on the vertical panel correspond in position to their control apparatus on the bench.

Other examples of this type of switchboards are benchboards for 11,000 volts set up by the Westinghouse Company in the Williamsburg power house of the Brooklyn Heights Company. A separate bench with instrument board is furnished for the generator control. Opposite the bench is placed a double switch-

Fig. 221.—Fifty-cycle alternating- and direct-current controlling and instrument board.

board for the control of all other high-tension machines and feeders. The space between the opposite panels in the feeder board is closed by two doors, one at each end. The front and rear board carry both instruments and control apparatus (front board), and also the necessary relays (rear board). The d.c. panels with the exciter, battery, station-lighting, and motor panels are mounted independent of the high-tension switchboard. Symmetry and artistic appearance are observed throughout, and by compact arrangement a minimum amount of space for controlling the entire large plant is taken up.

Figure 221 shows a combination of a bench and instrument panel with a panelboard in back of them. In the illustration the panelboard is not shown. The space between the boards is closed by two doors. A cast-iron framework holds the panels together, and the whole forms an independent closed compartment. The two boards at the extreme right and left control the exciters. Relays, controls for electrically operated field switches, wattmeters, switches for station lighting, and watt-hour meters are carried on the rear panel. The benchboard and instrument panel at the front carry the control apparatus and instruments for the generators, feeders, and transformer banks. When Tirrill regulators are used they are mounted on the vertical boards on either side of the bench. On the wall in front of the bench there are sometimes mounted the control apparatus for electrically operated rheostats and governor controls and for the electrically operated gates for the intake water for the power house. The separate parts of the bench and panels are held together by gas piping and clamps. The exciter buses and gas pipes connecting opposite panels should be mounted so that one may safely walk underneath them. Similarly the mounting of the exciter field rheostats and the copper connections to the exciter buses must be such that safe passage is possible.

In some industrial plants, and especially in mining collieries, where very little space for the installation and inspection of the necessary electric apparatus is available, a switchboard system consisting of single removable panels has come into use. This type may be adapted for d.c. as well as a.c. systems of low-, medium-, high- and extra-high-tension current.

Each movable panel of the ironclad switchboard stands upon rails in the base of a fixed framework, so that the panels may be withdrawn separately upon a portable temporary track or by means of a light carriage.

A high-tension switchboard has the single panels separated from one another and from the busbar compartment by means of fireproof barriers so that the front of the panels carries only the instruments and handles which control the high-tension apparatus mounted on the back of the movable panel. An interlocking arrangement combines the fastening latches with the handles of disconnecting and oil switches so that the switches must be opened before the panel can be removed or returned to position.

The adoption of this method renders it unnecessary to wait

for week-ends, or for other occasions when the plant is shut down, to effect repairs or alterations. Hence it permits periodical examinations to be made at reduced trouble and cost. If needed, a reserve panel possessing all the necessary apparatus can be quickly put in place of one removed for repairs.

Instead of mounting the instruments on vertical panels, they may be placed on posts with the corresponding control apparatus assembled on a bench or pedestal. Two of these instrument posts with a different number of symmetrically arranged instruments are shown in Fig. 222. They are often used as railing supports when they are placed on the edge of the gallery. Their upper parts are generally movable so that they can be turned by means of a handwheel at their base. Some of them carry a set of receptacles, with plug switches on the base, for testing the instruments. All instrument wires are carried inside the posts.

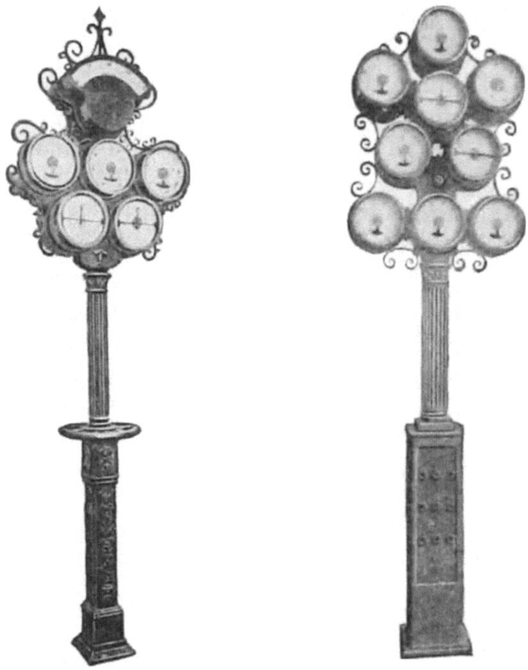

FIG. 222.—Instrument posts.

Figure 223 shows two pedestals with control apparatus, which constitute independent units, with the corresponding instrument posts. One controls a generator, and the other a bank of three transformers. The control of the former is for oil switch,

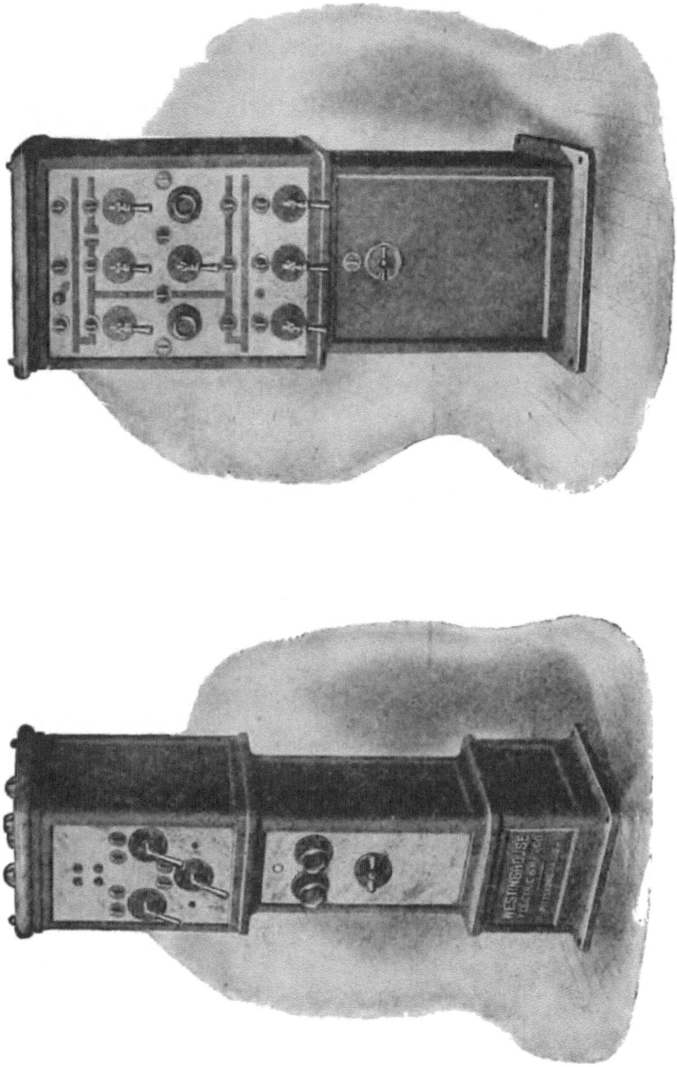

Fig. 223.—Control pedestals.

rheostat, governor control, and field switch. Handwheels for chain-operated rheostats are also sometimes mounted on them. The frame is of cast iron and the panels of blue Vermont marble. Control wires are led through the open base to the interior, and small doors are provided for inspection of the various connections. The pedestals are set up in a row in front of the instrument posts so that the attendant can overlook the entire station. On the second pedestal are mounted the mimic buses and indicating lamps.

By using such post and pedestal units for individual generators and transformers the chance of the operator mistaking the wrong control switch is greatly reduced.

CHAPTER XXII

CELLS AND COMPARTMENTS

As has been previously stated, high-tension apparatus and busbars must be guarded by fireproof insulating barriers. The purpose of such barriers, cells, or compartments is to protect apparatus, etc., against destruction by fire, which may be caused by sparks, arcing, or short-circuit, and at the same time to protect attendants from accidental contact with high-tension parts.

Two kinds of apparatus requiring such protection must be distinguished. The first includes appliances whose action is aecompanied by sparking or arcs, the duration of which is determined by the voltage and value of energy back of them, as well as by the construction and material of the instruments. Lightning arresters, disconnecting switches, fuses, etc., will serve as examples. Busbars, transformers, oil switches, etc., are appliances which will produce arcing accidentally. The causes for this may be either careless handling, damaged insulation, too high tension, improper spacing, or short-circuit. Arcing is detrimental not only to the apparatus itself, but also to adjacent machinery. In this kind of apparatus the destructive effect is mainly dependent upon the energy back of the arc.

In order to reduce the dangers mentioned above to a minimum the following points should be observed:

1. The system of connections should be laid out as simply as the control of the station will admit.

2. The connections from generators to transmission lines should follow the shortest and most direct path.

3. Apparatus of the first group should always be placed in compartments when under high or extra-high tension, while that of the second group at high and extra-high tension and large current, and those for smaller current, which are easily accessible, must also be enclosed in fireproof compartments.

4. High-tension parts must be separated from each other, and from inferior insulation, by the greatest possible distance.

The choice of material for compartments is somewhat restricted, for no material combining the qualities of cheapness, fireproofing, strength, and perfect insulation is at present available. Although the material most often used may possess some of the above qualities, it generally lacks the property of perfect insulation. High-tension parts must therefore be widely separated on account of the danger of grounding, and this in turn calls for considerable space. Concrete or brick 4 inches thick is most generally used for these compartments. The horizontal divisions are made of soapstone or slate. Glass, porcelain, and asbestos are less in use on account of their fragility and poor insulation respectively. For very high tensions, busbars and bus wires are mounted bare. They should therefore be placed near the ceiling in order to prevent accidental short-circuit through falling objects and contact of the attendants.

Busbar Compartments

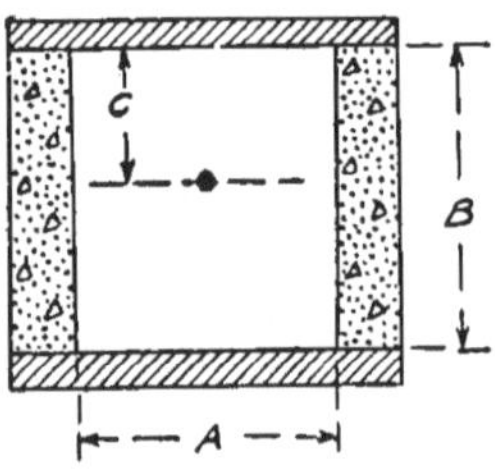

Dimensions of 22,000, 33,000, 45,000, 66,000-volt compartments are based on the assumption that bus wires are fastened to the insulator by means of tarred rope. If metal fastenings are used or metal caps are required for disk switches or other purposes, the dimensions should be increased so as to obtain the same distance between the compartment and the nearest live metal.

Standard size of brick, 8½″ × 4¼″ × 2¼″.

Volts	Bus	*A*	*B*	*C*
5 000 to 15,000	One—2″ × ¼″ bar	13″	12⅜″	5¼″
5,000 to 15,000	Two—2″ × ¼″ bar	13″	12⅜″	4¾″
5,000 to 15,000	One—3″ × ¼″ bar	13″	12⅜″	5¼″
5,000 to 15,000	Two—3″ × ¼″ bar	13″	12⅜″	4¾″
5,000 to 15,000	2″ or 3″ × ¼″ bar	12″	12⅜″ or 14⅞″	4″ or 5½″
22.000	Wire	15″	16½″	8½″
33,000	Wire	18″	19¾″	9¾″
45,000	Wire	2′ 1″	2′ 2″	13″
66,000	Wire	3′ 0″	3′ 1″	18″
100,000	Wire	4′ 8″	4′ 8″	2′ 4″

Where such mounting is not feasible, or where the connections are too complicated, the busbars are built in compartments. These may be entirely closed, save for the openings, or may be open on one side, in which case they would consist of a vertical wall with horizontal barriers between the buses.

The usual method of building the cell and compartment structure is by erecting forms and pouring the concrete, using reinforcing rods where necessary. Owing to the difficulty of securing a uniform and smooth job, sometimes an expanded metal framework with the concrete body plastered into it is used. The thickest portions plastered are 3 inches. The scratch coat contains a small quantity of hair, the remaining coats being pure cement and sand. The finishing coat is of neat cement, which gives a hard, white smooth surface. In the finishing coat a small quantity of lime is used to make the cement work more smoothly. Angles are used in the heavy barriers and small channels are riveted or tied, the expanded metal being wired to this skeleton framework.

The figure and table on page 336 show a typical cross section of one of these compartments, with the dimensions for different voltages for horizontal or vertical mounting of busbars and the use of bus wire. All these dimensions are for the same make of insulators (General Electric), which are built into the horizontal base. The figure and table below show these insulators for different voltages mounted on piping with the distances from center to center and to ground. With any change in dimensions in other makes for the same voltages the corresponding dimensions of the compartments must be changed.

BUS WIRE SUPPORTS

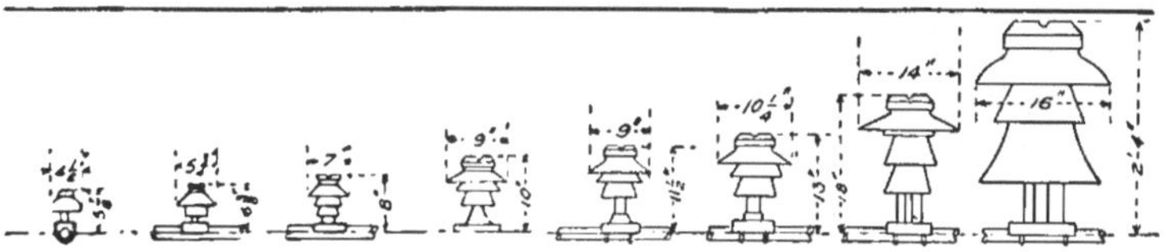

Number	E.M.F., volts	Spacing, inches	Ground distance, inches	Number	E.M.F., volts	Spacing, inches	Ground distance, inches
1	6,600	8	6	5	33,000	18	10 to 12
2	15,000	10	7	6	45,000	25	13
3	22,000	12	8 to 10	7	60,000	36	19
4	33,000	18	10	8	100,000	56	30

NOTE.—Ground distance (usually) = $\frac{\text{Wire spacing}}{2}$ + 1 in.

Spacing of series transformers (self-cooled):

600 volts..............	1 in. clear	6,600 volts..................	3 in. clear
2,300 volts..............	1.5 in. clear	13,200 volts..................	5 in. clear

In Fig. 224 there are shown the cross sections of compartments for 6,600 volts at medium rating and 6,600 volts at greater rating. This construction with insulators in the sides of the compartment is good up to 15,000 volts, but for higher tension openings must be left for the connection. In Fig. 225 are shown a number of different arrangements of busbar compartments, in connection with the oil-switch cells (form H3) for 13,200-volt

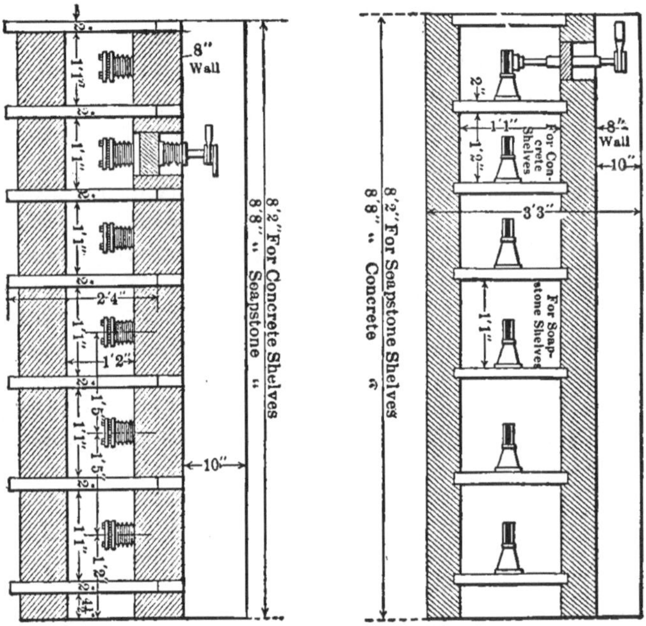

FIG. 224.—Bus compartments for 6,600 volts medium and high ratings.

plants. Note the mounting of the disconnecting switches between the barriers which are shut in by doors in the front. In the left-hand installation the lower stud of the disconnecting switch is connected to the buses. This arrangement is not to be recommended, because when the disconnecting switch is open, which presumes that the oil switch is also open, the handle is alive, although it should be dead in this position. The connections in this case are made of copper rods or tubing, but might just as well be made with copper wires. The space over the compartments is sometimes utilized for shunt transformer compartments. The busbar supports should be set up near the openings for the connections, or, if special openings in the back

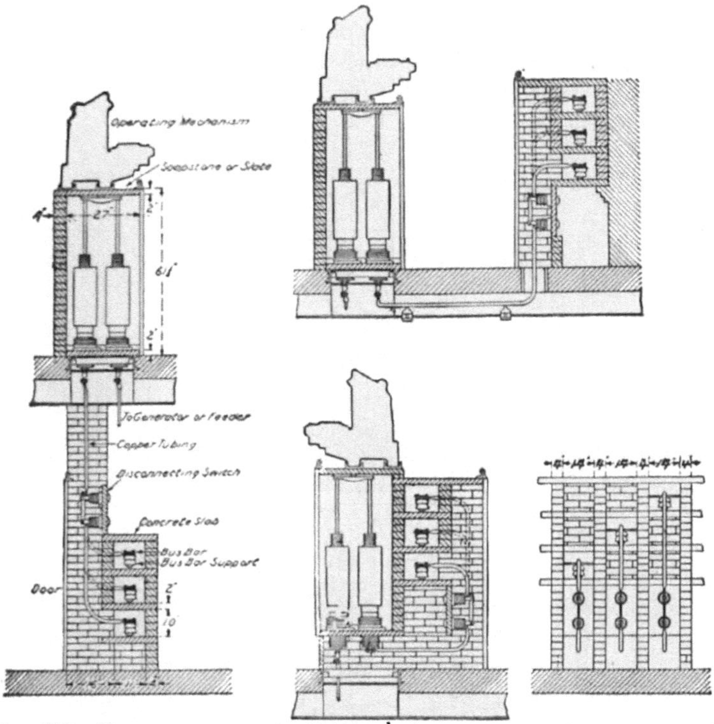

FIG. 225.—Bus compartments for 13,200 volts in connection with motor operated oil switches.

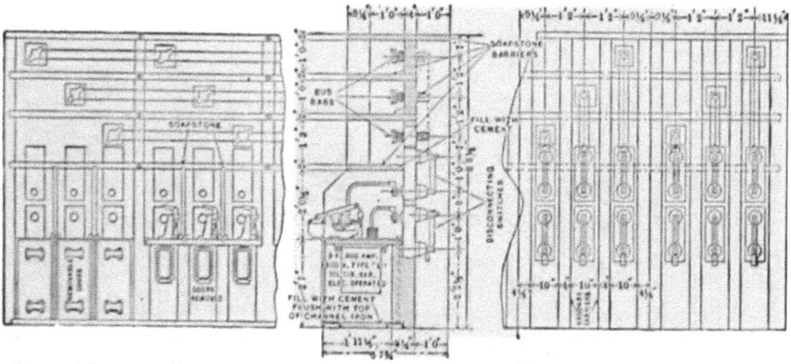

FRONT VIEW OF STRUCTURE **REAR VIEW OF STRUCTURE**

FIG. 226.—Busbar and oil-switch structure with solenoid-operated switches.

of the compartment are provided, near these, so as to admit of inspection. They are generally spaced about 4 feet apart. Figure 226 shows a busbar and oil-switch structure for a 6,600-volt, three-phase generating plant with type E solenoid-operated Westinghouse switches.

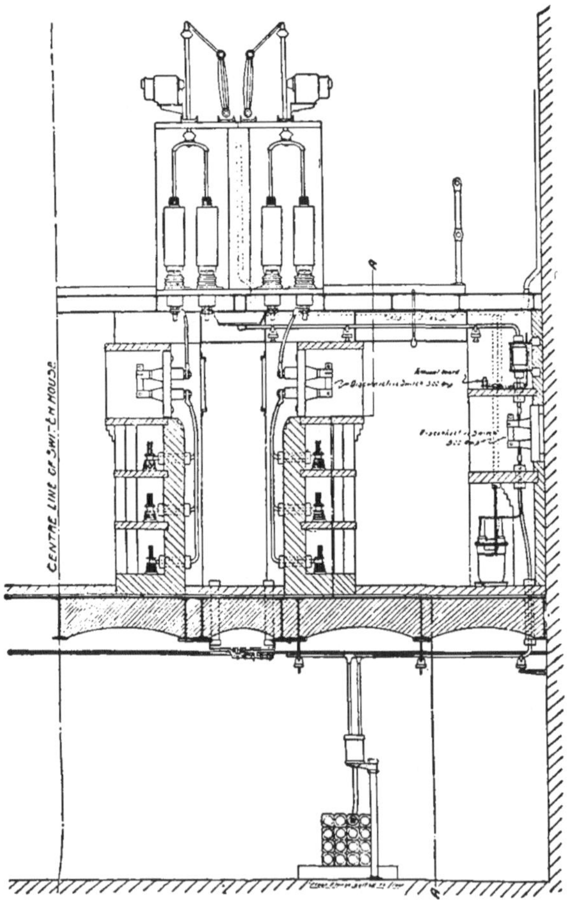

FIG. 227.—Cross section of bus and instrument compartments of the Boston Edison Company power house.

In the design of the enclosing structure it is of great importance, especially when the higher voltages (11,000, 13,000, etc.) common to structural mounting are employed, that openings at different parts of the structure, which may create a path for

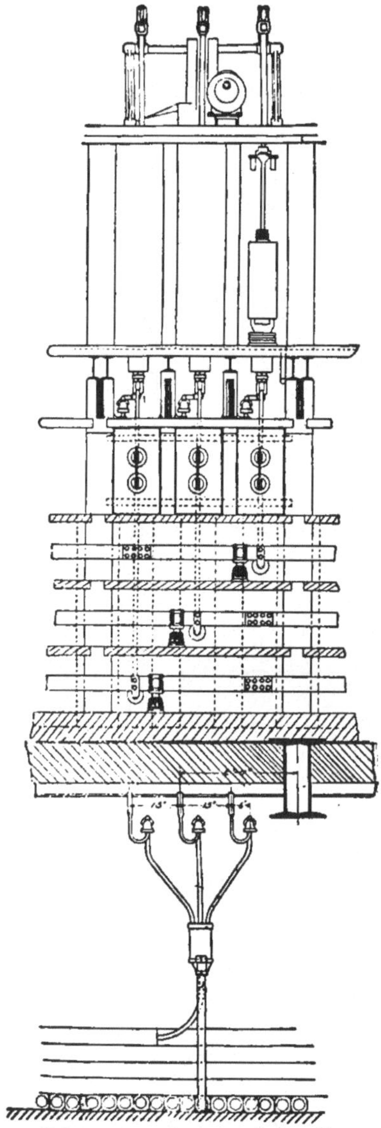

FIG. 228.—Elevation of bus compartments of the Boston Edison Company power house.

drafts, be avoided. For the lower voltages, especially 2,200 volts and below, the copper connections are usually very heavy and of large cross section. It is difficult to take such a connection through a wall without considerable expense and local heating of the conductor, except by means of an opening large enough to give sufficient insulation of air around the conductor.

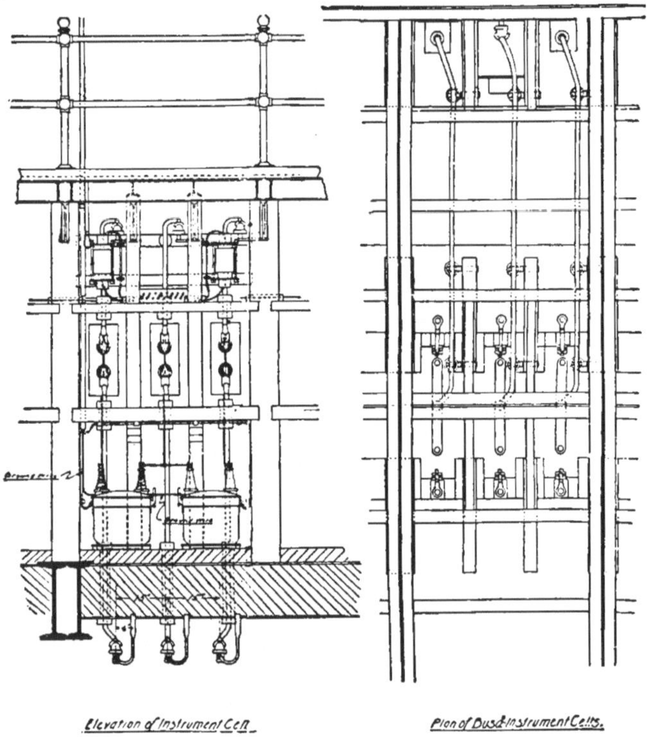

FIG. 229.—Plan and elevation of bus and instrument compartments of the Boston Edison Company power house.

Figure 227 shows a section and Fig. 228 an elevation of oil switch cells (form H3) and busbar and instrument compartments for a 6,600-volt plant whose control panels were shown in Fig. 219. It will be remembered that in that arrangement the generator was provided with one main and two selector oil switches, and the feeders and transformers with two selector switches each. The two oil switches of Fig. 227, mounted back to back, are the two feeder selector switches mentioned above. The two inner poles are connected together, and the feeders run out from this common connection. The feeders are run on the basement ceiling down the wall, where series transformers are built into the two outer legs. (See Fig. 229.) On the outer side of the transformers are inserted the disconnecting switches which

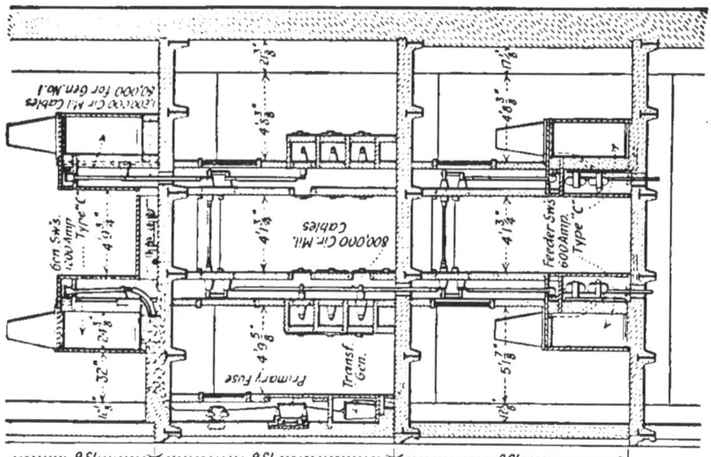

Section through Gen. Switches.

FIG. 231.—Cross section through electric galleries of Williamsburg power house.

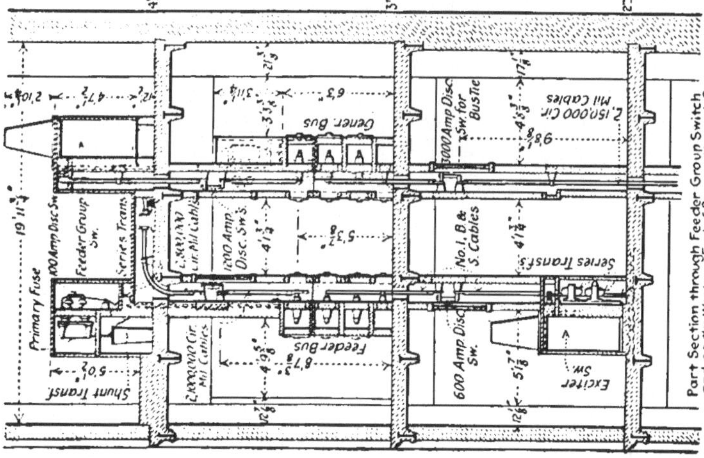

Part Section through Feeder Group Switch and at the Western End of Busses through Bus Tie Connections and Switch for Motor Exciter.

FIG. 230.—Cross section through electric galleries of Williamsburg power house.

join the transmission line to the station apparatus. The entire equipment on the wall is enclosed in brick compartments. The cables are conducted through the walls and floors in bushings, and the three-conductor cables are joined to the three single feeders by end bells.

In Figs. 230 and 231 the arrangement of oil-switch cells and bus and instrument compartments of the Williamsburg power station of the Brooklyn Heights Railway Company is shown. It is a four-story arrangement, with the generator and feeder group switches on the fourth mezzanine, the busbar compartments on the third, and the feeder, motor, and tie oil switches on

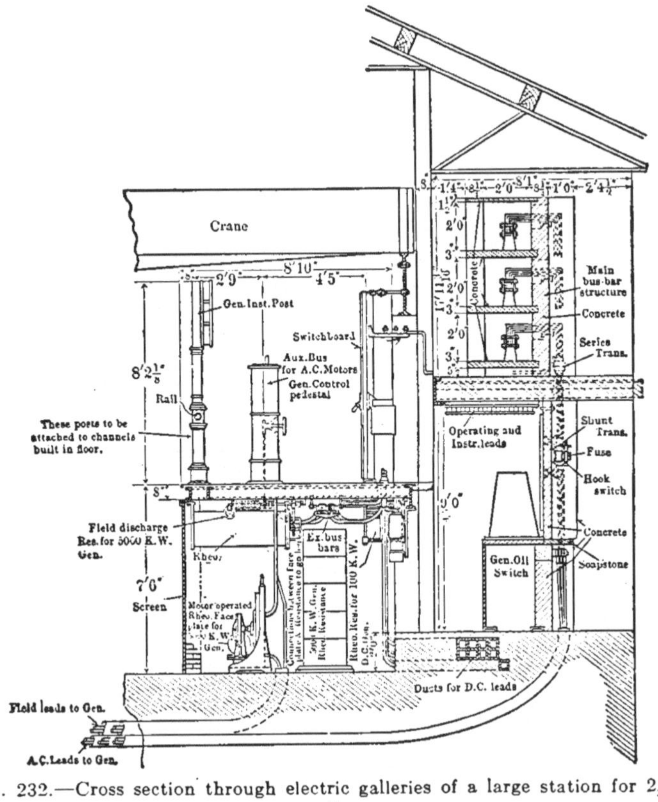

FIG. 232.—Cross section through electric galleries of a large station for 2,200 volts.

the second. The oil switches are type C Westinghouse, with terminals on the back of the cell. The series transformers on the fourth mezzanine are set up under a false floor, and those on the second in compartments on the back of the oil switches. The shunt transformers for the generators in Fig. 230 are set up in compartments on the wall of the third mezzanine, and those for the feeder group switches on the fourth mezzanine in cells similar to oil-switch cells. Primary fuses are mounted separately and are easily disconnected from the shunt transformers. Disconnecting switches are also provided between the shunt transformers and the high-tension leads. All disconnecting switches are set up in compartments and are operated through rectangular openings on the front side. Notice that the generator and motor cables, as well as the potential and instrument wires, are laid in

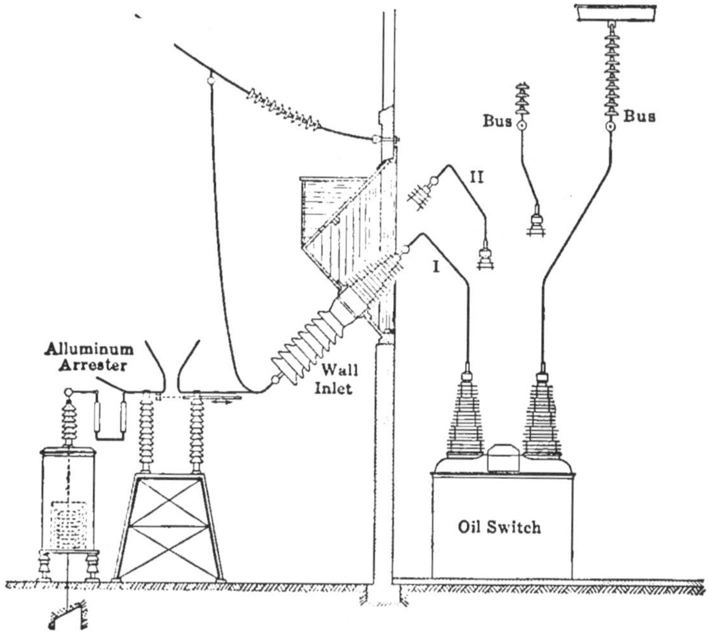

Fig. 233.—Arrangement of lightning protection outside the station building for a 110,000-volt system.

conduits built into the concrete floors. All of the galleries are entirely separated from the engine room, with the exception of the third, which has a glass-enclosed platform in front of the main switchboard facing the engine-room floor.

Figure 232 is an arrangement of cells for a 2,200-volt plant of large capacity. The arrangement is divided into two parts, one a control-operating division situated in the engine room, and the other a high-tension part in a separate room outside the engine room. Part of the first division is on the engine-room floor, and the other part in the gallery. That on the floor is screened off and carries the field rheostats, their operating motors, and the exciter buses. In the gallery are mounted the instrument posts, control pedestals, and the switchboard for controlling the exciters, motors, etc. The second division is also composed of a basement and gallery. On the part on the floor all the oil switches are found, and the busbar compartments are situated in the gallery. The connections are made by copper bars. The bus compartments are open on the front side, and the overhanging horizontal divisions are supported by concrete pillars.

It is recommended by some engineers that all instruments and apparatus requiring special cells and compartments on the different galleries be set up outside the station so as to do away with costly construction, at the same time preserving the requisite conditions for safety to station and attendants. Their argument is that as long as overhead transmission lines are left unprotected by barriers or special enclosures, even though they carry the same e.m.f. as the station appliances, this station apparatus may just as well be set up in similar manner, provided it is properly spaced and so arranged that accidental contact is made impossible. Such devices may therefore be installed overhead like transmission lines, with proper protection against snow and rain, etc. (See Chap. XXVI, Sub-stations.) Figure 233 shows such an arrangement with the lightning protection situated outside the station building. Note the mounting of buses on hanging insulators.

It will be noted that a more or less arbitrary distinction has been made throughout this treatise between high and extra-high tensions. Extra-high or highest voltages have been taken as those over 33,000 volts.

CHAPTER XXIII

WALL OUTLETS

Bringing high-tension lines into stations necessitates carefully constructed wall inlets which will successfully resist weather and electrical influences. When designing the building the installation of these inlets must be taken into consideration and proper provision for them must be made, for otherwise their location relative to the steel work and the position of the machines might be undesirable. It is impossible to state any hard and fast rules covering all cases, for local conditions generally determine the ways and means for solution. C. E. Skinner recommends that the following points be looked into before a definite arrangement is decided upon:[1]

1. Voltage of the transmission circuit.
2. The climate in which the plant is to be operated.
3. The size and insulation of the high-tension conductor.
4. The kind and height of building used.
5. The conditions of approach to the building and the location of the apparatus in the building to which the high-tension line is connected.

After these points have been investigated, the following requirements must be met by a proper form of inlet.

1. It must maintain proper insulation of the circuit under normal as well as abnormal atmospheric and electric influences inside and outside the conductors.
2. Snow, rain, cold air, and dust must be prevented from entering, since they weaken the insulation at the point of entrance, which may result in damaging the contents of the building.
3. The end strains of the line must be taken up and must not be transmitted to the inlets. Special line poles and supports with insulators outside and inside the station take up this strain and serve to center the lines in the inlet openings.
4. The construction must be reliable, simple, and cheap.

[1] C. E. Skinner: "Methods of Bringing High-tension Conductors into Buildings," *Proc. A. I. E. E.*, July 1, 1903.

The simplest form of inlet is a hole in the wall of sufficient diameter to allow enough open space about the wire to prevent any possibility of an arc striking across to the walls or surrounding material. The opening is protected against snow, rain, etc., by a terra-cotta pipe sloping outward or by sufficient extension of the roof above. Special steel or wooden hoods, or a gallery built around the opening on the outside of the wall, are also

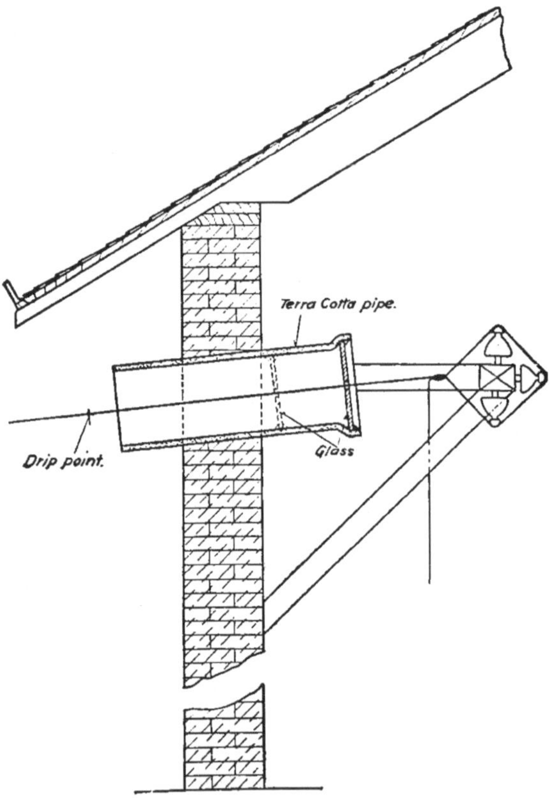

FIG. 234.—Wall outlet with a terra-cotta pipe.

sometimes used. The line must be held in the center of the hole or tube to keep it away from the wall or side of the tube, which is considered as ground. The above arrangement is applicable only for medium tensions up to 15,000 volts in dry and warm climates. For 15,000 volts or less the terra-cotta pipe is closed on the station side by one or two glass disks with holes. (See

Fig. 234.) In this case the pipe must be of a diameter which will allow a sufficient surface insulation of the glass plate against leakage, which might otherwise produce grounding. The insulation at the inlet should be somewhat greater than that of the line itself, as otherwise any electric disturbance might cause

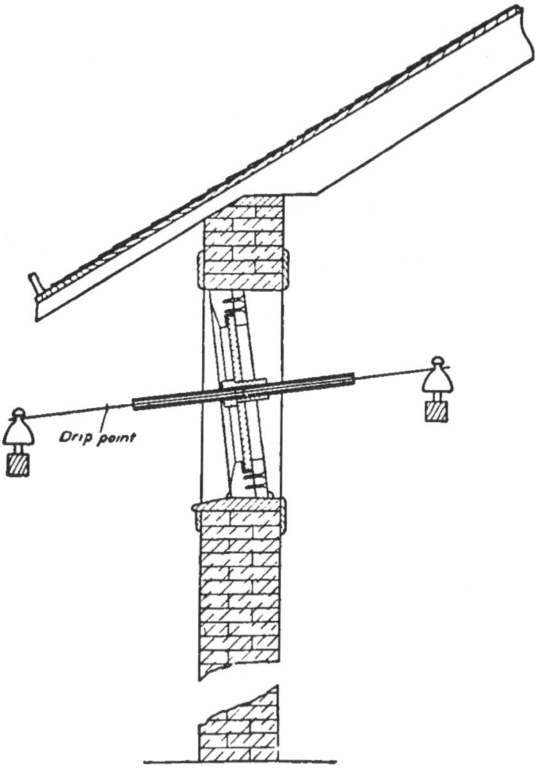

FIG. 235.—Wall outlet with a slab of insulation material and insulating tube.

an arc at this point which could prove disastrous to the building. The main disadvantages of the arrangement are fragility of the glass plates, opportunity for accumulation of dust and moisture, reducing the surface insulation of the plates, access for birds and insects in the outer part of the pipe, and, above all, the large diameter of the pipe which is necessary for higher voltages and unfavorable conditions.

A method used for high voltages up to 60,000 is shown in Fig. 235. A long insulating glass or porcelain tube of small diameter and very heavy wall is placed over the wire and passed through a slab of insulation set in the wall of the building, the whole being protected from driving rain by an extension of the roof or special hood. Both tube and slab should be of fireproof material. The chief difficulty is in securing the proper insulating tubes. Glass and porcelain are electrically the best materials for the purpose, but on account of their lack of mechanical strength even the difference between out-of-door temperature and that

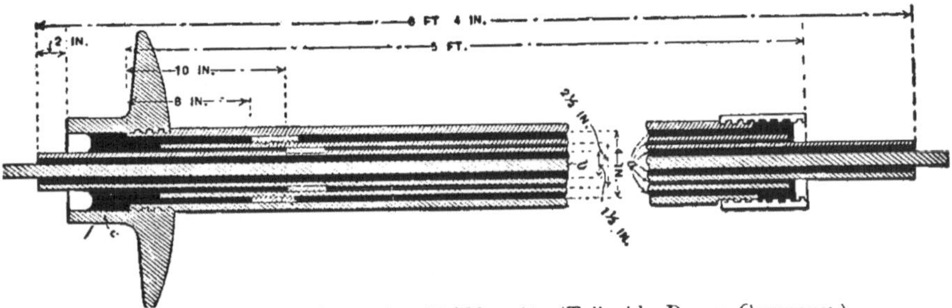

FIG. 236.—Wall bushings for 44,000 volts (Telluride Power Company.)

inside will weaken them. For this reason the end strain must be taken up outside the building by a suitable guide pole. Sometimes two or three glass plates are used and the tube is cemented through them. In other cases the glass tube is fitted into a wooden tube of paraffined wood and this is set into wooden panels. This arrangement is sometimes used in iron construction. As a rule, however, it is not advisable to make use of combustible material near the line.

If the building is too low and the line wires must be carried at a considerable elevation in the immediate neighborhood of the building, or the multigap lightning arresters require a high space for setting up, a tower construction may be necessary for the entrance of the line. Otherwise the inlet is the same as through the side of the building. There are any number of different forms of inlets in use, as every plant and manufacturing concern devises its own appliances so as to suit local conditions.

Figure 236 shows a wall bushing used by the Telluride Power Company, of Provo, Utah, for 44,000-volt lines. It consists of a set of concentric fiber-conduit tubes, the spaces between which are filled with ozokerite, while the ends are sealed for short consecutive spaces with chatterton compound, minerallac, and a very brittle asphaltum compound, to prevent the ozokerite

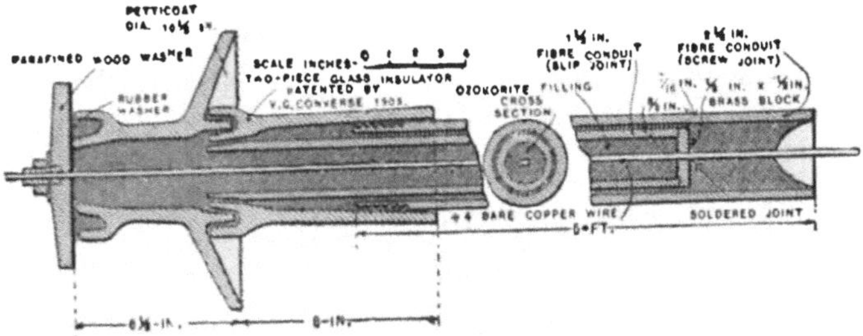

FIG. 237.—Wall bushings of the Telluride Power Company

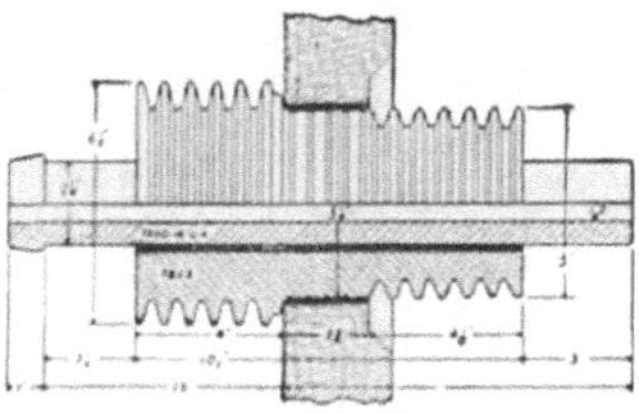

FIG. 238.—Wall insulator for 13,200 volts (R. Thomas & Sons Co.)

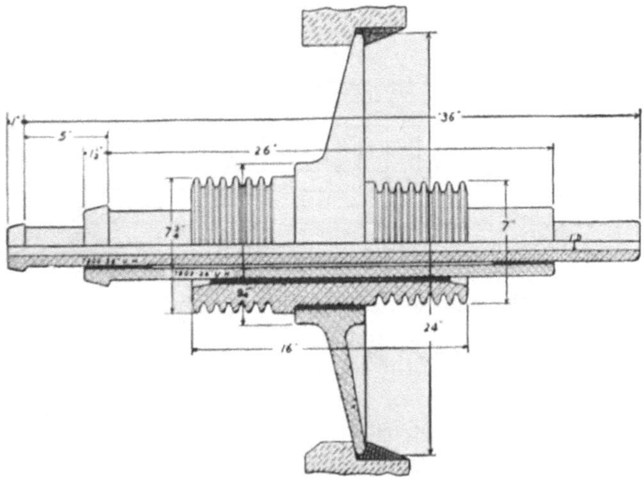

FIG. 239.—Insulator for 80,000 volts (R. Thomas & Sons Co.)

from oozing out. The outer end is covered with a porcelain sleeve, and the whole is fitted into a sewer pipe in the wall. It is protected against weathering by a steel hood. These hoods were found to be too expensive and cumbersome, and in

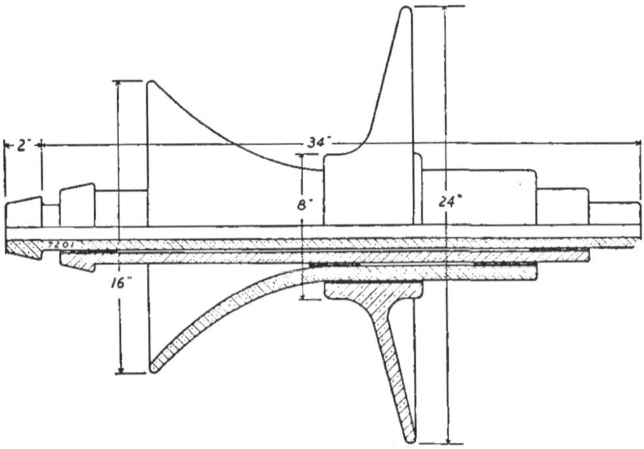

Fig. 240.—Wall insulator (R. Thomas & Sons Co.).

severe storms they were torn away together with lines and parts of the building. It therefore became desirable to obtain some other design of bushing of which one end might be directly

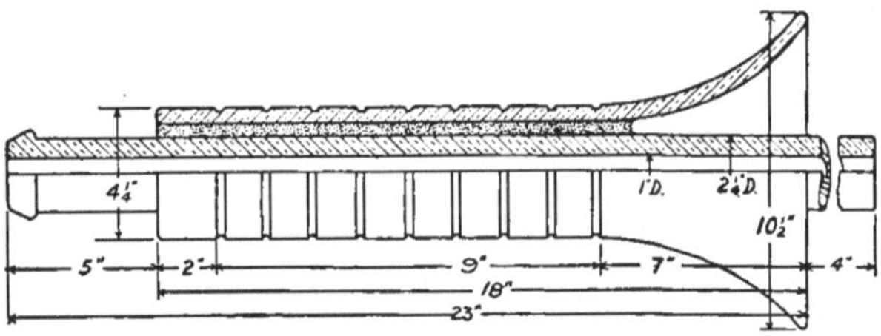

Fig. 241.—Wall insulator (R. Thomas & Sons Co.)

exposed to the weather, thus doing away with the necessity of a hood protection, and that might be adapted to the walls as well as to the roofs of the buildings, and might be of such insulating strength that even a building of the cheapest sheet-iron construction might be used without danger from break-down by puncture; and, finally, the cost of material and labor was to be less than for the devices previously used. The bushing in Fig. 237 was found to comply with all these requirements. It consists of a porcelain petticoat and a glass insulator with a fiber conduit, which together constitute the mantle. A second fiber-conduit cylinder is placed inside the mantle. The spaces are filled out with ozokerite, and in the center is laid the bare No. 4 B. & S. copper wire.

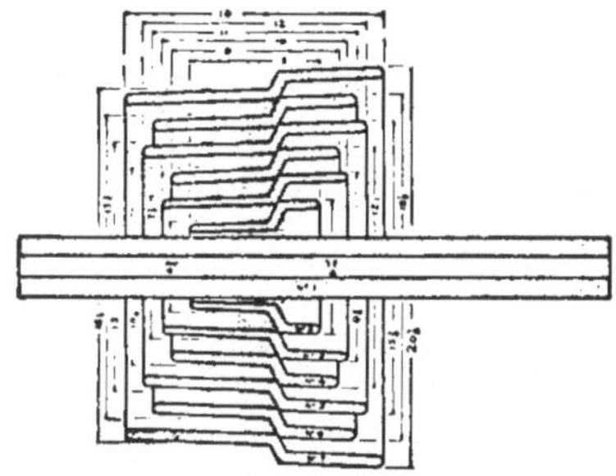

Fig. 242.—Wall insulator (Locke Manufacturing Company).

Figure 238 shows a wall insulator for 13,200 volts manufactured by R. Thomas & Sons Company. It is made of a porcelain tube cemented into a porcelain bushing which is fitted into the opening in the wall.

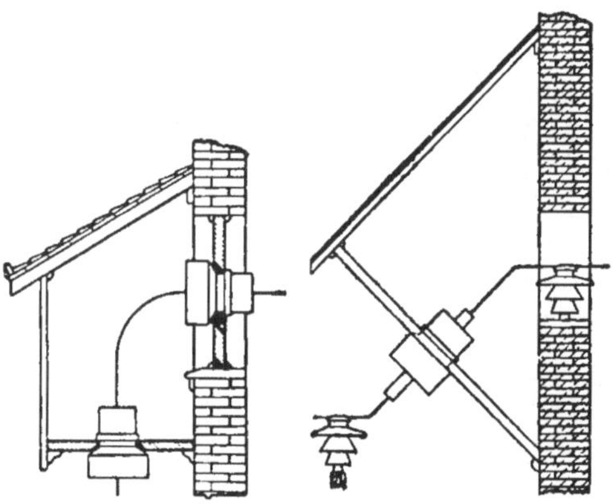

Fig. 243.—Location of wall insulators.

Figure 239 shows an 80,000-volt insulator of the same make. A long porcelain tube is cemented into a shorter one, and this into a porcelain bushing which sits in a tray-shaped plate fitted into the wall opening. The surface insulation of the bushings must be large enough to prevent leakage.

Other wall inlets for different voltages of the same make are shown in Figs. 240 and 241.

The Locke Insulator Manufacturing Company produces a bushing consisting of concentric porcelain rings, cemented together,

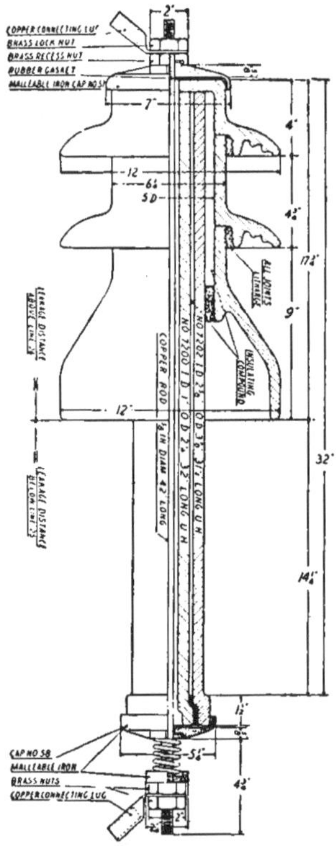

FIG. 244.—Roof insulator (R. Thomas & Sons Co.)

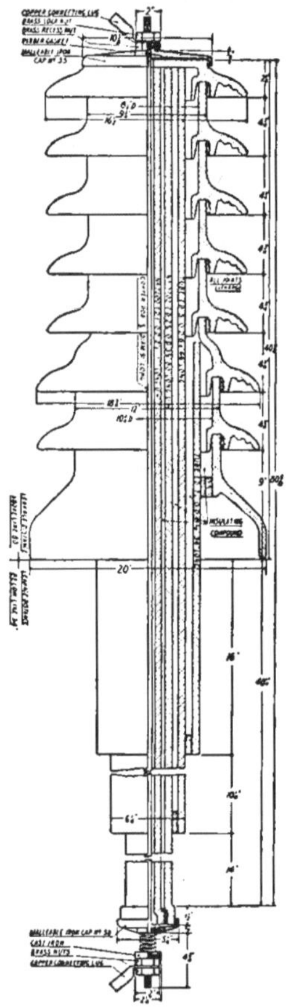

FIG. 245.—Roof insulator (R. Thomas & Sons Co.)

making them thoroughly substantial mechanically. The number of rings depends upon the voltage of the line. The bushing and settings are shown in Figs. 242 and 243. The bushing is set into a marble, slate, or thoroughly varnished wooden panel, whose diameter is twice that of the outside bushing diameter. A second insulator tube is often fitted into the bushing.

It sometimes becomes necessary to bring the transmission line in through the roof, in which case extra precautions must be taken in order to prevent the entrance of moisture. Figures 244, 245, and 246 show roof insulators for different voltages manufactured by R. Thomas & Sons. The outside line must be carried on the roof by specially built supports.

Figure 247 represents a floor bushing for 33,000 volts. The construction of floor tubes is comparatively simple since the whole insulator is kept dry and clean.

Where conditions admit, it is always best to bring the lines in on the gable end, as on the other ends ice and snow are apt to collect and damage the bushings. *D*rip points should be fastened on the line near the bushings, and extra guide poles should take up the end strain, so as to keep it away from the bushing or wall. The several feeder inlets should be placed far enough apart to secure adequate fire protection and easy orientation.

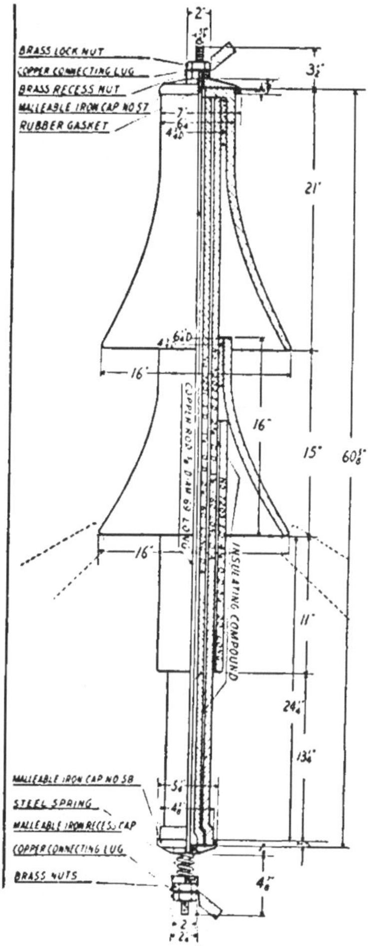

FIG. 246.—Roof insulator (R. Thomas & Sons Co.)

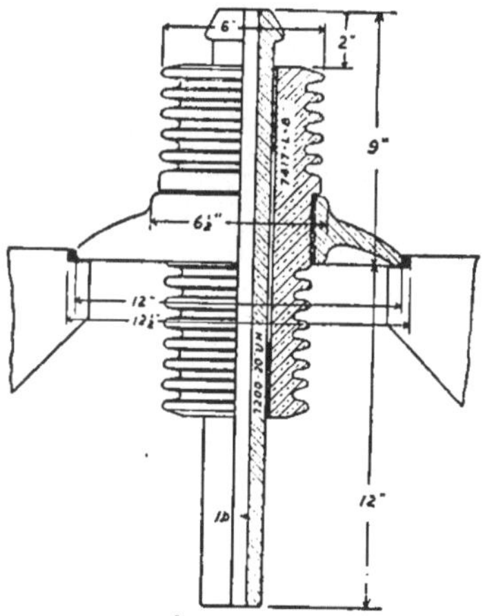

FIG. 247.—Floor bushing for 33,000 Volts.

CHAPTER XXIV

CENTRAL STATIONS

In distribution systems exceeding a certain size the use of a single d.c. central station is uneconomical. The reasons for this were stated in Chaps. X and XIII. With such systems either one of two methods is followed. The system may either be divided into independent districts, each with its own isolated d.c. plant, or the power generation may be concentrated in one large high-tension a.c. station and fed to sub-stations where it is converted to low-tension alternating or direct current. The choice is determined by local conditions, but in the majority of cases the latter method is adopted.

The advantages of a high-tension a.c. central station regardless of the kind of load may be summed up as follows:

1. The first cost per kilowatt of the large units entering into the construction of a large plant is lower than that of the small units of a small a.c. or d.c. station, thereby reducing the first cost of the apparatus necessary for the given service and reducing the annual fixed charges thereon.

2. A large plant inherently can be operated more economically than a small one, and further, the large plant can afford to introduce economies which would be out of the question in a small one.

3. The existence of a diversity factor whereby one kilowatt of capacity in a central station will serve a combined load that would take considerably more than one kilowatt if each part of that combined load were to be served separately.

Comparing a large central station suitable for delivering power for every purpose with a plant of the size that an individual customer of that central station would have to install, if he supplied his own wants, or comparing it with a small independent district plant, it is found that the central-station plant will be very much larger, anywhere from tenfold to a thousandfold, possibly even more. However, in comparing complete plants with such a difference in size, it will be found that the whole of the difference in cost that would be indicated by comparing difference in the cost of individual parts is not realized. The reason for this is that the large plant is inherently more elaborate than the small one. A large plant will naturally be equipped with such items as mechanical stokers, coal and ash handling machinery, coal bunkers and storage yards, superheaters, economizers, condensers with all their auxiliaries, double sets of busbars, distant-control electrically operated switches, heat-insulating lagging on all boilers and steam pipes, feed-water heaters, water-softening plant, etc., some or all of which items may be omitted from the small plant. Although such items cause an increase in the first cost of the large plant, they do not run its cost up as much per kilowatt as that of the small plant not equipped with these refinements. In the large plant the first cost of these items is more than offset by the operating cost which their use will save, and that fact accounts for their use. On the other hand, as the size of the plant is reduced, the point is approached where these cost-saving items may be eliminated one by one. It is only where power is reduced in relatively large quantities that these various cost-saving devices will pay.

Another saving which would be effected by concentrating power-producing apparatus at a single point instead of distributing it throughout the premises of the various customers served or throughout the independent individual small districts is the tremendous aggregate saving in space. As the size of the units increases, the saving in space occupied becomes marked.

The aggregate saving in space occupied in the housing of a plant and in the auxiliary equipment will amount to a large item, and the comparison of central-station power with individual plants will result in a very material reduction in the first cost of the former compared with the latter. The high-tension system, furthermore, makes it possible to locate the station outside of town limits, so that a greater number of suitable and cheap building sites are available.

The total cost of power may be divided into two parts: (1) Annual fixed charges (interest, depreciation, taxes, and insurance) that depend upon the first cost, and (2) operating costs (fuel, labor, repairs, supplies, and superintendence) that are almost entirely independent of the first cost. The proportion in which fixed charges enter total power costs is usually somewhere between 30 and 60 per cent., being smaller with small plants.

The first matter that requires attention as the size increases is usually the matter of fuel saving. In large plants the cost of fuel usually amounts to somewhere between 50 and 75 per cent. of the total operating expense. In plants of small output the fuel bill bears a smaller proportion to the total, not at all because of better economy, but because the other expenses, such as labor, superintendence, etc., go down more slowly than does fuel as the plant size increases. The use of condensing apparatus, superheaters and economizers affects the fuel economy to a very considerable extent.

The purchase of fuel under specifications is a refinement in fuel practice to secure a uniformity of supply. It is only such plants as use fuel in large quantities that can avail themselves of this method of fuel purchase.

In the large plant it is possible to introduce labor-saving devices that would be out of the question in the small plant. The mechanical stoker is one of the greatest labor-saving devices. Not only does it save labor, but it also gives ability to force boilers to a point not possible with hand firing. The amount of boiler equipment necessary during peaks is therefore affected.

The use of automatic machinery to handle the coal from the coal pile to the grates and the application of ash-removal appliances as labor-saving devices are also particularly limited to large power plants.

Another item of economy which can be exercised by the large plant, but is prohibited to the small one, is that of adjusting the apparatus to the load to be carried. A large plant usually is equipped with a comparatively large number of units and can operate them so that the units are run near their maximum economy point. On the other hand, a small plant is usually equipped with a relatively small number of units and must often run a unit considerably larger than necessary to take care of the load and consequently operate it at a relatively low point on its economy curve.

The large plant can afford to insure itself against a coal miners' strike by installing a coal storage yard, and to insure itself against electrical breakdowns by spare units, spare cables, and a double set of busbars, refinements that would be entirely out of reach of the individual small plant. The maintenance expenses in general are lower, and when repairs must be made these may be accomplished with greater safety and reliability and at a smaller cost.

A diversity factor might be defined as the advantage in capacity which is secured by the large plant serving many different kinds of loads over the aggregate capacity of the many plants that would be required to serve each individual part of this load separately. Not only is the large plant cheaper per kilowatt than the small one, but also the large station does not have to have as many kilowatts of capacity installed to take care of a given aggregate load as would individual plants for taking care of the same service.

There are two arguments that may be urged against the central-station method of supply. The first is that the central station requires the addition of a high-tension transmitting and low-tension distributing system with the necessary sub-stations before its customers can be supplied, and the cost, upkeep, and losses in this transmitting and distributing system must be considered to make the central station comparable with the individual small plant. It is, of course, possible to imagine a condition where the cost and losses of transmission and distribution would make the central-station supply more expensive than that of an isolated plant, but this would mean that it had gone beyond its proper radius. The advantages of the central station are so great that the utmost additions in the way of transmission and distribution, which still allow good regulation are not so costly as to permit isolated plants to compete.

The second argument in favor of the isolated plant is that during the season heating is required the steam may be used twice, once for power and the exhaust later for heating. In general, however, a separate heating arrangement coupled with the supply of the central station will secure the advantages of the centralization of power production.

The location of an a.c. station is governed by the same considerations that were found to apply to d.c. plants, with the exception that not so much attention is paid to the location with respect to the load center. The main points to be investigated are as follows:

1. Accessibility, (*a*) for delivery of coal and removal of cinders; (*b*) for delivery of heavy machinery; (c) for connections to mains· (*d*) for officials and workmen.

2. Proximity of water for condensing and boiler feeding.

3. Stability of foundations. If an artificial pile or concrete foundation must be made, the initial cost is materially increased.

4. Isolation to secure freedom from causing nuisance by noise, vibrations, smoke, dirt, fuel, carting, etc.

5. Facility of extension, which includes the possibility of purchasing adjacent ground or future extensions.

The method of operation of a hydro-electric system and the lay-out of the system depend upon many factors, some of which are the relative importance, the number and the relation of each of the different power stations in the system to the whole system, the location and character of the load, the number and capacities and locations of the steam reserve stations, and, in general, whether the high-tension line is a purely transmission line, a high-tension distribution circuit, or both.

With water-power stations the choice of building site is determined by the topography, which must be such as to give inlet and exit of water with minimum loss of head, the shortest feasible penstock, and the greatest security from variation of head. It must also afford security from floods, possibility for extensions, and easy access for overhead lines.

The normal conditions of operation are determined largely by the capacities of the different generating stations belonging to the same system, by the water conditions, and by the characteristics of the load curves to be met. Practically no hydro-electric development of great size with the capacity of installed apparatus above rating at minimum stream flow is attempted without a steam station on the same system. This steam station may serve any one or all of three purposes: First, as an auxiliary station to be used at periods of low water; second, as a reserve in case of interruptions; and third, as a regulating station to take care of the variations in the load with the hydro-electric plant running at constant output.

The normal problems of operation, such as starting up the whole system, paralleling the generators and the power stations of the system and properly dividing the load among them, putting units into service previous to the demand of load conditions, regulating the voltage for the proper distributing points, connecting the high-tension lines, and, in general, so manipulating the generating, transforming and switching apparatus as to deliver the desired load at the distributing centers, with the desired characteristics at the maximum degree of reliability, are met only after a careful study of what is wanted, by so laying out the power-station system of connection, the apparatus, and the transmission lines as to accomplish the desired object.

In deciding upon the number and capacity of the units in a station the combination of the water-wheel and the generator must necessarily be considered together. Besides hydraulic conditions and the limitations of the water-wheel design it is governed by the load factor, the nature of the load, the reserve capacity, the reliability and flexibility of the service, etc. The units should be operated as near full load as possible, and new units preferably should be started as the load increases instead of utilizing overload capacities. Where sudden overloads of considerable magnitude come on the system for short periods it is necessary to have wheel capacity sufficient to care for them. Single units are never desirable except for multiple-plant systems, in which case the necessary reserve can be obtained from other stations. For a single-plant system the number of units preferably should not be less than four, but above this number should be governed by the limit in design, considered both from a technical and an economical point of view. With a small number of large units the first cost, the maintenance charge, and the necessary floor space are reduced, and the efficiency is also usually better than for a large number of smaller units.

There are some cases where the development of hydro-electric energy has attracted industries to such centers, a notable example of which is Niagara Falls. The cause, however, which has been active in locating these industries at the Falls has not been exclusively cheap electric energy, but it has been the combination of this essential element with unexcelled transportation facilities by both land and water and the presence of a large part of the consuming population of the United States within a comparatively short radius. The position held by the developments at Niagara is, of course, unusual. For the most part, hydraulic power is found in remote localities which, owing to topographical and other adverse conditions, do not lend themselves well to the establishment of ideal manufacturing sites. It is apparent, then, that water power generally is subject to limitations as to location, and if energy generated from waterfalls is to become available for industrial uses, it usually means electrical transmission over a greater or lesser distance. With a given percentage of profit the distance that may economically intervene between the point of generation and that of utilization depends principally upon the total amount of energy that may be marketed, considered in connection with the unit price that may be obtained through its sale.

A comparison drawn between a steam-driven power plant and a hydro-electric plant shows the limitations and advantages of one or the other system.

While the reliability factor of the generating and receiving machinery of a water-power plant is equal to or better than that of its competitors, and while transmission construction has attained a substantial basis, reliability of the latter being much enhanced by the use of duplicate lines, yet where it is necessary to transmit energy uninterruptedly over long distances reliability is a characteristic which should receive careful attention. To insure this reliability large expenditures in reserve equipment at the receiving end of the transmission system are regarded as a necessary adjunct to any well-planned enterprise.

If energy from remote waterfalls is to compete successfully through the medium of electrical transmission with energy locally generated, one of the chief essentials is that the former shall be generated a nd transmitted in large quantities. The nature of power loadas measured by load factor forms quite an important element in determining the value of hydro-electric energy as compared with energy generated in steam-electric plants. In large steam-driven electric plants operating at load factors with a value in the neighborhood of 20 per cent. or below it is the rule to find that this method of energy production outstrips its hydro-electric competitor in the matter of cost. However, as the load factor increases above a certain value hydro-electric energy shows up to better advantage. This is due to the fact that the cost of hydro-electric energy is practically fixed and independent of the amount used, whereas in the case of steam-produced power a much larger proportion of the operating expenses depends upon the consumption.

Owing to the comparatively small number of employees required for the operation of hydro-electric plants, together with their freedom from difficulties of fuel supply incident to strikes, freight congestion etc., steam-electric energy suffers in comparison with hydro-electric energy in the ability to supply non-interrupted service. Large investments in fuel are therefore necessary to tide steam-power plants over an emergency like a fuel famine. This need of steam plants for comparatively large amounts of working capital as compared with hydro-electric plants gives the latter quite a strategic position as compared with the former and is a point of great importance, particularly in times of financial stringency.

Quite an advantageous operating characteristic of a water-power plant compared with a steam-engine plant is the relatively large increase in gross receipts that is possible without a corresponding increase in operating expenses.

Owing to the relatively large proportion of total investment in hydro-electric enterprises devoted to construction of more or less permanent nature as compared with the equivalent steam plalets the amount that is necessary to set aside annually to cover depreciation of plants of the former type ranges probably from not more than one-third to one-half of that required for plants of the latter class.

The measure of success which hydro-electric energy will meet in competition with steam energy locally produced will depend largely upon the price of steam fuel in the locality in question, together with the possible economies attending its use. Our principal fuels are wood, oil, and coal. Wood is no longer taken into consideration, and oil, although used quite extensively in some parts of the country, is of limited occurrence and may cease to be an important element as a source of energy. Coal is therefore the chief remaining steam fuel, and as it advances in cost hydro-electric energy furnished under proper conditions promises well to hold its own with energy generated from this most important combustible.

One of the most important parts of a generating station is the exciter system. The capacity of the exciter units designated for a complete station, the proper division of the required exciter capacity into several units, as well as the method of drive, the arrangement and connections of the different units and their voltage regulation, are all factors which should be given careful consideration and which have an important bearing upon the successful operation of the plant.

The capacity of the exciters should be sufficient to excite all the synchronous apparatus in the station when these machines are operating at their maximum load and at the true operating power factor. It is not enough to provide for the excitation when the machines are operating at unity power factor, because the excitation required at lower power factors is considerably higher than at unity power factor. Exciters are generally designed for a 25 per cent., two-hour overload rating, and by providing exciters with a combined normal capacity equal to the excitation required for all the synchronous machines in the station when operating at the maximum load, sufficient margin will be left for operating auxiliary apparatus.

A 125-volt excitation pressure has been considered standard for moderate-size installations, while for larger plants a 250-volt exciter system will generally be found more economical.

Different means are accepted to produce the exciter current. Almost all exciters are now of the direct-connected type. They may be driven coupled with the main generator, or by separate motive power (steam or water-turbine), or by motors.

There seem to be a number of advantages obtained from the use of direct-coupled exciters with water-wheel or steam-turbine generators, especially where generator speeds conform to standard exciter speed. Among them may be cited simplicity of operation and the elimination of a considerable amount of wiring and exciter switch connections. In some cases additional building facilities will be required to accommodate the greater length of the complete generating unit. But as we seldom know the ultimate capacity of the generating station, and therefore the number and probably the future size of the generators, there is a serious objection against the use of a large number of direct-connected exciters, because the initial desire for simplicity may result ultimately in complication. Furthermore, trouble with a direct-connected exciter may involve the shutting down of a large unit at a time when its capacity is greatly needed. With a few generating units in the station it is advisable, therefore, when using the direct-connected exciter to give each exciter a rating equal to twice that required for one generating unit, or additional excitation arrangements must be provided.

The system which seems to be most favored from an operating point is one in which the excitation is obtained from a common source consisting of as few exciters as possible. Three units are then generally provided, two of these being used normally and the third one being held in reserve. The two exciters used for normal operation are usually motor-driven, while the third one is driven either by a water-wheel or by a steam turbine as the case may require.

Until now the motor-driven exciter has been supplied from the a.c. generating system. Great objection, however, has been raised against taking feeding current from the system, on account of the fact that any short-circuit may cause the driving motor to drop out of step and thus cause a shut-down. Another objection against feeding the motor of the electrically driven exciter from the main line has been brought out in the operation of some of the large systems. Where several large machines are operating in parallel, with exciter and other small auxiliary machines operated from the system, the loads are necessarily small, the current transformers operating the relays are of small capacity, and consequently it is difficult to obtain adequate protection against trouble on these auxiliary circuits.

To avoid these objections C. W. Stone[1] recommends for a steam-driven power house the use of a non-condensing turbine unit with a low-voltage generator, which should supply, independently of the main system, all the electrically driven auxiliary machines such as circulating pumps, exciters, etc. In the case of larger stations it may be advisable to use two of these low-voltage generator units. As an emergency connection with this exciter system, transformers could be installed so that the auxiliaries could be operated from the main system if necessary. This connection would probably be of a capacity equal to that of the other feeder going out from the station. The protective device, therefore, could be made as reliable as those on any of the feeder circuits.

In some of the latest hydro-electric developments each of the generator units is provided with a small motor-driven exciter set of a capacity corresponding to that required by each generator. The terminals are connected directly to the generator field. The motors of the various exciter sets are fed, in line with the proposition of C. W. Stone, by one or two low-voltage generators driven by independent water-wheels, with the additional emergency connection through transformers to the main line.

Where separate exciters are operated for each generator and these exciters are not operated in parallel, the regulators can be used for preventing wattless cross-currents flowing between the generators. This is accomplished by installing current transformers for each generator and connecting them 90 deg. out of phase with the potential transformer, so that cross-currents which

[1] C. W. STONE: "Some Problems in Central Station and Sub-station Operation" *General Electric Review*, June, 1912.

tend to flow between the generators will be reduced by the regulator action which tends to strengthen or weaken the excitation of the units as required.

Storage batteries are very often used to supply exciting current in emergencies. They are controlled through a booster by means of an automatic regulator which keeps the exciter bus at constant potential.

In any large up-to-date installation a great number of automatic appliances will be found. From boiler room to feeders, the control and handling of large outputs implies the use of labor-saving devices to cut down operating expenses whenever possible without impairing the efficiency of the service. In very large plants the appliances of this sort will include circuit breakers and oil switches, potential regulators, boosters and batteries, motor-operated circulation pumps, etc. Another class of devices serves to indicate the conditions existing in the station apparatus. Signal lamps, for instance, are used to show whether the oil switches are open or closed, which of the exciters is in service, or what set of busbars or feeders is in circuit. In the same way automatic telltales serve to attract the attention of the operator to the temperature of the transformers or to the height of the water level in the boilers. When an overload switch opens that fact is announced in the same way. Automatic steam-engine cut-offs and speed-limiting devices are important protective appliances. When it is desired to start or stop an engine, generator, exciter, etc., communication between switchboard attendant and boiler engine room is facilitated by whistles, gongs, telephones, speaking tubes, and illuminated letter signs. Sluicegates, engine and turbine governors, steam and water inlet valves, as well as field rheostats and generator field switches, can be operated electrically, from the switchboard.

Where many different power plants, as in a hydro-electric system, are attached to a common high-tension transmission line or to the low-tension distributing circuits, the direction for operating the different stations and apparatus come from a central source, where the chief operator has before him a diagram of the station and information regarding the capacities of the generators in use and the magnitudes of the loads at the different places of distribution. The successful operation of such a large and complicated system is therefore intimately connected with the means which are provided for communication.

The use of the telephone service is especially important at times of electric disturbances, or when there is the greatest danger that communication may be interrupted, and for this reason the greatest care must be exercised in the routing and erection of these circuits.

High-tension switching under load should be entirely abolished, but where necessary, special precautions should be taken to avoid the disturbances resulting therefrom as explained in Chap. XX. The apparatus should be so arranged as to have as large a condenser capacity as possible left in position behind the switching point to absorb the electrical disturbances. As a rule low-tension switching is preferable as this avoids steep-wave-front phenomena.

For instance, in energizing a line, it is preferable to connect the dead line to the dead step-up transformer and then connect, by low-tension switches, the line and transformer to the generator rather than to connect the transformer alone to the generator and then switch the line onto the high-tension winding of the transformer. Of course, in this case it would be still better to connect the line, the step-up transformer, and the generator together while the whole system is dead, and then bring it up gradually to full potential by the excitation of the generator.

In case it becomes necessary to open a high-tension switch in a loaded line, the circuit should if possible first be paralleled with another before opening the switch.

When the switching is done on the low-tension side of the transformer both in the generating station and the sub-station the switches in the former are equipped with inverse-time-limit relays and the switches in the latter with reverse-power relays.

In order to make the right selection, it is absolutely essential that the final installation be carefully considered as well as the ultimate capacity which must be controlled. It is also very important that the essential features of switchboard design and arrangement be well worked out before the station design is settled whenever the switchboard control involves ducts, conduits or other arrangement for cables, special provision for floor beams and supporting structures or masonry, line outlets or any other structural details.

In alternating-direct current systems the cost of sub-stations and feeders is an item of considerable importance. Under the discussion of d.c. plants it was pointed out that an increase in the service voltage materially reduces the expense for the copper transmission lines. This applies to an even greater extent to a.c.-d.c. systems such as traction systems when the d.c. service voltage of the system supplied requires to be raised. A larger service voltage, therefore, with the same line drop will give a greater length of line; that is, if the line drop is fixed, any increase in the service voltage makes possible a corresponding increase in the distance between sub-stations. If the load increases directly as the length of the line, then the distance between sub-stations and consequently the length of the transmission line will vary directly as the increase in service voltage, the line drop having the same value in all cases. If, on the other hand, the line is extended and the load remains constant, then the distance between the sub-stations will vary as the square of the voltage increases. In actual practice, however, the true value is the average of the two extremes; that is, if the voltage is doubled the distance between sub-stations is tripled, and with three times the voltage the distance will be five times as great. It therefore becomes evident that if in traction systems a high d.c. voltage is used, the number of sub-stations is reduced and a great saving in the cost of installation is afforded.

In Chap. XIII there are enumerated a number of considerations which apply to the design of switching arrangements. Point 5 states that the choice of units is determined by the size and kind of service required of the machines. Under the heading of "kind of service" we have the following classification:

1. Direct-current feeding for street or interurban railways. This implies an a.c. central station with sub-stations. The service voltage may be either low (500 to 600 volts) or high (1,000 to 1,200 volts), and the a.c. may be converted either by synchronous converters or motor-generator sets.

2. Alternating-current single-phase traction service with a.c. central station and sub-stations. The high tension is transformed to a service voltage of from 2,100 to 2,300. Motor-generator sets are sometimes used to change the frequency, phase or voltage of the supplied current to suit the required conditions.

3. Alternating-current series lighting systems for arc and incandescent lamps, with central-station and constant-current transformers. The primary voltage of the transformers is 1,100 to 2,200, and that of the secondaries depends upon the number of lamps connected to them in series.

4. Power distribution at 110 to 220 volts or 500 volts, alternating or direct current. Direct-current power distribution is analogous to case 1. The direct current may also be supplied through mercury rectifiers. Alternating-current distribution is analogous to case 2.

5. Three-wire system, requiring central station and sub-stations. Motor-generator sets or synchronous converters are used for converting the a.c. into d.c. The service voltage is 110–220 or 250–500.

In any of the above cases transformer banks may be used in the central station to step up for high tensions for long-distance transmission. Where direct current is delivered, the sub-stations are equipped sometimes with storage batteries which are kept as reserve or to be used in service during a part of each day. (See Chap. V.) The converter and transformer stations may be located in the central station itself or in separate buildings, or they may be portable. The transformers for a.c. service are often put on poles or the sides of buildings near the places of consumption. (See Chap. XXVI.)

The matter of size of units for a given total station rating is decided by the kind of load and the cost of reserves. In the case of a generating station feeding into an underground system of a given length there exists a minimum value as regards the output of the smallest generating unit. It is unsafe to drop below this value, lest we produce resonance of the fundamental frequency. The following data were obtained from a number of installations in actual service:

Size of Units for Given Size of Station

Required kw.	Units		Reserve	
1,000	Two	500	One	500
2,000	Three	800	One	500
3,000	Four	800	One	800
5,000	One Three	800 1,500	One	1,500
7,000	Two	3,500	One	3,500
10,000	Three	3,500	One	3,500

Still larger stations employ units of 5,000, 8,000, 10,000, 15,000 up to 20,000 and 25,000 kw., and even in the near future there will

24

be units of 30,000 kva. in use. For all of the material of a high-tension distribution system, viz., generators, transformers, motors and cables, there are proposed as safety coefficients the values given in the following table:

Volts	Coeff.	Volts	Coeff.	Volts	Coeff.	Volts	Coeff.
2,000	4.7	4,500	4.0	7,000	3.5	12,000	3.0
2,500	4.5	5,000	3.9	8,000	3.4	13,000	2.9
3,000	4.3	5,500	3.8	9,000	3.3	14,000	2.8
3,500	4.2	6,000	3.7	10,000	3.2	15,000	2.7
4,000	4.1	6,500	3.6	11,000	3.1	17,000	2.6
.....	....	....	....	....	....	19,000	2.5

These coefficients represent the relation between normal pressure and that which may be applied instantaneously to the material without breaking it down. This does not refer to the break-down resistance between the winding and the frame, but between the wires of the windings. For cables these safety coefficients are understood to apply between the conductors or between one conductor and the armor.

The minimum pressure for a given system below which it is impossible to go is determined by the distance of transmission and the drop of potential allowed in the line. On the other hand, the choice must also be guided by the maximum surge tension produced by a sudden change in pressure. The maximum surge tension produced theoretically in the alternators can be shown to be almost independent of the normal pressure of the machine and to approach the value of 150,000 volts. It is obvious that such values can be obtained only by a combination of the most unfavorable conditions, viz., an absolutely instantaneous break of a short-circuit at the exact moment when the current is at its maximum; but nevertheless it is certain that even under less difficult conditions there may be built up surge tensions which are much higher than the pressure which the machines are able to withstand. This mentioned fixed value of the surge tension, whatever the nature of the machines, justifies the conclusion that it is far less dangerous to employ high-tension alternators than generators of medium voltage.

If we adopt the foregoing values as coefficients of safety, it is easy to show that for each voltage there is a certain output of the generating station which must not be exceeded to prevent the surge tension from rising to a dangerous value; or conversely, that for a given output of the generating station there exists a definite minimum value of the working pressure. The following table is the result of a combination of voltages and outputs per phase arrived at by this method:

Volts	Value	Volts	Value	Volts	Value	Volts	Value
2,000	70,000	4,500	285,000	7,000	580,000	12,000	1,370,000
2,500	104,000	5,000	345,000	8,000	730,000	13,000	1,500,000
3,000	140,000	5,500	400,000	9,000	890,000	14,000	1,700,000
3,500	186,000	6,000	465,000	10,000	1,050,000	15,000	1,800,000
4,000	235,000	6,500	520,000	11,000	1,200,000	17,000	2,200,000
.....	....	.....	....	......	....	19,000	2,600,000

The absolute value of the surge tension decreases in inverse ratio to the voltage, but the relative value is inversely proportional to the square of the voltage, and it is obvious that we should employ as high a voltage as possible. On the other hand, alternators constructed with windings from which the air has not been expelled very quickly deteriorate through the formation of ozone and are also less safe if the voltage is too high.

The voltage of the system determines the spacing of the various members of the arrangement and influences the size and the type of the oil switch chosen. Moreover, it has certain influence on the arrangement in that concrete or masonry structures are not strongly recommended for voltages above 13,200. This is due to the fact that for higher voltages concrete or masonry structures must be considered as dead ground and therefore, since the tendency toward leakage and corona increases as the voltage increases, safe spacing distances would necessitate a very large and expensive structure.

The tables on page 372 and 373, showing the rating of some of the more well-known large stations, will give a general idea of the development of modern plants.

In general, the central-station building should be of fireproof construction. It is usually built of brick and steel. The foundations are of concrete resting on solid subsoil (or piling with concrete). The machine foundation should be independent of the building foundations. It should support the machine by distributing its weight over sufficient soil to carry it. Then it has to absorb the vibrations of the running machine, and at the same time it must withstand any tendency to topple sideways due to the pull of belts or the settling of any soft spot in the soil beneath the foundation. The less depth given to a machine foundation

RATING OF WELL-KNOWN STEAM POWER PLANTS

	Rated kw. capacity	Overload kw. capacity	Ultimate kw. capacity
Commonwealth Edison Co. (Fisk St., Quarry St., and Northwest plants)........	....	244,000	492,000 (overload)
N. Y. Edison Co (Waterside No. 1 and No 2 and 140th St. plants)...............	216,000	249,000	366,000 (rated)
Interborough Rapid Transit Co. (59th and 74th St. plants)....	....	171,750	475,000 (overload)
Boston Edison Co. (L St. and Atlantic Ave. plants)......	100,000		
N. Y. C. & H. R. R Co. (Port Morris and Yonkers plants).	40,000	60,000	80,000 (rated) 120,000(overload)
Pennsylvania R. R. Co. (Long Island and service plants)...	40,000	75,000	86,000 (rated)

the better. Therefore, should great depth be required in order to support the machine at a considerable distance above bedrock or firm soil, it will be better to put in a system of pillars, piling, or arches, to carry a low, flat foundation to which the machine can be attached, and the posts, columns, or arches should be calculated independent of the fact that a machine foundation is to be supported. A traveling crane which will safely carry the heaviest machine parts is essential. Ample space should be allowed for the machines for easy and safe inspection and handling of parts while repairing. With vertical steam turbines sufficient space between the outer casing and the gallery or engine-room floor should be left to allow both expansion at high temperature and removal of the sections of the casing. In designing the building future extensions should be taken into account with respect to ultimate symmetrical arrangements of the machines and to continued running while building annexes or while installing new machines.

RECENT HYDRO-ELECTRIC INSTALLATIONS
With Transmission Voltages Exceeding 100,000 Volts

Name and location	Transmission voltage in kv.	Present capacity of plant in kw.	Ultimate capacity of plant in kw.	Frequency	Total length of transmission in miles
Eastern Michigan Pr. Co., Michigan	140 △	10,000	....	60	125
Mexico Northern Pr. Co., Mexico	110 Y	31,200	46,800	60	125
Mississippi River Pr. Co., Keokuk, Ia	110 Y	135,000	270,000	25	150
Georgia Power Co., Tallulah Falls, Ga	110 Y	30,000	60,000	60	160
Ontario Power Co. (Hydro-El. Comm.)	110 Y	78,800	175,000	25	280
Sierra-San Francisco Power Co., California	104 Y	34,000	34,000	60	100
Yadkin River Pr. Co	103.9 Y	27,000	....	60	150
Great Falls Water Pr. & Townsite Co., Montana	102 △	21,000	21,000	60	135
Southern Power Co	100 Y 50 △	66,000	....	60 60	500 400
Great Western Pr. Co., California	100 △	40,000	100,000	60	157
Central Colorado Pr. Co., Colorado	100 △	20,000	20,000	60	183
Grand Rapids-Muskegon Power Co	110	......	10,000	30	
Shawinigan Water Pr. Co	100	......	40,000	60	80
Commonwealth Power Railway Co	140	......	15,000	60	220
Pacific Light & Power Co	150	......	120,000	50	280

CHAPTER XXV

TYPICAL CENTRAL STATIONS

Coney Island and Brooklyn Railroad.—This system controlled previously three independent d.c. central stations, conveniently located with respect to the three main lines of traffic. One was situated in De Kalb Avenue, the second in Smith Street, and the third in Kings Highway. The latter was used only during the summer, when traffic to the pleasure resorts at Coney Island is heavy. The equipment of all three stations was largely composed of old machines, which were not equal to the increase in traffic, and the company therefore decided to convert the d.c. installation into an alternating-direct current system. A change of this kind in a service like that between New York and Brooklyn could only be accomplished gradually. It was also planned to keep some of the oldest d.c. machines as reserves and to maintain two of the latest types in service. The following arrangement resulted from these considerations: An extension had to be built to the largest of the three stations, at Smith and Ninth Streets, making an 11,000-volt a.c. central station with two 2,000-kw. Curtis turbines. The extension had to contain a sub-station with two 1,000-kw. converters and the d.c. reserve and the 800-kw. d.c. machine which were to be kept in service and remain in the old part of the power house. The other two d.c. central stations were replaced by sub-stations. The boiler room and water and coal supply arrangements were evidently adapted to the larger a.c. central station. The author's designs for the electrical part of the extension, which were adopted, are given herewith.

The plan, cross section and outline of the engine room and gallery are shown in Figs. 248, 249, 250, 251, and 252. From the two turbo-generators the lead-covered, three-conductor cables are passed through the foundations in clay tiles to the middle column supporting the gallery. They then lead up this column to the engine-room floor, where they diverge and run half way up the columns on opposite sides.

This arrangement has been somewhat changed, so that now there are three separate cables leading from a terminal box on the turbine to the basement. These then run along the basement ceiling to the columns.

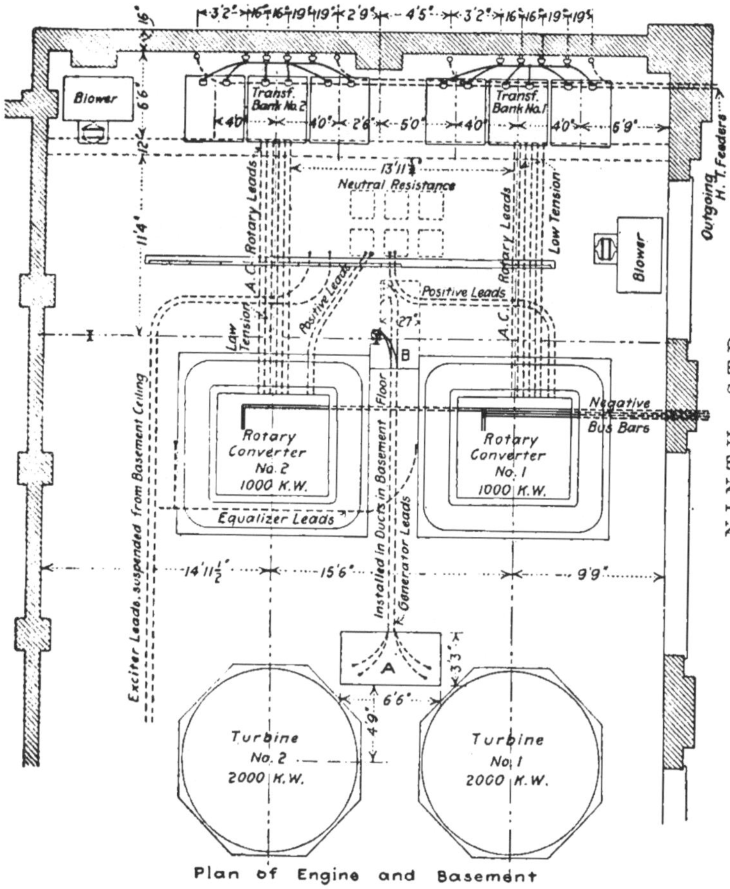

FIG. 248.—Plan view of engine floor and basement of the Coney Island and Brooklyn R. R. Co. power house extension.

The three phases are separated in an end bell, at which point they diverge, and after passing through the series transformers reach the terminals of the generator oil switches (type H3). Note that these switches are set up in the front part of the gallery. The gallery is not large enough to allow setting all

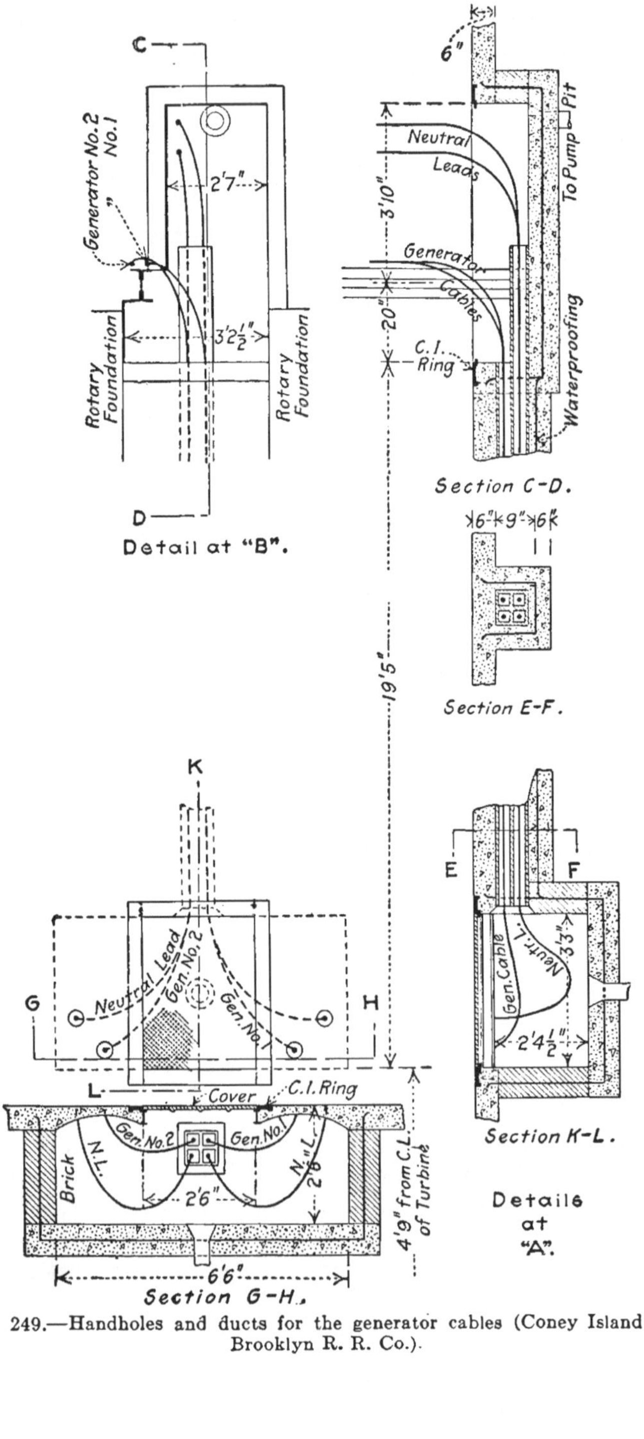

FIG. 249.—Handholes and ducts for the generator cables (Coney Island and Brooklyn R. R. Co.).

the necessary oil switches in one row and still keep sufficient space for operating. The series transformers are placed on the gallery ceiling.

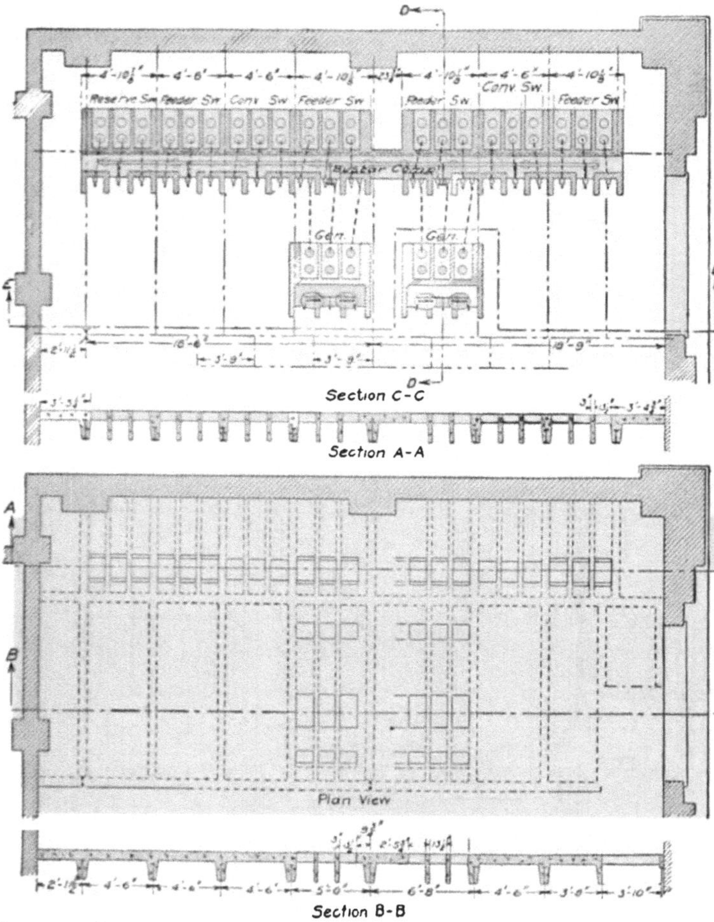

FIG. 250.—Plan and construction of switchboard gallery (Coney Island and Brooklyn R. R. Co.).

The second change that was made was to place the generator oil switches on a compartment at the same height as the other oil switches, and the series transformers were put into these compartments.

The shunt transformers are in a cell back of the generator oil switch cell, and their fuses are mounted on a separate slate base

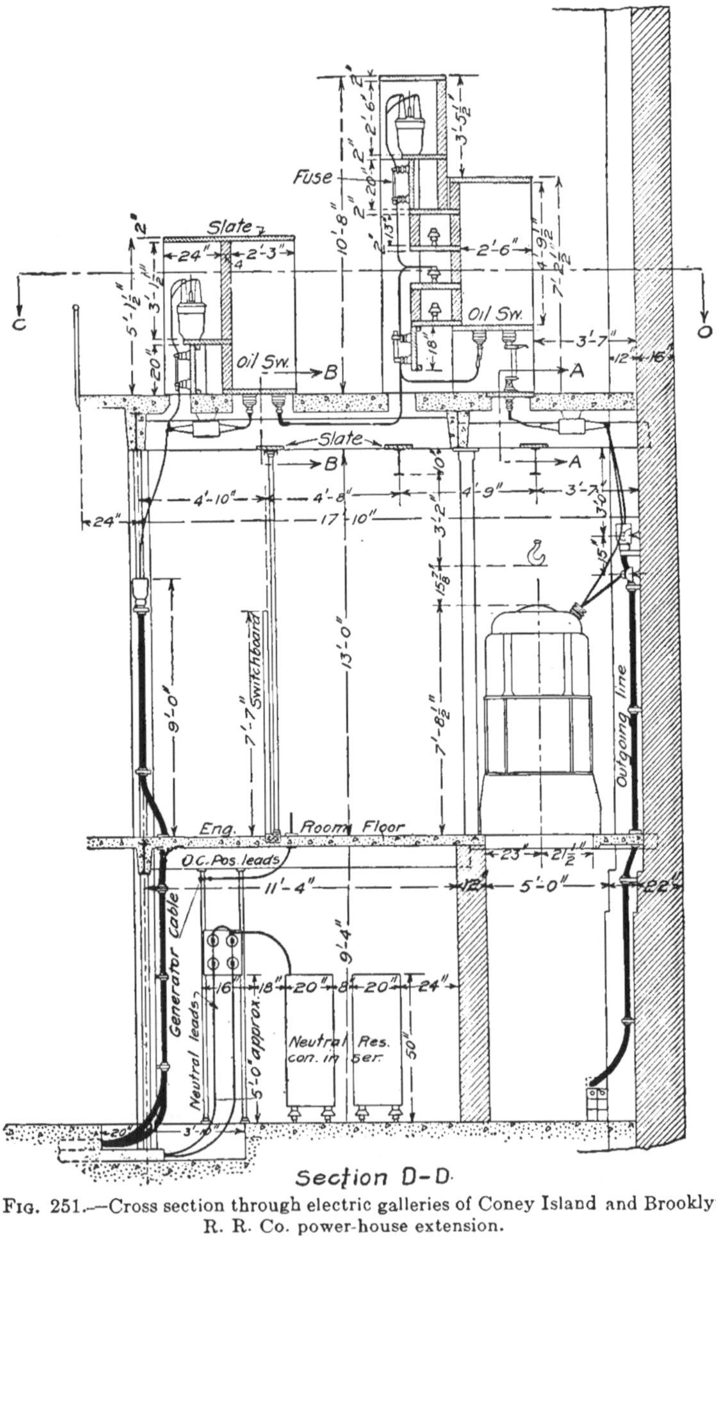

FIG. 251.—Cross section through electric galleries of Coney Island and Brooklyn R. R. Co. power-house extension.

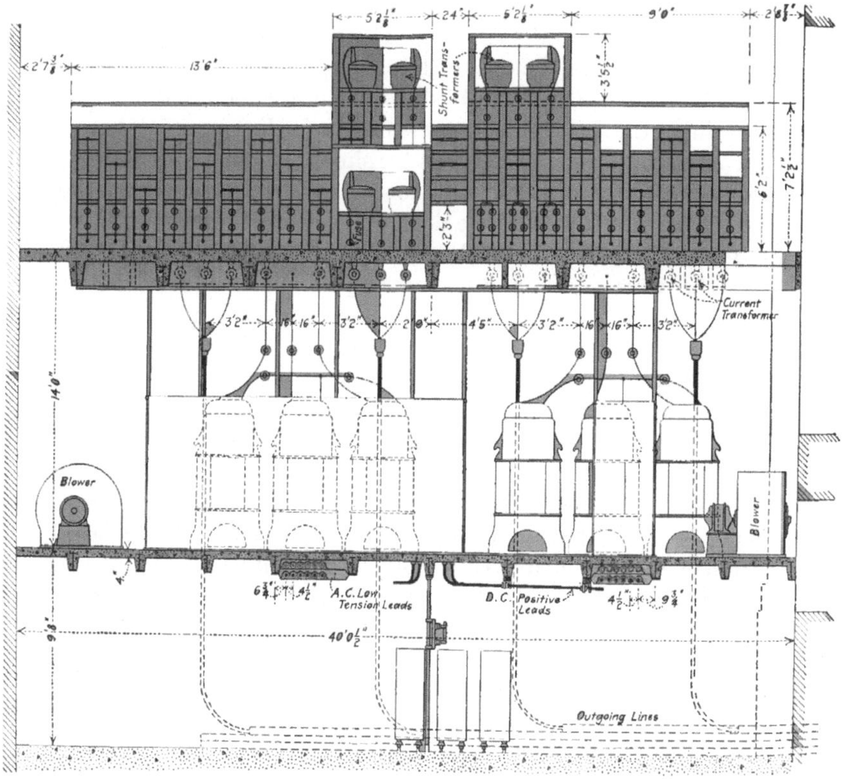

Fig. 252.—Elevation of electric galleries (Coney Island and Brooklyn R. R. Co.)

(Facing page 379.)

and may be used as disconnecting switches. There are no disconnecting switches on the generator side of the oil switches, for it is assumed that the oil switch will be accessible when open, i.e., when disconnected from the busbars and when its corresponding turbine is not running. From the generator switches, the cables run to the disconnecting switches through which they are connected to the busbars. The buses are enclosed in compartments back of the feeder oil switches. The walls are of brick and the cover of slate, and the openings for the cable connections are on the front side. Over the busbars there are mounted the cells containing the shunt transformers for the feeder instruments with the necessary fuses. Their cells are similar to those for the generator shunt transformers. The feeder and converter oil switches are set up in a row with the door side toward the rear wall.

The plan of the gallery shows that the rows of oil-switch cells and the corresponding disconnecting-switch compartments are broken at about the center. These breaks are necessary to afford room for operating the bus-sectionalizing switches, which are inserted for the purpose of dividing the entire equipment into two distinct parts. On each side of the section switches there are connected to the buses one generator, one synchronous converter, and two feeders. The two sub-stations are connected to both sides of the buses each by two sets of feeders, and a reserve switch is provided on the side toward the old building. The feeder oil switches are provided with disconnecting switches on both sides. Those on the busbar side are mounted in a row of compartments facing the front of the gallery, and those on the feeder side are on the terminals of the oil switches in separate compartments under the oil-switch cells. Note that all disconnecting switches are located on the same gallery together with their oil switches. This has the advantage that the attendant may easily assure himself that the oil switch which he is inspecting or repairing is dead and will remain so while he handles it. The feeders are run out of the station underground in tile ducts as lead-covered, three-conductor cables. The converter oil switches, as mentioned above, are located in the same row as those for the feeders, but are furnished with only one set of disconnecting switches and these on the busbar side. They are omitted on the transformer side for the same reason that they are left out on the generator side of generator oil switches, namely, that the oil switch is inspected when it is open and disconnected from the busbars and when the converter is not running. On the engine-room floor under the gallery there are set up two sets of three single-phase 375-kw. air-cooled transformers. The transformation ratio of these Y-connected transformers is 11,000 to 430 volts. The twelve low-tension cables run along the basement ceiling in two rows to their starting panels, whence six cables lead through the foundations of the converters to the brushes of the collector rings. Figure 253 is a detail drawing of the foundations of one of the converters, showing the relative positions of the ducts and openings for the cables and busbars.

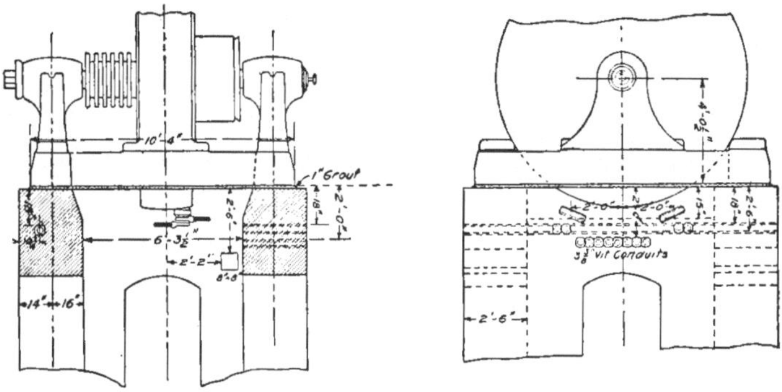

FIG. 253.—Concrete foundation with ducts installed for a converter.

A wall shuts off the rear part of the basement, and the air circulation for the air-cooled transformers in the resulting chamber is produced by two blower sets placed on the engine-room floor. The blower on the right side toward ninth Street is joined to the chamber by a sheet-metal air passage. The other, which is located directly over the chamber on the left side, will have to be removed in case of any further extension. The transformers are set up over the chamber on air-tight frame supports of channels and I-beams. In order that the transformers may be removed, two traveling differential pulleys running on tracks fastened overhead to the gallery beams are provided. If a transformer is to be moved, therefore, it is first raised by the pulley directly over the transformer row and a small carriage is run underneath while the opening in the air chamber is closed. The carriage is then run under the second hoist, which carries the transformer between the back of the switchboard and the row of transformers to the left wall, where it can be handled by the main crane.

The plan and elevation of the gallery show the construction of the separating barriers between the different phases and the series transformers on the gallery floor. All of the concrete work is reinforced. The floor is 7 inches thick owing to the conduits for the secondaries of the instrument transformers which are built into it. The switchboard is made up of two generator panels, two feeder panels controlling two feeders each, and one synchronous-converter panel equipped with the controlling apparatus and instruments for the two present converters and having space for the third machine eventually to be installed. Besides this it carries the starting switches for converter 2, in front of which it is located. Following these come the two exciter panels and three panels for the d.c. side of the converter. In front of converter 1 a starting panel is set up which is in line with the main board. The space between this panel and the main switchboard is reserved for five 16-inch d.c. feeder panels ultimately to be transferred from the other side of the building. The main board is held on pipe supports running up to a channel on the gallery, and rests on a wooden strip 1 inch above the floor. The positive cables are led from the converters along the basement ceiling to the d.c. board. The negative buses are laid in the foundations, in openings 8 inches by 8 inches. The generator neutrals are led from the turbines to the resistance in the basement floor in the same set of ducts as the generator cables. The resistance for the neutral is grounded. Manholes and ducts are made water-tight and are provided with sewer connections for drainage since they lie below the level of the extreme high-water mark of the adjacent Gaudemus Channel. These plans were laid out before designing any of the structural work, so that the latter may be fitted to the former.

Waterside Station No. 2 of the New York Edison Company.—This company delivers direct current to motors and lamps mainly in the Borough of Manhattan, New York, City. High-tension alternating current is generated in two central stations whose busbars are interconnected; low-tension direct current is distributed from twenty-four sub-stations, while three a.c. substations deliver low-tension alternating current.

Figure 254 is a diagram of the high-tension connections between the two power houses. The main and auxiliary buses are connected together by two sets of cables each. Note that each of the connections contains two non-automatic H3 oil switches in series in each of the stations. The reason for making this connection, as well as others to be described later, specially secure is the fact that as the company supplies the most densely populated portion of the city, it is imperative to maintain the service absolutely without interruption under all circumstances. In the figure there are shown the subdivisions of the buses of the two stations. The busbars of the new Waterside station have four subdivisions, which give greater flexibility to the generator-feeder group units.

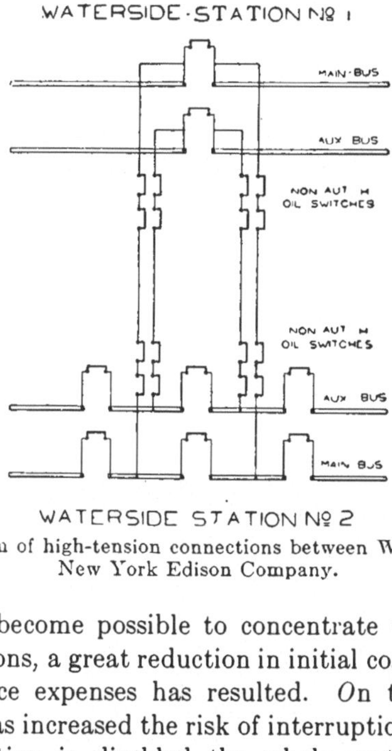

FIG. 254.—Diagram of high-tension connections between Waterside stations of New York Edison Company.

Since it has become possible to concentrate the output of a number of stations, a great reduction in initial cost and operating and maintenance expenses has resulted. On the other hand, however, this has increased the risk of interruption of the service, since if one station is disabled the whole system is crippled. To reduce this danger as far as possible the different parts of large stations are divided into groups of units which can be worked together or independently as desired. By subdivisions in the boiler room, by independent installation of the turbo-generators with their accessories, by separate laying of cables and subdivisions in busbars, a group composed of the above parts can be run as an independent unit.

This is the method adopted in the Waterside Station No. 2.

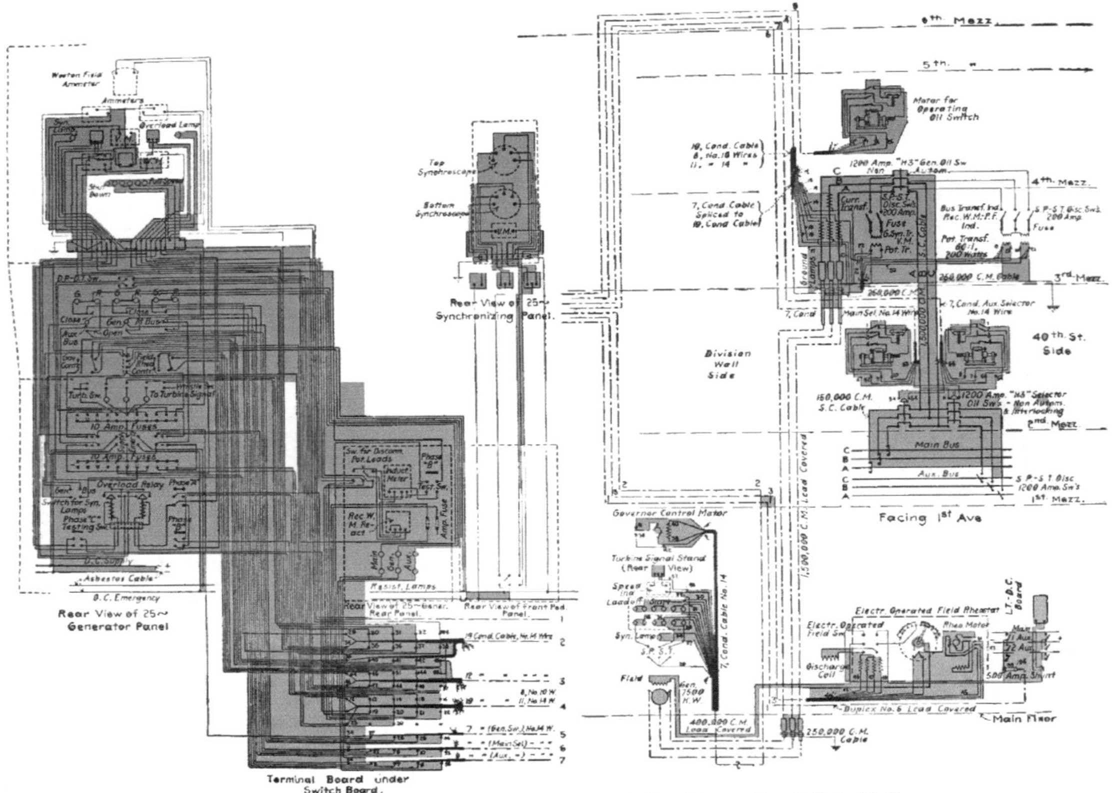

FIG. 255.—Wiring diagram of high-tension 25-cycle generator (New York Edison Company, Waterside Station No. 2).

(Facing page 383.)

At present the station equipment consists of two Westinghouse-Parsons steam turbines on whose shafts are mounted two generators of 7,500 kw., 6,600 volts, 25 cycles, and 7,500 kw., 7,500 volts, 60 cycles, respectively. Only one of the generators is run at any one time. Besides this there are six Curtis turbines with generators of 8,000 kw., 6,600 volts, 25 cycles, two additional Curtis turbines with 14,000-kw., 25- and 60-cycle generators respectively, and one 20,000-kw. turbo-generator. Four exciters of 150 kw., 280 volts each driven by a 220-hp., 6,600-volt, three-phase, 25-cycle induction motor, and two 500-kw. motor-generator sets supply the excitation. Two storage batteries are used as a reserve for the exciters. Sixty-four high-tension feeders at 25 cycles and eight at 60 cycles are supplied by the station.

Figures 255, 256 and 257 show the connections of a generator and feeder and a cross section through the galleries. The cross section is taken at the feeders, so that Figs. 256 and 257 will be described first. It was mentioned above that each busbar set is divided into four parts. (This applies to the 25-cycle buses. The 60-cycle buses have an independent subdivision.)

From corresponding divisions of the buses cables are led to two automatic H3 oil switches. They are connected to the buses by means of disconnecting switches. (Fig. 256.) The buses and disconnecting switches are located in compartments on the first mezzanine. (Fig. 257.) The selector switches mentioned above are on the second mezzanine and are interlocked to prevent closing both of them at the same time. The main feeders which are connected to either set of buses through the selector switches diverge on the fourth mezzanine into two outgoing feeders connected to separate automatic H3 oil switches and to the two end bells to which the three-conductor cables are joined. On the third mezzanine are the shunt transformers, mounted in compartments with their disconnecting switches. The three-conductor cables are laid in the division and main walls and are led out of the station underground in tile ducts.

Figure 257, first mezzanine. The busbar compartments are separated from each other and from the walls by corridors, the inner one of which is used for operating the disconnecting switches and the other one for inspecting the busbars and insulators, which are accessible through small doors in the back of the compartment. The disconnecting switches are double-break switches, first disconnecting the cables and then the busbars. Tie oil switches are inserted between the divisions of the busbar sets. (See Fig. 254.) These are set up in line with the busbars in separate divisions, so that the whole mezzanine is divided into a series of chambers containing alternately sections of the busbars and the bus-tie oil switches. The chambers are connected with doors. The tie oil switches are connected to the buses through disconnecting switches as indicated by the bottom switch in the first mezzanine.

On the second mezzanine the selector switches are mounted back to back. The outer corridor is for inspection of the oil switches, and the main feeders are led up in the space between them. The series transformers are also placed on this mezzanine, being built into the feeders as shown.

On the third mezzanine there are set up the shunt transformers with their disconnecting switches. The two outer corridors are moved over toward the middle of the gallery to make room for the end bells and three-conductor cables and the series transformers. The secondaries of the shunt transformers are laid in tiles embedded in the concrete floor.

The fourth floor is similar to the first, with the addition of a false floor for the outgoing feeders. All control and instrument wires are led up through the wall to a frame support on the sixth mezzanine, whence they run to the switchboards.

Figures 255 and 257. The generator cables are led in through the floor in conduits up to the division wall, whence they pass up the wall to the electrically operated non-automatic oil switches in the fourth floor. From here they are led back to the second floor in the same way as the main feeders, to the non-automatic type H3 selector switches which join them to their respective buses. These switches are interlocked so that only one can be closed at a time. The connections between selector switches and busbars are made through double-break disconnecting switches. The generator shunt transformers are set up in line with those for the feeders, and the series transformers are mounted on the division wall. On the fourth mezzanine there are also located the station-tie oil switches, which tie feeders are led along the division wall to the ducts in the basement floor. There are, therefore, two non-automatic oil switches in series on each high-tension generator. A relay lights a signal lamp to indicate over-

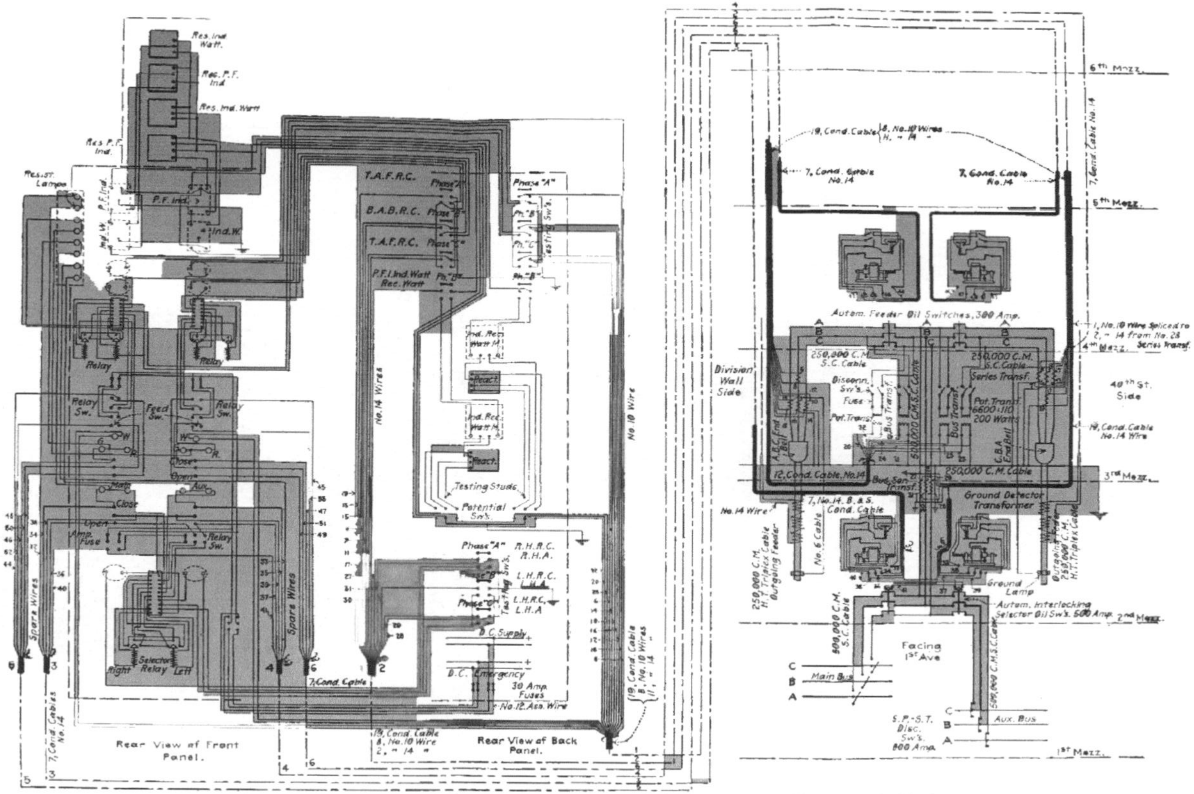

Fig. 256.—Wiring diagram of high-tension 25-cycle feeder (New York Edison Company, Waterside Station No. 2).

(Facing page 384.)

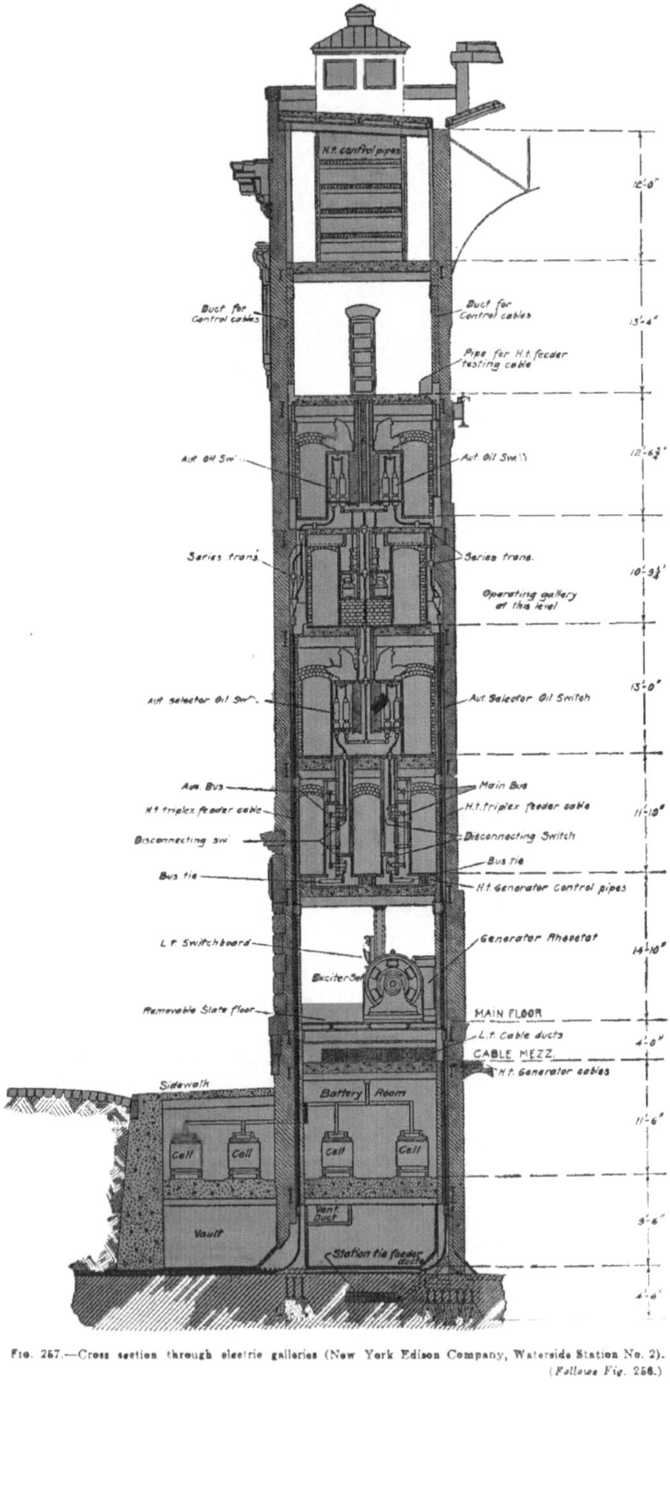

FIG. 257.—Cross section through electric galleries (New York Edison Company, Waterside Station No. 2).
(Follows Fig. 256.)

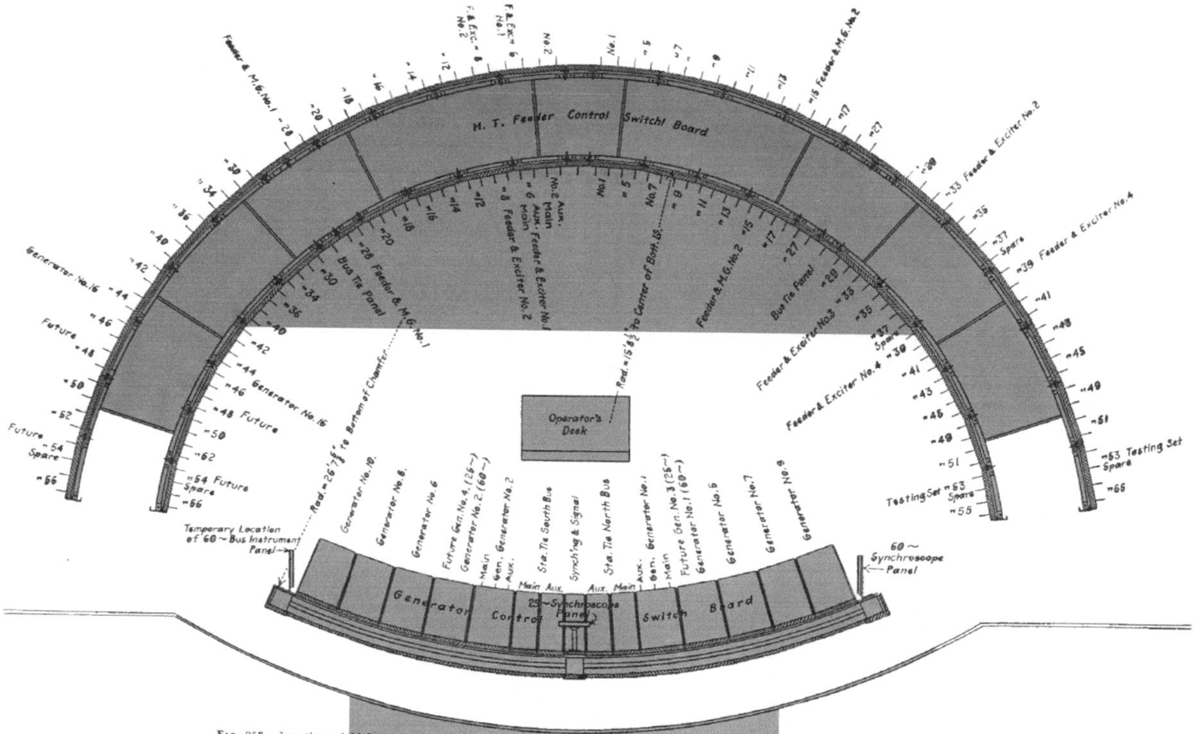

FIG. 258.—Location of high-tension feeder and generator control switchboard (New York Edison Company, Waterside Station No. 2).

(Follows Fig. 257.)

load or short-circuit on a generator, which may then be disconnected by the operator at his discretion.

The outgoing feeders are connected to the high-tension buses through two automatic oil switches in series. An overload or short-circuit opens these switches automatically by means of an

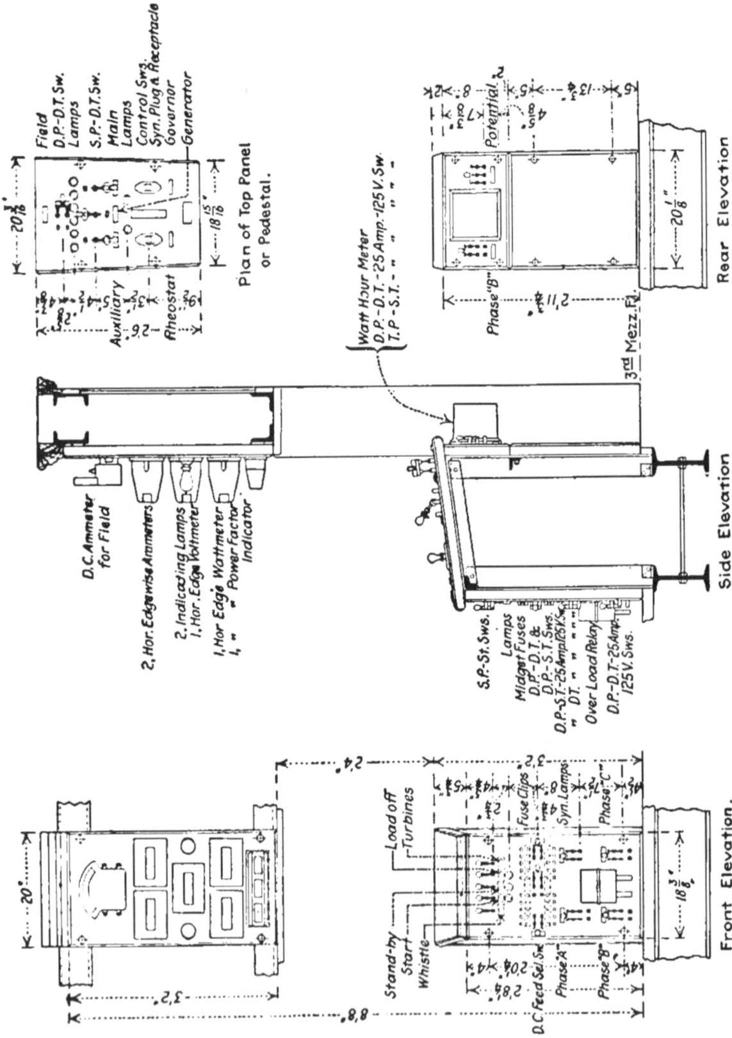

FIG. 259.—Generator control 25-cycle pedestal and instrument panel (New York Edison Company, Waterside Station No. 2).

inverse time-limit relay, disconnecting the feeder. All high-tension feeders are also equipped with a selector ground-indicating device, which indicates but does not disconnect a grounded feeder. This device frequently makes it possible to locate and disconnect a grounded feeder before the trouble develops into a short-circuit. The exciters, the field rheostats with their motors, the d.c. switchboard, and the exciter buses are located on the main floor of the high-tension division. The annex is really nothing more than a part of the main power house shut off by a division parallel to the north side of the building. It is completely separated from the engine room and is a complete fireproof building in itself. The main operating room with the main switchboards is situated on the third mezzanine on the west side of the building, on a balcony facing the engine room. The generator and feeder panels are grouped in two independent switchboards. (See Fig. 258.) On the engine-room side facing the inner part of the balcony are set up the control benches and instrument boards for the generator, and station-tie and bus-tie oil switches. The benchboard is arc-shaped. (Figs. 259, 260, and 261.) Opposite to this board there is set up a double semicircular row of panels for the feeder control. This makes a very compact arrangement, and the entire system is very economically and safely controlled from a comparatively small space and with a small number of attendants. The instrument boards over the benches are set up so as not to obstruct the view over the station. Ample light is obtained through the engine-room skylight and the windows on the west side.

Figures 255 and 259 show the connections and location of all the control apparatus and instruments for a 25-cycle generator. Every generator has its own bench and instrument board made of blue Vermont marble, the general dimensions of which are given on the drawings.

Station instruments for the synchroscope are placed on a swinging panel in the center of the benchboard. (See Fig. 260.)

Figure 260 also shows the equipment for the station tie control. Only that part located in Waterside Station No. 2 is shown.

Figures 256 and 261 show the control apparatus and instruments for a feeder board, consisting of two panels placed back to back. Each board controls four outgoing feeders.

The generator and feeder for 60 cycles are similar to those described above for 25 cycles with corresponding changes in the ratings of the instruments.

Besides carrying the equipment for the generator and bus and station ties, the benchboards are provided with signal apparatus for use as a means of transmitting signals between operators on

Panel	Mark	No. of Panel	Instruments (Total)
Synchroscope	f	1	2, Synchroscopes 1, V.M.
Bus Instrument	k	2	8, Frequency Indicators 8, V.M. 8, Lamps
Sta. Tie Instrument	h	2	4, Ind. W.M. 4, V.M. 8, A.M.
Top Sta. Tie Control Pedestal		4	16, Lamps 8, S.P.-D.T. Sw's. 8, Name Plates 4, 4 P.T. Syn. Plugs & Receptacles 4, Name Plates
Front Sta. Tie Control Pedestal	m	4	16, D.P.-S.T. 125 V.-25 Amp. Sws. 16, Name Plates 4, T.L. Relays
Rear Sta. Tie Cont. Pedestal	d	4	12, D.P.-D.T. 125 V.-25 Amp. Sws. 12, Name Plates
Top Cent. Ped.	c	1	
Front Center Pedestal	n	1	2, D.P.-D.T. 125 V.-25 Amp. Sws. 4, Name Plates 1, Fuse Panel (24 Midget Fuses)
Top Cent. Ped.	c	1	10, Sp. D.T. Sw's.-20 Name Pl's 2, Name Pl's. 1, Sn. Oil Sw.

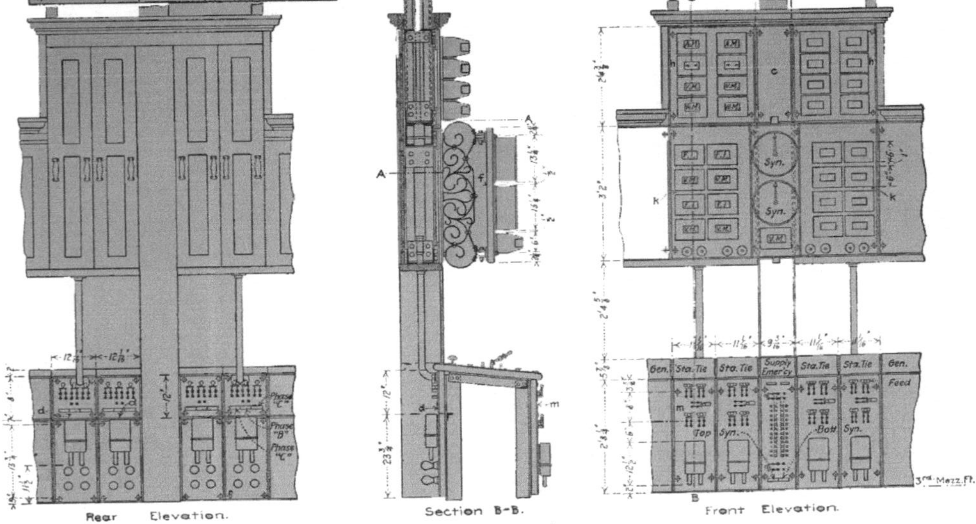

FIG. 260a.— High-tension tie control pedestals and instrument panels.
FIG. 260b.— Bus instrument panels and synchroscope panel (New York Edison Company, Waterside Station No. 2).

(Facing page 386).

the high- and low-tension switchboards and the engineer stationed at the turbine throttle. Provision is made for three sets of signals. One is for regulating the turbine governor and is transmitted through illuminated signs located on the west wall of the engine

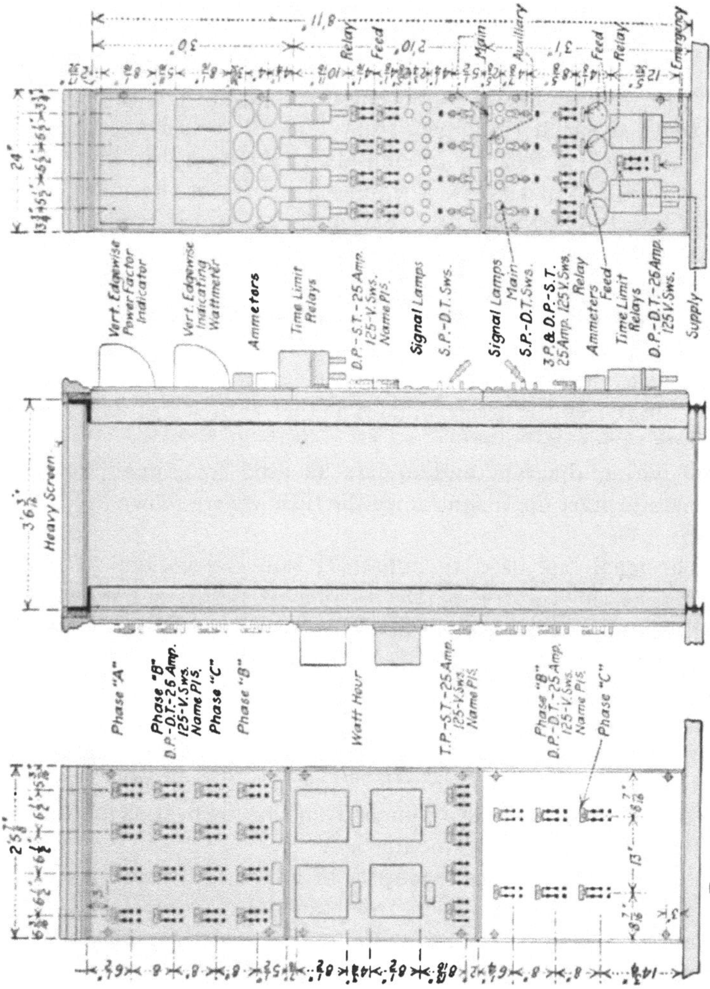

FIG. 261.—High-tension feeder control panels for 25 cycles (New York Edison Company, Waterside Station No. 2).

room. It is used in conjunction with the signal whistle located on the south division wall. The other signals relate to the starting or stopping of the turbines and are transmitted to the engineer direct through the signal stand located near the turbine throttle.

The third signal is a call whistle from the d.c. control board on the first-floor electrical gallery.

There may be flashed individually the turbine number and the red or green engine signal of the illuminating sign by means of small knife switches located under the synchronous panel of the turbine board.

This sign consists of a substantial sheet-iron box with angle-iron framework set flush with the engine-room wall and displaying the stenciled turbine numbers from 1 to 10 inclusive. Above each numeral is a separate ground-glass signal panel to indicate red for "raise," and below a similar panel indicating green for "lower," as the case may be. Within the sheet-iron box are the lamps and circuits for illuminating the turbine numbers and the colored signals mentioned.

The operator at any one of the generator pedestals is able to blow the signal whistle and flash the turbine number. He is also able to blow the signal whistle from the center pedestal and flash either the red or the green signal, together with any of its corresponding turbine numbers.

The wiring diagram and apparatus used for signaling the engineers to start up or shut down the turbines are shown in Figs. 255 and 259.

Each signal box used to impart certain instructions to the engineer at the throttle of the turbine possesses a frequency indicator and a small ground-glass illuminated plate containing the following legends: "Start," "Stand by," "Load off," "Full speed," "O. K.," and "Shut down."

This apparatus enables the operator at the high-tension generator control pedestal to flash the following signals to the engineer standing at the turbine throttle :"Start," "Stand by," "Load off," and to receive in turn signals from the engineer as follows: "Full speed," "O.K.," "Shut down."

The boiler signal system consists of an electric carriage call, used for transmitting numbers from the system operator to the foreman in the boiler room. The number flashed indicates to the foreman the number of boilers that will be required within the next fifteen minutes. Working in conjunction with the carriage call is a registering time clock, so that the signal and the time it was transmitted are automatically recorded. In addition to these systems there is a complete intercommunicating telephone system connecting the high-tension and low-tension switchboards, system operator, foreman, superintendent, sub-stations, etc. A pneumatic-tube service is also maintained between the system operator's desk in Waterside Station No. 1 and the high-tension control board in Waterside Station No. 2. In order to insure the greatest possible safety for the workmen the system operator transmits all orders in writing. A telautograph is employed for transmitting messages between the high-tension control board and the d.c. switchboard, so that a written record is kept of all the orders passed between the two switchboards.

Every precaution is taken to render communication between the system operator and the various sub-stations by telephone reliable and instantaneous. The telephone lines to the sub-stations are routed through two telephone exchanges. The lines and switchboards are in duplicate, and the trunks within the station are carried in separate conduits. While the company has therefore depended entirely upon the telephone for transmitting orders between the system operator and the sub-stations, it has appreciated that this system would be too slow in case of emergency. The Edison Company has, therefore, installed a signal system similar to a fire alarm to convey simultaneously to all of the sub-stations signals which will reach the operator no matter in what part of the sub-station he may be. A series of ten signal buttons is connected in multiple to private telephone trunk lines, so that by pressing any button its signal is transmitted to all of the stations. Thus, should the system operator anticipate any trouble which would need prompt attention, all sub-stations would be apprised of the condition at once. The distribution of all energy from both Waterside stations is under the exclusive control of a system operator. Feeders and rotary converters are put into or out of service only on his order, and at frequent intervals he receives reports from all sub-stations giving the load on each high-tension feeder and the reserve held in the storage batteries. He keeps before him a complete map of the system, including stations, sub-stations, and feeders. By means of suitable markers the system operator can tell at a glance what feeders are connected, their routes and their loads. Any feeder or machine held for repairs is indicated also.

Long Island City Power Station.—The western lines of the Long Island R. R. (N. Y.) were the first of this system which were electrified, the tunnel under the East River connecting the Long Island lines with the new depot of the Pennsylvania

Railroad, which was at that time in course of erection. The power station and its chain of sub-stations therefore were at first

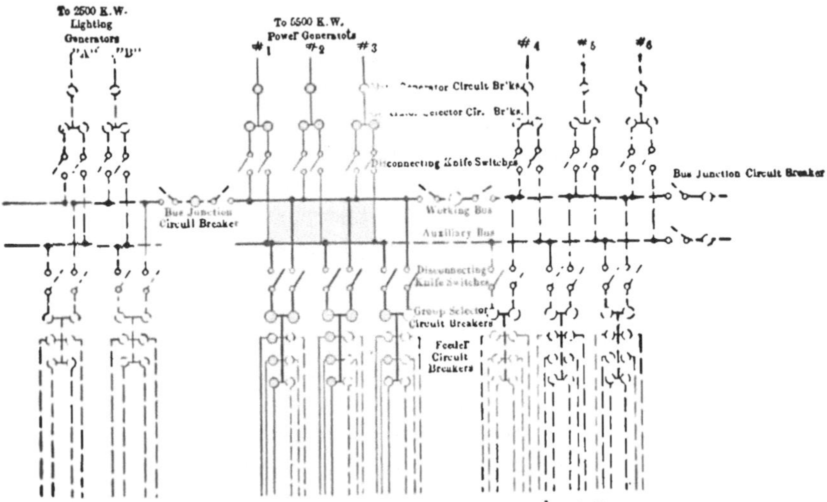

FIG. 262.—Diagram of high-tension wiring, Long Island City power station.

FIG. 263.—Feeder gallery, showing oil switches for feeders and generators.

erected and installed to suit the demands of the electrified western road; nevertheless, means were provided for increasing the power capacity.

The present power station contains a machine overload

capacity of about 72,000 kw. There are installed three steam turbines of the Westinghouse-Parsons type, driving 5,500-kw.,

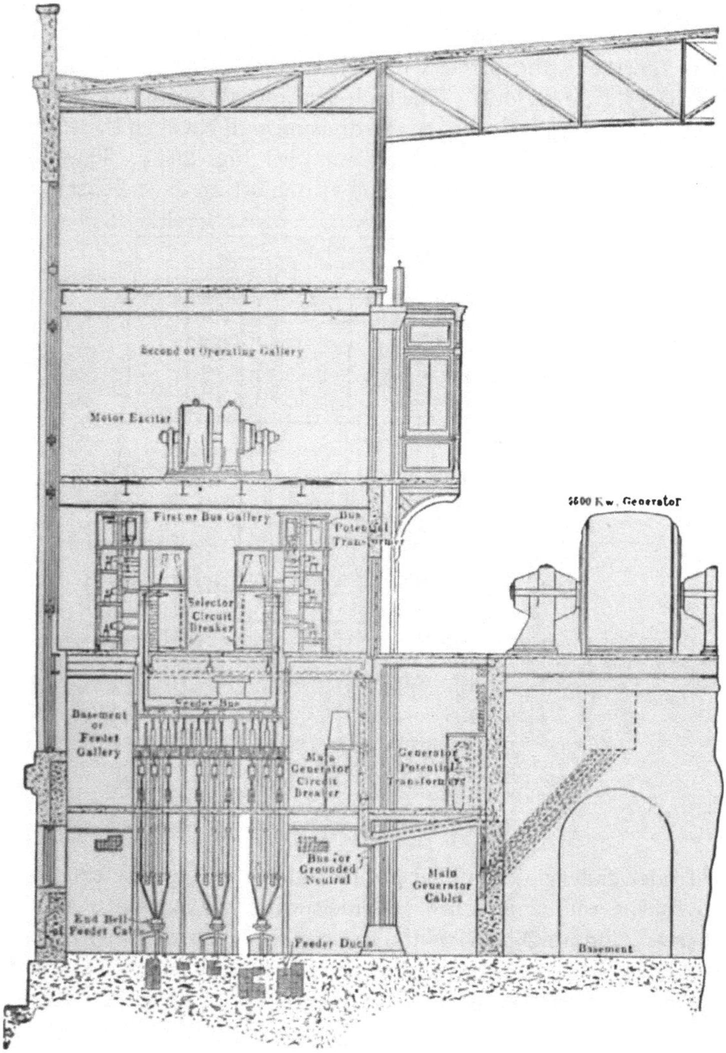

FIG. 264.—Cross section of electric galleries, Long Island City power station.

11,000-volt alternating-current generators; two 8,000-kw. steam-turbine units, two 3,000-kw., 11,000-volt units, and one 1,000-kw., 240-volt unit. The armature windings are star-connected and the neutrals of all the machines are grounded through a common resistance.

Figure 262 shows the wiring diagram of the high-tension cables, indicated in simple lines. The full lines and a part of the dotted lines represent the present installation, the rest of the dotted lines the future lay-out. The high-tension cables are run through ducts in the foundation up to the division wall between the engine room and the switching galleries. (See Fig. 264.) They are then run up this wall to a main oil switch set up on the basement

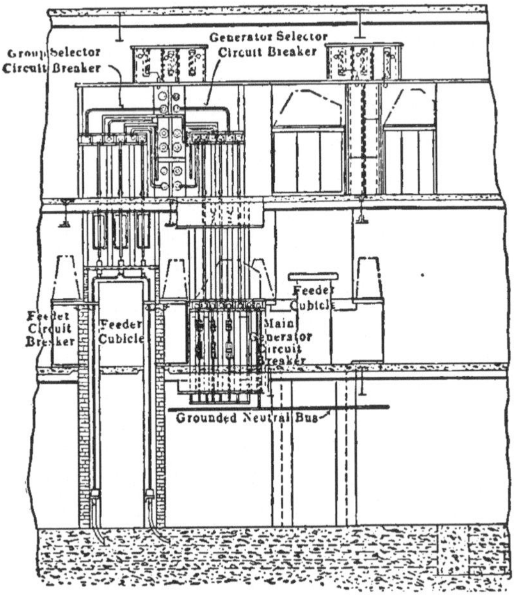

FIG. 265.—Longitudinal section between buses.

or feeder gallery. The backs of the oil-switch cells are extended up to the ceiling and are provided with barriers between the phases. Figure 263 shows the three main oil switches with their extensions in the background on the right-hand side. The shunt transformers for the generators are set up in special compartments near the division wall on the same floor. From the main oil switches the cables run along the basement ceiling to the selector switches installed on the first or bus gallery.

Figure 267 shows a plan of the engine room with the three steam turbines first installed and the bus gallery, showing two rows of cells. Each row, as planned in the first installation, did consist of three groups of two oil-switch cells each. The smaller are the selector switches for the generator. Two opposite oil switches belong to the same machine.

A compartment containing two sets of disconnecting switches is inserted in each group of oil switches. The cables lead from the selector switches in compartments on the back of the cells to one set of the disconnecting switches. (See Fig. 265, longitudinal section between buses, and Fig. 266.) They then run across the space between the oil-switch structure and the busbar compartment to the corresponding busbars.

FIG. 266.—Bus structure and selector switches.

The generators are designed to run in parallel on either of two sets of main busbars, called the working and auxiliary buses, only one set of which is generally in use. The three busbars of the working bus are disposed in the three-story bus structure of brick and alberene stone along the north side of the gallery, the auxiliary bus being disposed in a similar structure along the south side directly opposite the main bus and toward the division wall.

As shown in Fig. 262, there are installed three main oil switches in the basement and six selector switches on the bus gallery.

Three groups of six feeders each are supplied from the busbars, each of the groups being connected to the two sets of buses through two group switches. The large oil switches shown in Fig. 266 are the group switches, of which there are two for each feeder group. As noted above, each disconnecting switch compartment contains two sets of switches. One connects the generator selector switches to the busbars, and the other joins the buses to the feeder group switches. The cables are led to the back of the group switches in similar compartments or barrier chambers. (See Fig. 265.) Corresponding pairs of group switches are connected under the floor by busbar sets. The latter are contained in brick compartments divided into elongated chambers by a number of cross barriers. These compartments are supported on the basement ceiling and run across the switching room. The six feeder switches for each feeder group are set up in two rows on either side of the auxiliary busbar compartments, three on each side, and running across the room parallel to the buses. From the small busbar sets cables are run down to the feeder switches. They then lead down the backs of these and down the wall, which extends to the foundations, until they finally reach the end bells of the three-conductor cables. The feeders are led out of the building underground in tile ducts. Figure 268 shows the connections of the three-conductor cables to the phases. The brick barriers between the phases extend to a level with the tops of the cells, above which they are replaced by asbestos barriers. As each group of feeder switches is set up directly below the corresponding set of group switches, and since the generator main switches are on the same vertical plane with their selector switches, the generator main switches are placed on the front side of the corridor between the groups of feeder switches. All oil switches thus become easily accessible. The switches employed are all type *C* Westinghouse electrically operated three-pole switches with the exception of the main generator switch, which is four-pole. The fourth pole is for the neutral connection of the machine. All oil switches are built for 600 amp. except the feeder group switches, which are for 1,200 amp. The pole pairs are enclosed in separate brick chambers, with doors on the front side and closed in on the back by a wall, into which are built the bushings for the incoming and outgoing cables. All cables and connections, including those of the same phase, are separated from one another by brick, soapstone, or asbestos barriers. Over the busbar compartments on the engine-room floor are

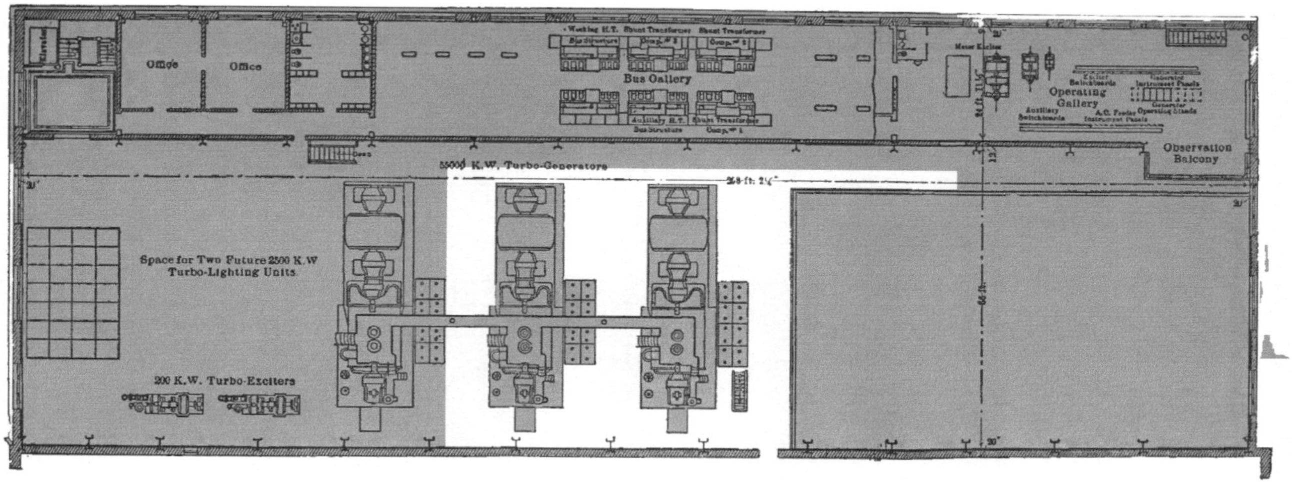

Fig. 267.—Plan of engine room and bus gallery.

(Facing page 394.)

set up the cells for the necessary shunt transformers and fuses. The busbars are composed of 3-inch by ¼-inch copper bars supported in separate compartments on heavy insulators. There are a number of openings in the backs of the compartments so that the bus insulators and connections may be inspected. The compartments for the small auxiliary buses in the basement are similarly built with the openings toward the outer wall. The

FIG. 268.—High-tension feeder cables in basement entering conduits.

main connections between buses and oil switches are made of strong copper bars mounted on porcelain insulators and enclosed in brick or asbestos compartments.

The four main cables of each machine are 600,000 cir. mils cross section with 10/32-inch varnished cambric insulation, and those between the group switches and small buses are 1,318,000 cir. mils sectional area. The feeder switches are connected with the three-conductor cables by 250,000-cir. mil cables, and those between the feeder switches and the transformers which supply the induction motor of the motor exciter set are 73,000 cir. mils. The neutral busbar which connects the neutrals of all the machines to the resistance is a 600,000-cir. mil cable. All of these cables are covered with $^{10}\!/_{32}$-inch varnished cambric insulation with double protective braiding. None of the station cables of either high or low tension are lead-covered. The three-conductor cables leading to the high-tension transmission line consist of three 250,000-cir. mil. conductors with $^{7}\!/_{32}$-inch paper insulation. A further $^{7}\!/_{32}$-inch insulation encloses all three cables, and the whole is covered with jute and $^{9}\!/_{64}$ inch of lead. These cables leave the building in ducts laid in the foundations. All of the high-tension three-conductor cables are connected to brass end bells 7.75 inches in diameter and about 5 inches deep, soldered to the lead covering and filled with an insulating compound.

Excitation for the generators in the first installation was obtained from three different sources. These are a set of two exciters each driven by its own steam turbine, a motor-generator set, and a storage battery. The two Westinghouse-Parsons turbines for the exciters are direct-connected to two 200-kw. d.c. generators which deliver 910 amp. at 220 volts. The motor-generator set consists of an induction motor of 290 hp. driven by three-phase alternating current from the low-tension side of the transformer bank at 440 volts. This bank is composed of three 175-kw. oil-cooled transformers, delta-connected on the high- and low-tension sides. Their high-tension side is connected to the high-tension buses by a feeder of the feeder group. The exciter transformers are set up on the east side of the switching-house basement.

The generator of the motor-generator set is of the same rating as the turbine-driven machines and is mounted on the shaft of the induction motor. The turbo-exciters are set up in the engine room and the motor-generator set on the operating gallery. The storage battery is a very reliable reserve for the machine excitation and for the oil-switch solenoids. It consisted of 110 cells, each containing seven type R″ chloride accumulator plates, as delivered by the Electric Storage Battery Company. There is sufficient space in each cell for eleven plates. The discharge rate of the battery is 366 amp. per hour at a normal pressure of from 180 to 220 volts. It is charged by a 15-hp. induction motor driving a 12.5-kw. booster. This machinery is also set up in the operating gallery, and the battery is placed in a separate chamber in the engine-room basement.

Figure 269 shows the wiring of the exciter and battery. There are two sets of busbars on the d.c. side, installed under the operating room together with the equalizer buses. The battery or either one of the exciters can be connected through double-throw

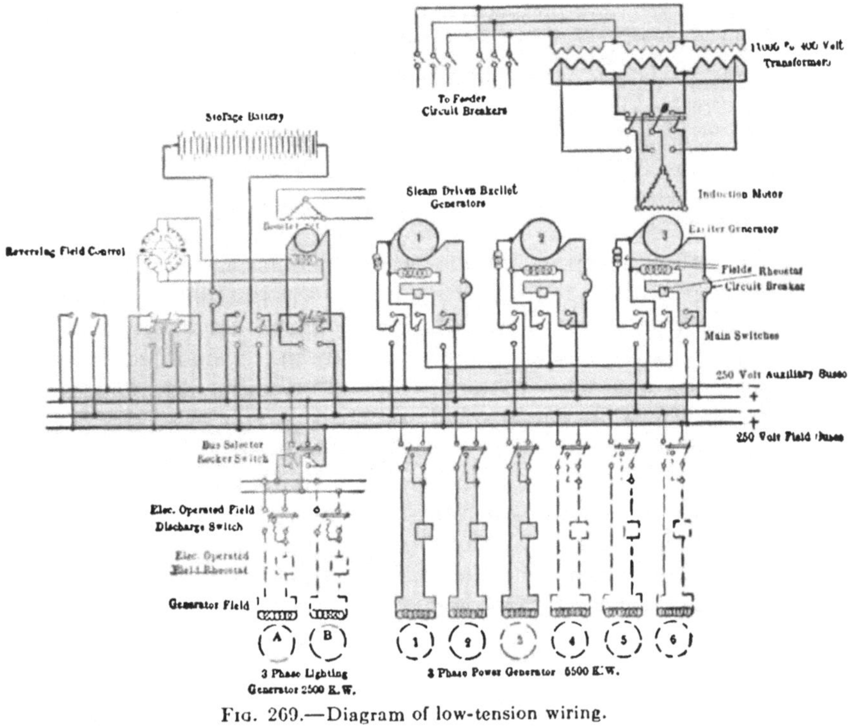

FIG. 269.—Diagram of low-tension wiring.

switches to the main or auxiliary buses. The generator field windings are connected to the main buses by double-pole switches, and the busbar sets can be connected together with lever switches, so that a generator field may also be excited indirectly from the auxiliary buses.

The entire switching system is controlled from the operating stand at the east end of the second gallery. The present position of the operating room corresponds to what will be the middle of the station when the contemplated addition on the east side is built. (See Fig. 267.)

The gallery is about 13 feet over the main engine room and has a small overhanging observation balcony which gives a comprehensive view over the entire station.

Fig. 270.—General view of gallery showing exciter buses, generator, rheostat, and auxiliary wiring.

The switchboard arrangement consists of (a) generator control bench; (b) greater instrument board directly in front of the bench; (c) feeder control board; (d) exciter switchboard; (e) auxiliary switchboard for station lighting and other purposes.

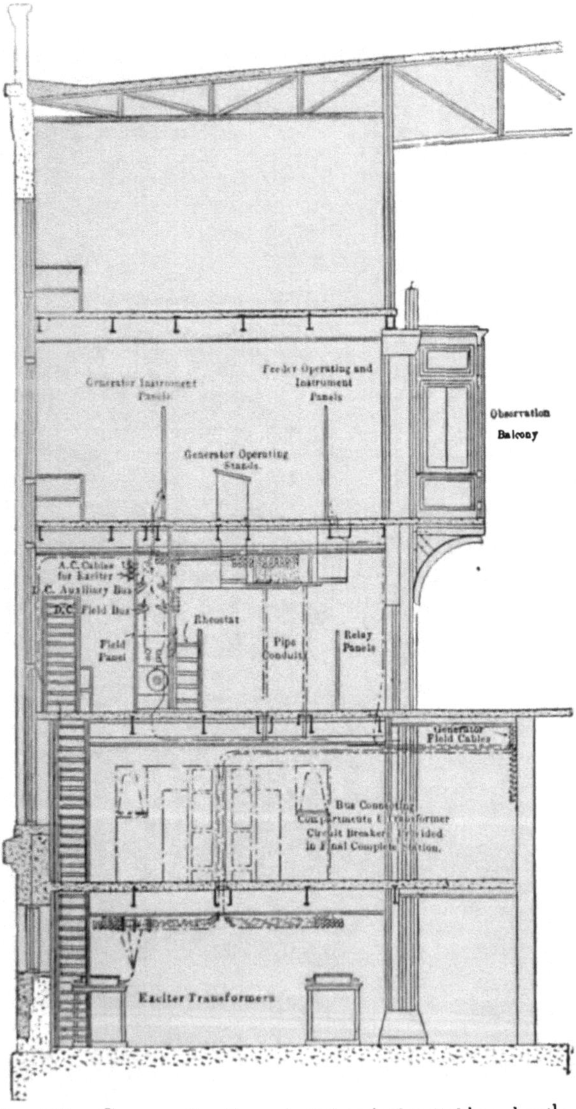

FIG. 271.—Cross section through east end of switchboard gallery.

All of the panels are of marble and the panels and instruments have a dull black finish. On the top of the generator bench are mounted all the necessary control switches and signal lamps for the generator oil switches, and also the control devices for the sectionalizing switches. Directly opposite the generator control panel is the corresponding instrument board to which the secondarics of the series and shunt transformers are connected. These leads are enclosed in fiber conduits. On the left side of the instrument panel is a board carrying three a.c. voltmeters, giving the e.m.f. in each leg of the busbars, and also a frequency indicator and two synchronizing lamps, one for each bus. To the right of the instrument boards there is a panel with one differential a.c. voltmeter, two synchroscopes, two synchronizing lamps, and one a.c. ammeter indicating the current in the grounded neutral bus from the generators.

Figure 270 shows how the rheostats, motors, and electrically operated field switches are set up. Operation is from the bench. This equipment is installed directly under the operating room. (See Fig. 271.) The two last-mentioned illustrations also show the mounting of the d.c. exciter busbars. The feeder board did consist of three panels, each one of which controlled two group switches and six feeder switches. The relays for the automatic oil-switch control are on a separate panel on the first gallery near the field rheostats. An auxiliary panel which is set up in line with the feeder board controls the supply for the d.c. motors, for the station lighting, and for the electrically operated oil switches. The exciter board also carries the necessary lever switches for the positive, negative, and equalizer buses for the circuit breakers and instruments. A number of empty panels for future use were set up between the instrument and exciter boards. The disposition of the fiber ducts containing the control and instrument wires and of the generator field rheostats is shown in Figs. 266 and 267.

A complete signaling system is provided for communication between operating and engine rooms. It consists of a set of letter signals set up in the engine room which are illuminated by pressing push buttons in the operating room after the engineer's attention has been attracted by means of a signal whistle. A set of return signals is set up in the operating gallery and is actuated from the engine room, the operator's attention being attracted by a gong. Ready intercommunication is thus possible. A large synchroscope is placed over the signals in the engine room. Opposite each turbine there is an opal globe inside of which are a red, a white, and a green lamp, and on a bracket outside of the globe a plain incandescent lamp. The lamps in the globe are used to denote the position of the weight upon the governor control of the turbine. The red lamp signals speeding, the white lamp dead center, and the green lamp slowing. These signals are pilot indications of the position of the governor control weight, which is controlled from the operating room.

The other apparatus is used as follows: (The letters O.K., *I*N, OUT, S.B., and the numbers are illuminated on the board in the engine room and in the operating room as stated above.)

1. To *C*ut in a Generator

Operator	Engineer
(The number of the generator) IN-whistle.	(*I*f O.K.) from the corresponding generator. IN-gong.

When the turbine is ready for load the engineer sounds the gong again. The operator connects the synchroscope and synchronizing lamp and when switched in parallel disconnects the synchroscope and synchronizing lamp, signaling O.K.

2. To *C*ut out a Generator

Operator	Engineer
(The number of the generator) OUT-whistle.	(*I*f O.K.) from the corresponding generator. OUT-gong.

Operator connects the synchroscope and synchronizing lamp.

When the generator is cut out the synchroscope commences to revolve and the synchronizing lamp to pulsate.

Operator signals O.K.-whistle and disconnects the synchroscope and lamp.

Engineer signals O.K.-gong.

All signals disconnected.

3. To *C*ut in a Turbine

Engineer	Operator
From the turbine, IN-gong. Operation then continues as in first case.	From corresponding generator, IN (if O.K.).

4. To Cut out a Turbine

Engineer	Operator
From the turbine, OUT-gong. The operation then continues as in second case.	From the corresponding generator, OUT (if O.K.).

5. To Change Over

First and second or third and fourth to be followed in sequence, depending upon whether the change is made from the operating room or from the turbine room. In case of trouble of such a nature as to require one unit to be cut out before the other is cut in, the operator or the engineer will show both signals, as (number of generator)—OUT (number of generator)—IN, and the whistle gives a long blast or the gong is sounded several times rapidly to call attention to the illuminated signs.

6. To Cut in an Exciter

Operator	Engineer
(Number of the exciter) IN-whistle.	Brings the exciter up to speed, when ready signals O.K.-gong; from No. 1 turbine.
Operator cuts in and signals O.K.	
All signals disconnected.	

7. To Cut out an Exciter

Operator	Engineer
(Number of the exciter) OUT-whistle.	O.K.-gong; from No. 1 turbine.
When exciter is cut out operator signals O:K.	
All signals disconnected.	

8. Stand-by Signals for Engineers

Operator	Engineer
S.B. (number of generator)-whistle.	S.B.-gong.
When over	Engineer signals O.K.-gong.
Operator signals O.K.-whistle.	
All signals disconnected.	

9. Stand-by Signals for Operators

Engineer	Operator
From turbine, S.B.-gong.	Number of generator, S.B.-whistle.
When over	
Engineer signals O.K.-gong.	Operator signals O.K.-whistle.
All signals disconnected.	

10. When whistle is given one long blast, or gong is sounded several times rapidly, it is to convey the meaning that the operation, whether a single or a double one, is to be performed as quickly as possible.

In starting a turbine the fields are built up gradually to give full voltage after the machine has attained its full speed. The generators are then synchronized and the main generator switch is thrown in. This switch is so wired that it cannot be closed until the synchronizing plug has been inserted into its socket in the operating board. In cutting out a machine the load is taken off the generator by reducing the field excitation and then the main switch is opened. Afterward the field circuits are cut out. With the field circuits out, a turbine will continue to revolve about forty-five minutes before coming to a standstill, but it will stop in about ten minutes with the fields excited.

The switchboard operator gets his orders to cut current off or throw current on the high-tension lines from the sub-station at Woodhaven Junction. (See Chap. XXVII.) He does not, however, change the switches until he is assured that the person giving the order has authority to do so.

60,000-Volt Station.—Figures 272, 273, and 274 (by courtesy of Stephen G. Hayes) are the designs for a hydraulic power station to be located on a hillside. The generators are each rated at 7,500 kw. and 6,600 volts, which is stepped up for transmission purposes to 60,000 volts through twelve 5,000-kw. transformers. The entire switching plant is in a separate building. In the engine room there are only the operating switchboards mounted on a gallery and the field rheostats for the generators. The generator cables lead through the foundations and walls to the oil switches on the lower floor of the switching house, whence they lead to the 6,600-volt busbars, which are installed in the same room over the oil switches, being connected to the latter through disconnecting switches. (See Figs. 273 and 274.) On the next floor are the single-phase transformers, set up in compartments separated from each other. Each set of three transformers is delta-connected on the high- and low-tension sides, the connection on the low-tension side being fed from the 6,600-volt buses. A pair of disconnecting switches is inserted on the high-tension side of each transformer so that the apparatus can readily be disconnected, and the high-tension deltas are joined to reactance

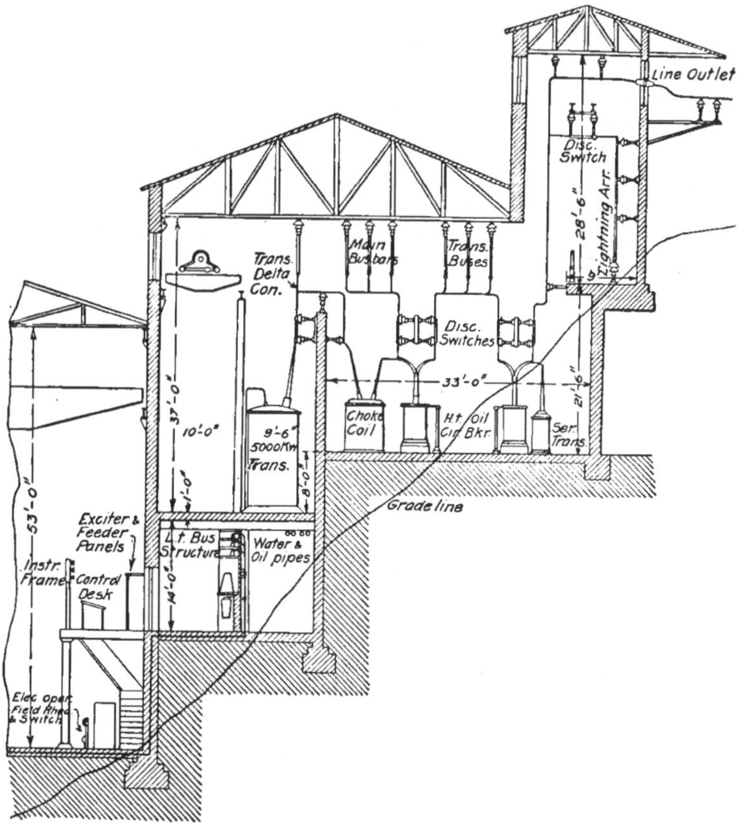

FIG. 272.—60,000-volt station; section through switching house.

coils contained in separate oil vessels. From the coils the cables lead through oil switches and disconnecting switches to a set of transformer busbars. There is a set of buses for each bank, which can be connected through oil switches and disconnecting switches, to the main buses or to the outgoing feeders through a similar set of switches. The lightning arresters, line-disconnecting switches.

and wall bushings for the line outlets are contained in two tower-like structures on the uphill side of the building. The high-

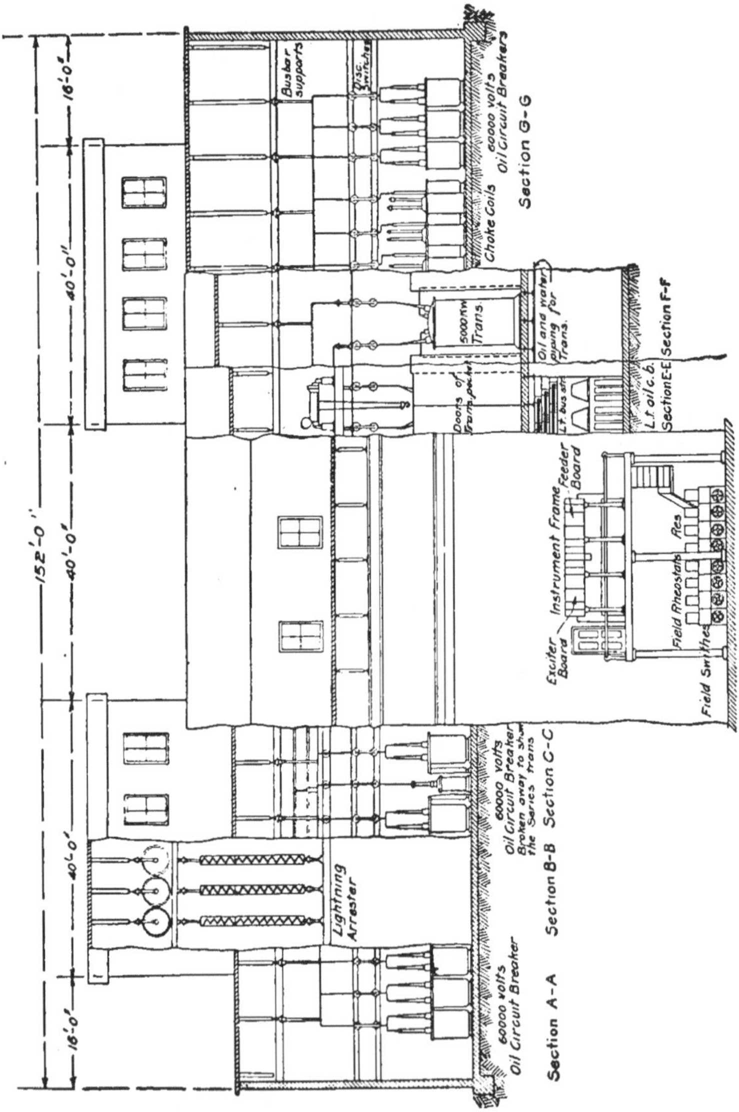

FIG. 273.—Sections taken in lines A, B, C, D, E, F, G, of Fig. 274.

tension oil switches are type G Westinghouse solenoid-operated, and are set up in two rows. (See Fig. 274.) The end switches

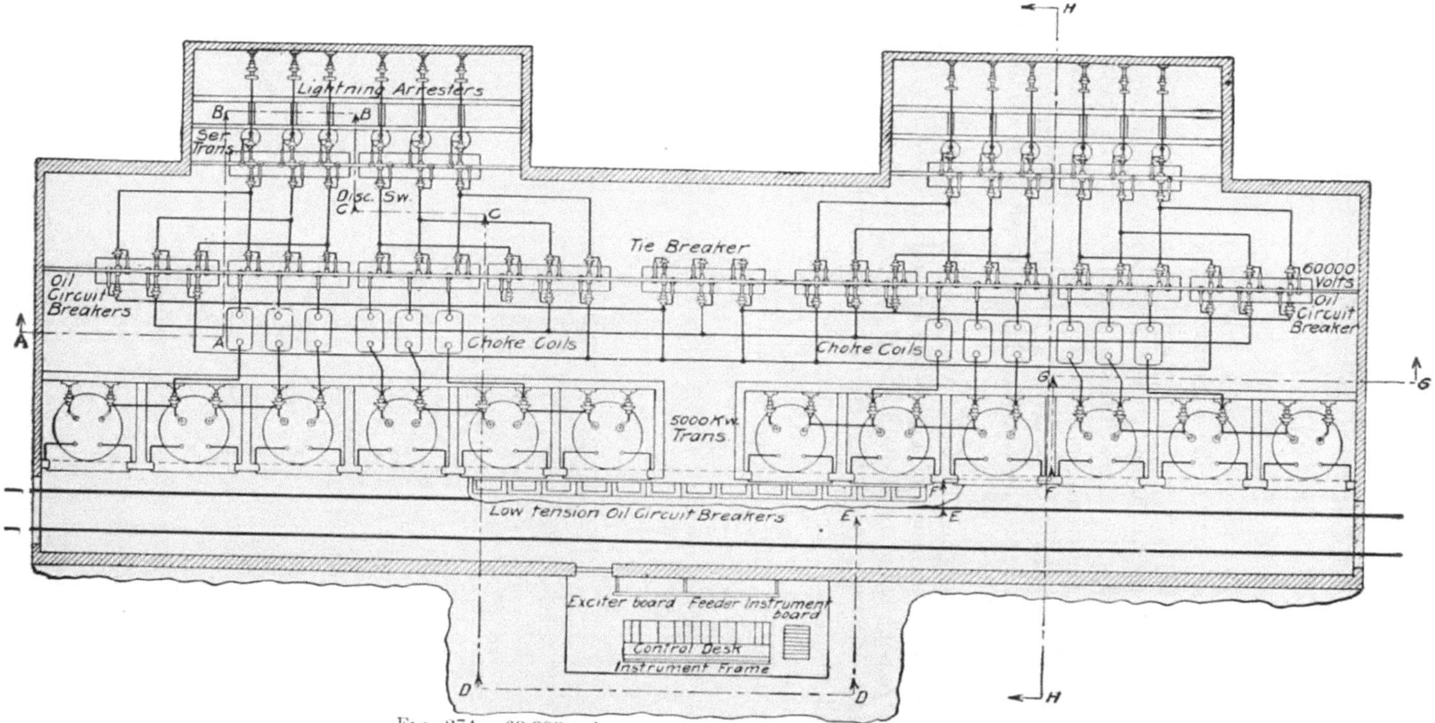

Fig. 274.—60,000-volt station; plan view of switching arrangement.

of the first row as well as the fourth switch from each end are for connecting the transformer buses to the main buses, and the second and third from each end connect the reactance coils to the transformer buses. The central switch is the bus sectionalizer. The first row therefore consists of nine oil switches each built up of three single-pole vessels. In the second row there are four oil switches, two in each tower, which connect the transmission lines to the transformer or main buses. Back of this row the instrument series transformers are set up. A traveling crane is provided to carry the large and heavy transformers up to the compartments.

On the gallery in the engine room the required switch board is mounted. It consists of a generator control bench with vertical instrument board back of the bench and mounted on supports which also form the gallery railing. It also includes the feeder and exciter boards located near the back wall. The motor-operated field rheostats and the electrically operated field switches are under the gallery. (See Fig. 273.)

CHAPTER XXVI

SUB-STATIONS

Modern systems have come to assume such large proportions that it has become necessary to find some method of feeding other than those heretofore employed. Formerly the feeding capacity was increased by raising the d.c. voltage, by using boosters, or storage batteries with or without boosters, or by increasing the number of units in the independent generating stations. The method now almost universally adopted for large systems is to generate a high-tension alternating current in a central station and distribute it to the net through a number of sub-stations. (See Chap. XXIV.) The determination of how this method should be applied depends (1) upon the manner in which the system has expanded; (2) upon the kind of service, and (3) upon the reliability of the machines and apparatus:

1. The first point divides itself into three cases. (a) In the first case the system for which the generator station was originally designed has expanded to such an extent that isolated auxiliary plants become necessary to assist in supplying the excess demand in the various stations. This problem may be solved by dividing the systems into sections and supplying each section from its own independent station. This case arises most frequently in lighting or street railway systems in large cities. (b) A large system may also be formed by combining a number of independent, widely separated smaller systems, each having its own station. Suburban and interurban electric railways combined with street railways will be included under this head. (c) Electrifying railroads or designing large hydraulic plants for electrical transmission purposes necessarily involves large service systems from the very start.

Lately the development of gas-engine and steam-turbine design has made such progress that it is safe to predict that in the future central stations of enormous power will be built near the sources of coal supply to feed systems of wide radius.

2. The kind of consumption of electrical energy materially affects the ways and means for distribution and supply. Under

this head there will be included: Street and interurban railways or trunk lines, arc or incandescent lighting systems, and power distribution for motors, or for electrometallurgical or electrochemical purposes. (See Chap. XXIV.)

3. The method to be used for power distribution will depend upon the cost and reliability of machines for converting the energy from the central station into energy suitable for the desired purposes. It will also depend upon the adaptability of the consuming apparatus to the varying conditions to which it is subjected in service.

A study of the three above-mentioned considerations will lead to a determination of which course to pursue in order to obtain the best results with regard to cost, reliability, and adaptability.

Three systems are then available: (1) Alternating-current central station with sub-stations in which the high tension is stepped down and is converted by means of converters. (2) Alternating-current central stations with sub-stations in which the converting to direct current is done by means of motor-generator sets. Transformer banks may be employed to step down the high tension when necessary. (3) Low-tension alternating-current feeder service. This requires either sub-stations with transformer banks or transformer banks alone located at the points of consumption. In some cases it becomes necessary to make use of motor-generator sets to change the voltage, frequency, and phase of the high-tension supply.

Sub-stations, like independent d.c. stations, supply districts in which they are located, and in order to save copper they must be located at the load centers of these districts. Since they perform the same service as the former d.c. stations which they replace, and which are located at the load centers, it follows that the new sub-stations may be located in the old buildings or on adjacent property. The equipment of the old stations is often used as a reserve or for peak loads. The number and size of the sub-stations will naturally increase with the growth of the system and with the intensity of the load. The cost of real estate, buildings, machines, operation, and attendance will increase with the number of stations. It must therefore be considered whether it would be cheaper to use a large number of stations near together or a smaller number farther apart but having a correspondingly increased power rating. Moreover, it must be remembered in this connection that sub-stations with large units necessarily involve greater cost for reserve machines and for feeder systems. High-service voltage allows greater distance between sub-stations, as noted in previous chapters. In direct-current service the line drop may be compensated and the efficiency of the machines may be increased by using storage batteries with boosters. (See Chap. V.)

Synchronous Converter Sub-stations.—Since *synchronous converters* first came into practical use in 1897 their development to the present day has tended to simplify a great many problems arising in large installations.

The voltage ratio between the a.c. and d.c. sides of a rotary converter is practically constant and cannot be materially altered by varying the field strength. With sine-wave current the conversion ratios hereinafter given are closely approximated. The d.c. potential being taken as unity, the potential at the collector rings will be:

For single and two-phase machines 0.71, for three-phase 0.61, for six-phase 0.71 or 0.61, depending upon whether a diametrical or double-delta connection is used for the transformers. For example, a two-phase rotary receiving alternating current at approximately 426 volts will deliver direct current at 600 volts, while a three-phase rotary receiving alternating current at 366 volts will also deliver direct current at 600 volts.

The 150-, 200-, and 300-kw., 25-cycle converters and the 100- and 200-kw., 60-cycle converters are wound for three-phase operation, all the larger machines being usually wound for six-phase.

The direct-current voltage of a converter can be changed only by a change in the impressed alternating-current voltage or by a change in the shape of the magnetic field set up by the field winding of the converter. For the first method, until recently, resistance coils and variable field excitation were in use. Taps from the lowering transformers or from separate regulating transformers and induction regulators, the last mostly in connection with shunt-wound converters, were applied. More recently the method using a synchronous alternating-current booster mounted on the same shaft as the converter armature, and connected electrically in series with it, has been introduced. Another method of direct voltage variation based on the second of the two principles mentioned above, i.e., by a change in the shape of the magnetic field, has also been introduced recently. This is the "split-pole" or "regulating-pole" converter, of which each pole is split up into two or more sections and the field form varied by varying the excitation of the different sections.

The generating and regulating apparatus used in a synchronous converter sub-station, therefore, is divided as below.

- Synchronous converter
  - Commutating-pole type with
    - Induction regulator
    - Reactance
    - Transformer taps
  - Non-commutating-pole type with
    - Induction regulator
    - Synchronous booster
    - Reactance
    - Transformer taps
  - Regulating-pole type

The use of resistance either in the lowering transformers or in special coils in connection with series windings on the field poles of the converter is based on the principle that an increase in excitation causes the phase relation of the current to be leading with respect to the line voltage, and leading current through resistance gives a rise in the resultant voltage. This voltage variation is automatic and therefore is best suited for railway work in which the load fluctuation is large and rapid. On the other hand, the variation is obtained at the expense of a varying power factor.

As to the use of taps from transformers with dial switches and the application of induction regulators for d.c. voltage regulation of synchronous converters see Chap. XVII.

The use of a synchronous booster involves the addition of an alternating-current generator of special characteristics to a standard rotary converter. The generator has the same number of poles as the converter, and the two revolving parts are carried by the same shaft. The booster, therefore, generates a voltage of the same frequency and phase as the induced voltage of the converter, and by properly proportioning its parts the wave shape of the booster may also be made the same as the wave shape of the converter voltage. The booster armature is connected in series with the converter armature, so that the voltage delivered to the converter armature is the sum of the line voltage and the booster voltage. The delivered voltage is varied by varying the field current of the booster—in one direction to raise, and in the other direction to lower, the converter voltage. The booster is usually placed between the collector rings and the main armature of the converter, and the taps from the collector rings are connected to equidistant points of the booster armature. Similar points on the booster armature are connected to the synchronous converter armature, thus placing the two in series. Separate and distinct windings are used for each phase.

In the split-pole converters the variation of the d.c. voltage is secured by varying the distribution of the flux under each pole, i.e., by varying the field form. For that purpose each pole, instead of being in one piece with one field coil, is made in two or three sections with two or three separate exciting coils.

It is a known fact that the voltage generated in any d.c. machine is proportional to the average value of the flux, or the average ordinate of the field form, between adjacent brushes. A variation of the field form which increases and diminishes the average ordinate will increase and diminish the generated voltage. This is done by means of the regulating poles with their separate exciting coils. On account of the presence of a.c. line voltage, the variation in field form must be accomplished without varying, to a corresponding degree, the wave form of the alternating-current voltage generated in the converter armature. The split-pole converter, therefore, consists of two essential elements: first, a means of varying the field form, and second, a means of eliminating the variation in alternating-current wave form resulting from the variation in the field form. Only that part of the change in field form which does not produce a change in the alternating-current wave form is effective in changing the direct-current voltage, and any change in field form which does change the wave form is neutralized by the consequent change in the alternating current.

In a great many interurban lines the load factor is very low, and at certain times the momentary peak loads are extremely high on account of the necessity for starting and accelerating a number of cars at the same time from a given sub-station. This momentary condition ordinarily would require rotary converters of much larger capacity than are suited for normal operating service. To meet such conditions railway rotary converters have been developed, having a momentary overload capacity of 300 per cent. of normal. The application of commutating poles has made it possible to meet this condition.

With higher speeds of rotary converters it has been possible to replace existing units with units of twice their capacity without occupying additional floor space, a most economical arrangement where space is usually limited, as in city sub-stations.

The greatest units of rotary converters now in service have a capacity of 4,000 kw., 600-volt, 25-cycle, for railway use, and for low voltage 3,500 kw., 270-volt, 25-cycle, which can deliver 13,000 amp. continuously and nearly 20,000 amp. for two hours. This type of machine is a combination of the booster and commutating pole construction (New York Edison Company). There are also. in use on lighting and railway systems 60-cycle converters with commutating poles of 1,000 to 1,500 kw. capacity. These

A Comparison of Motor-converters with Synchronous Converters and Motor-generators[1]

Characteristics	Compared with rotaries and their transformers	Compared with motor-generators
Efficiency:		
Full load	Within 1 per cent. of rotary.	Better by 2½ per cent. than motor-generator.
Half load	Within 1 per cent. of rotary.	Better by 4 per cent. than motor-generator.
Quarter load	Equal to rotary.	Better by 8 per cent. than motor-generator.
Power factor	Better than rotary with same variation of d.c. volt.	Better than induction motor-generator
Starting	Simpler than rotary.	Equal to induction motor-generator.
Synchronizing		Better than synchronous motor-generator
Regulation	Simpler than rotary.	Simpler than synchronous motor-generator.
Simplicity	Better than rotary.	Equal to motor-generator
	Simpler than rotary and transformers, which require more complicated switching.	Equal to induction motor-generator. Simpler than synch. motor generator.
Reversal of polarity	Motor-converters will not reverse their polarity. Rotaries are liable to reversal	

[1] N. E. L. A., thirty-sixth convention at Chicago, 1913.

machines are built either of the synchronous-booster or split-pole types. A new type of the converter, called *motor-converter*, has been developed lately. This machine consists of an induction motor and a rotary converter rigidly coupled together, with the rotor winding of the induction motor and the armature winding of the rotary converter together in parallel. In operation, the converter runs at a speed equal to half the frequency of the system, which is exceptionally favorable to commutation, and in fact the entire machine is of materially smaller proportions than would be a synchronous or induction motor-generator of the same capacity.

The machine is stated to be capable of performing all the services for which rotaries or motor-generators are used, including its application to three-wire direct-current systems. The usual methods of starting from either a.c. or d.c. are employed under greatly simplified arrangements. Synchronizing in particular is stated to be more easily accomplished than with any other form of converting apparatus.

The transformer has been the principal factor in the development of alterating-current distribution and long-distance transmission of electrical energy. Such energy cannot be transmitted economically for long distances at low voltages, and the transformer is the only means by which sufficiently high voltages can be obtained. With the rapidly increasing demand for electrical energy, the vast water-power resources of the country are becoming more and more valuable as the center of power generation for a large territory. As the distance between generation and consumption of energy increases economical considerations require that the voltages also increase.

With our present modern high-voltage systems, where it is undesirable to parallel the transmission lines on the high-tension side or to carry out any high-tension switching, it has become general practice to install the transformers in groups, each having a capacity corresponding to a line, the transformer group and the line being considered as a unit. Transmission lines generally have a current-carrying capacity ranging from 15,000 to 40,000 kw., and as the most economical size of high-voltage transformers is from 6,000 to 10,000 kw., it is obvious that it is entirely feasible to provide one bank of single-phase transformers for lines up to about 25,000 kw. capacity, while above this it becomes necessary, as a rule, to provide two banks in parallel for each line.

The most important requirement in connection with modern transformers of large capacity is that their design be such as to limit the current output of the system which they feed at times of short-circuits, and besides this to successfully withstand the tremendous mechanical strains to which the transformer windings are subjected by short-circuits. Such transformers are for this reason now generally designed with a considerable higher reactance than was formerly the case, 4 to 6 per cent. reactance now being quite common. (See "Reactive Coils" in Chap. XX.)

The transformers should always be provided with taps in the windings. In many instances it is found desirable to begin the operation of the system at half the ultimate voltage (in the generating water-power station), and in order to accomplish this the transformers are often provided with 50 per cent. taps in the high-voltage windings. It is also customary to provide the high-voltage windings with four 2½ per cent. taps below the normal operating voltage so that these taps can be used in compensating for the line drops as the load increases. Such taps are, of course, only necessary in the step-down transformers of the sub-stations, but for the sake of interchangeability they are usually provided in the step-up transformers as well.

The number and size of the transformers and whether they should be of single- or three-phase type depends entirely on the nature of development of the power system and on the condition to be met. In most long-distance power transmission systems the amount of energy transmitted is large, and in central-station design the modern tendency is toward larger and larger capacities, so that in most cases a small number of transformer units of large capacities can be used. On the other hand, the different classes of consumers' service require different smaller units to suit special cases.

Step-up transformers of the generating station as well as the step-down transformers used in sub-stations may consist, as mentioned, of banks of either three single-phase unit or one three-phase unit.

The advantages of polyphase transformers over the single-phase type used in a group are as follows:

1. The initial cost is less. This is due in part to the saving in active core material between adjacent phases on account of magnetic phase relation. It is a reduction of about 15 per cent. in active magnetic iron. Other savings are the use of one larger case (requiring less radiating surface than three small cases on account of higher three-phase efficiency) in place of three small ones; one set of end frames in place of three; one cooling system

(of forced-cooling type) in place of three; six terminals in place of twelve, and less oil. There is also less material to be handled, and it is usually less expensive to build one large unit than several small ones of the same total capacity. The first cost of small three-phase delta-delta-connected transformers should be, in general, 5 to 10 per cent. less than that of three single-phase units having the same aggregate capacity. The cost of three-phase air-blast transformers approximates the cost of equivalent single-phase air-blast units at about 3,000 kva., while for the oil-insulated, water-cooled type, the cost is equal at about 7,500 kva. For sizes larger than these three-phase transformers usually cost as much or more than three single-phase units.

2. Since there is less active material in the three-phase transformer, the losses (at the same densities) will be less, and therefore the efficiencies should be higher. A three-phase shell-type transformer requires just three times as much copper and approximately 2.36 times as much iron as a single-phase transformer having one-third the capacity. Consequently the copper losses would be three times and iron losses 2.36 times as great. The three-phase efficiencies should therefore be slightly higher than those of three single-phase units.

3. The actual floor space covered by a three-phase transformer will always be less than for three single-phase units, and on account of the necessity of leaving ventilating spaces and passageways between single-phase units, the total floor space required for single-phase groups is very much more than that required for three-phase. The economy of floor space for the three-phase transformers may be of great advantage in cities where real estate is expensive. Usually one bank of transformers is placed directly opposite the rotary converter or motor-generator it feeds, and often the dimensions of apparatus fed by the transformers are such as to provide ample space for the bank of single-phase transformers. However, high-speed rotary-converter and motor-generator sets, owing largely to the employment of commutating poles, require very much less floor space than did the older types of these machines. As the speed increases the available transformer space should proportionally decrease and the lesser space requirement of three-phase transformers becomes more important. In the case of vertical rotary converters the floor space becomes a minimum, a point especially favorable to three-phase units.

The weights of three-phase units should be less than those of single-phase type, except in the excessively large capacities, when the weights would be greater on account of the very strong and heavy mechanical design necessary to support the heavy three-phase transformer.

4. Wherever a high-voltage lead passes through the transformer tank or case there is a weak point. In three single-phase units six high-tension leads must be carried through the case (unless the transformers are star-connected with neutral grounded to the cores, in which case only one high-tension lead would be carried through each case. This latter method, however, is very seldom used). In three-phase transformers there are only half as many terminals and consequently only about half the chance of terminal trouble.

It is cheaper to install one larger unit than three small ones, particularly in case it is necessary to dry out the transformers at destination by the vacuum process. Less cable and fewer connections are also required for the three-phase units.

In the three-phase transformer, the star and delta connections are easily made inside the case, requiring no outside crossings, while with single-phase units these connections must be made outside of the case.

5. The higher total weight and less space required for three-phase units will, in general, result in reduced transportation charges. This advantage may, however be neutralized in the case of long hauls over mountain roads or poor facilities for handling heavy pieces. In this case the smaller weight and bulk of single-phase units may even become a necessity.

The disadvantages are as follows:

1. Large cost for reserve units. Although the cost of a three-phase transformer is larger than that of a single-phase, the rating of the former is three times that of the latter. Where a large number of transformers are used the relative cost of reserve is small. In some places in Europe it is customary to hold in reserve only the coils of the core type instead of the entire transformer, as these coils are easily replaceable. For the shell type the entire units must be used as reserves.

2. Transformer insulation is subjected to three different strains—those due to the normal working pressure; those due to disturbances, and those due to lightning. The normal working pressure strain is constant; the static strains due to disturbances in the system are more or less frequent, depending upon local

conditions; the strains caused by lightning are less frequent and do not exist in cases of underground transmission systems. In some localities lightning strains are unknown.

Either three single-phase delta-delta transformers or one three-phase delta-delta-connected transformer of the shell type may be operated temporarily at 58 per cent. of the combined capacity of the three windings, provided only one phase has been disabled. With the core type of polyphase transformer the service is entirely crippled under these conditions. In case of star-connected shell-type transformers either single- or three-phase a breakdown in one phase will disable the entire service unless the transformers are operated with a grounded-neutral transmission system. In such a case both single- and three-phase star-connected transformers may be operated at 58 per cent. of their original capacity by short-circuiting the damaged windings, both primary and secondary. Polyphase core-type transformers cannot be operated with a short-circuit on any phase.

The time required to replace or repair a polyphase transformer is greater than for a single-phase unit, therefore the time of duration of crippled service in the first case is longer.

3. In the shell type there is no more liability of insulation breakdown in the three-phase transformer than in three single-phase units; in fact, the arrangement of coils and iron is the same in both types. In the core type there is greater liability of insulation breakdown and of a burn-out in one phase being communicated to an adjoining phase than in the shell type since the phases are closer together. In the case of three single-phase transformers it is probable that in event of a breakdown in one transformer, it would be cut out before the other transformers in the group were damaged. With a three-phase unit, where the phases are so near together, there is a possibility that a breakdown in one phase may damage one or more of the others. In such an event there are two or three phases in place of one to be repaired. However, with oil-insulated transformers, oil-switch protection, and barriers between adjacent phases, there should be little likelihood of trouble being communicated to adjacent phases. In general, the repair of a polyphase transformer will be more costly and more difficult than that of a single-phase type of similar construction. This is particularly true in extremely large capacities on account of the difficulty of handling the large parts.

4. In some cases it is highly desirable to use only self-cooling

transformers. At present the capacity of the largest three-phase self-cooled transformer is approximately 2,000 kva., 60 cycles, at 110,000 volts, or 1,500 kva., 25 cycles, at 110,000 volts. These capacities could be increased for lower voltages. A group of three single-phase self-cooling units can be built with an output of nearly three times this amount. Hence if very large outputs of the self-cooling three-phase type are desired, it might be necessary to use two or three smaller three-phase units, a more expensive arrangement than three single-phase transformers.

5. For fairly low-voltage transformers there would be no difficulty in arranging for a number of tap voltages in polyphase transformers, but with extremely high voltages this may become objectionable on account of the limited space in a three-phase unit. In the air-blast type this would become still more serious on account of the greater spacing required when terminals are not placed under oil.

On comparing the above advantages and disadvantages it is evident that the three-phase transformers are applicable in large stations where the relative cost of the reserve is small, where there are sufficient handling facilities, and where saving in floor space is essential.

The transformers may be connected either in delta or in star, but it is now generally conceded that the isolated delta connection is preferred on account of the increased reliability which such a system affords. Under normal operation the voltage stress of the apparatus is the same whether the isolated delta or star connection is used, but in case of grounds on one line wire the isolated delta system will naturally be exposed to a higher voltage above ground than would be the case with the grounded star system. The disadvantage of the grounded star system, however, is that any ground of the line wires will cause a short-circuit and thus a shutdown, which is not the case with the delta system, where in certain instances it has been found that the operation could be kept up for a whole week with one of the line wires grounded. Usually the transformers used for 11,000 to 19,000 volts have the high-tension side connected in delta, and for 33,000 to 66,000 volts in star, with the grounded neutral. The secondaries are connected in the synchronous-converter sub-station as follows: For two-phase diametrically, for three-phase delta, star or T, for six-phase diametrically, double-delta, star or T.

Transformers are built self-cooled, air-blast, oil-cooled, or water-cooled, the manner of cooling depending upon the power and voltage.

In all transformers the first step in the dissipation of the heat generated by the losses is its conduction from the interior of the core and coils to the surface exposed to the cooling medium. In the oil-cooled type the oil heated by contact with the exposed surfaces of the core and coils rises to the surface, flows outward, and descends along the sides of the transformer case, from the outer surface of which the heat is radiated into the air. In the water-cooled type the same general process takes place, with the exception that most of the heat is carried away by water forced through a coil of pipe so placed that the heat is taken from the latest oil at the top of the transformer. The circulation of oil is completed after its descent along the sides of the case by again rising around the transformer core and winding and passing through ducts arranged for that purpose. In the air-blast type cooling is effected by a forced draft of air, which is directed through spaces provided between coils and in the core structure.

For installations where initial cost is of less importance than minimum attendance and where very large units may not be asked for, the oil-cooled type of transformer is recommended. For plants requiring the largest size of transformers, either air-blast or the water-cooled types must be used, if the choice depends on voltage, cheapness of water supply, fire risk, and first cost.

In the air-cooled type ventilators driven by induction motors must be furnished, and the transformers provided on the bottom with dampers must be set up over an air-tight chamber which receives the air blast from the ventilators. The diagram in Fig. 275 gives the amount and pressure of air required for single-phase transformers. Approximately 150 cubic feet of air per minute are required for each kilowatt lost. For the three-phase type the volume required is equal to that for three single-phase transformers each of one-third the power of the three-phase apparatus, but the pressure should be that required for one single-phase corresponding in power to the three-phase transformer. Turns in the air chamber should be avoided, and the cross section should be such that the velocity of the air blast will not exceed 500 feet per minute. It should consist of smooth fireproof material and should have drainage and be easily accessible. As the transformers are open on the top, they should not be located where there is much dust or dirt nor where water from any source is liable to fall on them. The blower should be so situated as to obtain clean dry air of a temperature not greater than 25° C. The air-blast transformers are used only up to a potential of 33,000 volts. Water- and oil-cooled transformers have external piping to carry the cooling liquid to the apparatus. In the former type the water is forced to circulate, an extra pump being necessary for this purpose. The oil-insulated types have the primary and secondary lead terminals set on top of the transformer, while the air blast have the primary on top and secondary on the bottom. Traveling cranes must often be provided to handle the larger units, or the transformers are constructed with rollers or are handled with roller wagons on tracks.

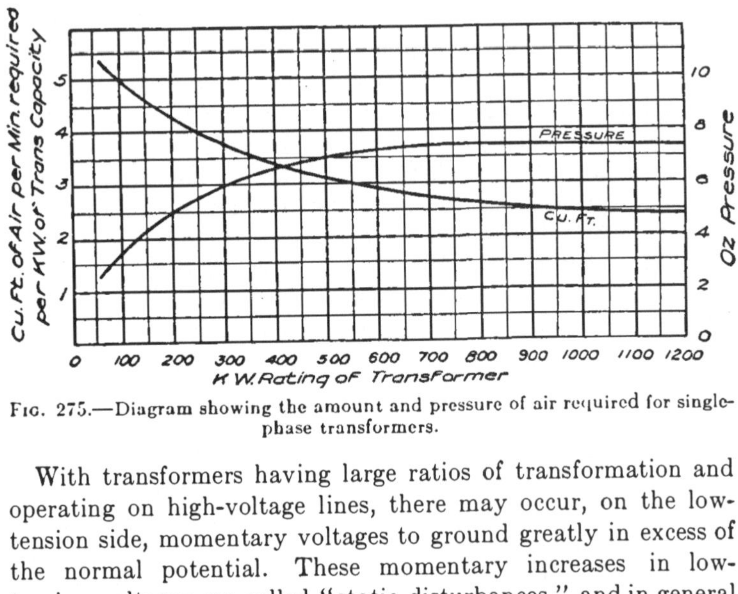

FIG. 275.—Diagram showing the amount and pressure of air required for single-phase transformers.

With transformers having large ratios of transformation and operating on high-voltage lines, there may occur, on the low-tension side, momentary voltages to ground greatly in excess of the normal potential. These momentary increases in low-tension voltages are called "static disturbances," and in general are the result of a change in the static balance of the high-tension side and its connecting circuits. In transformers with a high ratio of transformation this static disturbance on the low-tension side may cause serious strains in the insulation. It is more serious in high ratio transformers because its insulation is less able to withstand it, the induced static voltage being independent of the ratio of transformation. A method of relieving this disturbance is to join a discharge gap between a middle or neutral point of the low-tension side of the transformer and the ground. Any voltage very much in excess of the maximum normal will cause a discharge to ground over the adjusted opening of the spark-gap. The low tension side will be thus tied to ground during such a discharge, while at other times it is ungrounded. The commonly employed spark-gap connections to banks of transformers for different transformations are shown in Fig. 276. The low-tension windings only are shown as the connection of the high-tension windings is in general immaterial. One gap is used in all groups except in two-phase independent circuits.

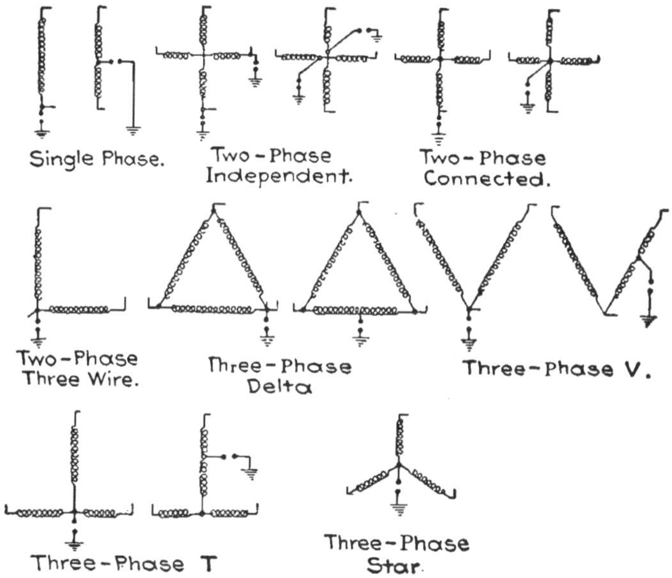

FIG. 276.—Spark-gap connection to transformer banks of different transformations.

In laying out a sub-station it should be borne in mind that even the normal operation of the machines will considerably increase the temperature, so that some provision should be made for good ventilation. This is particularly important where oil-cooled transformers are used, and ample ventilators should be provided in the roof so that the temperature will not become unbearable in the summer time. Where air-blast transformers are used the walls of the air-blast chambers should be waterproofed and the sub-station built at such an elevation that water will not stand on the floor of the air-blast chamber. If this is not done, the transformers may be damaged by the warm air from the blowers picking up moisture and depositing it in transformers which are not in service.

With air-blast transformers care should be taken to open the dampers when the transformers are put in service. Such transformers should also be cleaned by a high-pressure (about 40 pounds per square inch) jet of air at least once per month.

Faults in transformers, such as breakdowns between coils or turns, usually result in considerable smoke before the fault becomes sufficiently heavy to bring out an automatic switch Smoking transformers should, of course, be cut out at once to avoid further damage.

Where any cables come into the station underground the entering conduits should be sealed and suitable drainage provided so that water cannot break into the cellar through these openings. When oil-cooled transformers are used it is well to install a pit of sufficient capacity to hold the oil from several transformers, and to provide drainage piping.

An air compressor with a piping system or rubber-hose attachment should be provided to keep the machines clean from dust or metallic particles.

Sub-stations may be classed with regard to their location in the high-tension net as intermediate stations and end stations.

Intermediate stations are those which lie between two or more other sub-stations, and in their case the high-tension incoming lines feed the machines in the station and also the high-tension out-going feeders, which in turn feed the next sub-station. In case the intermediate station is located at a junction of several lines and feeds a number of other sub-stations, it is called a distribution sub-station. End sub-stations are the last stations on the system and may eventually become intermediate stations when the system expands. In case a load on a system is subject to shifting to certain sections during certain seasons, so that the load at these points requires additional sub-stations, portable sub-stations are made use of. A station of this kind consists of a car containing a complete sub-station equipment, which may be moved to any point along the tracks of the system and there joined to the high-tension line and the low-tension d.c. feeders. This form of sub-station is used only in railway service.

In case a high-tension short-circuit occurs during operation the usual result is the opening of many or all high-tension oil switches on the incoming or outgoing feeder and converter equipments. If power goes off and all oil switches do not operate automatically in such a case, as may happen on a fault not constituting a solid short-circuit, the operator should open at once the oil switches not brought out automatically, noting which ones they are and beginning with those of the sub-station machines. After a suitable length of time to allow the sub-station attendants to clear their oil switches, the main-station attendant should close in the line (or lines successively) to the first sub-station. It will be assumed that means are provided for indicating at the sub-station end when the power is restored. After receiving power from the generating station (assuming duplicate incoming high-tension lines) via both lines, the sub-station operator should close his incoming line switches and then proceed to close (one at a time) the switches which supply the outgoing lines to outlying sub-stations. If the short-circuit is on a line and has not cleared itself, power will vanish at the instant of closing one of the outgoing line switches in one of the sub-stations. Usually this will again clear out all switches back to the generating station if the short-circuit is solid. The generating-station attendant should then restore power at once and the sub-station attendant should pass it as before, leaving the faulty line open and reporting the same to the generating station as soon as all apparatus is in motion. To avoid confusion in the case of duplicate lines, an operator in a sub-station so supplied should not pass on power to outlying sub-stations until both incoming lines are alive or until he has received notice from the generating station that one is faulty.

In case there are several outgoing lines in different directions from the main station to the sub-station, or groups of sub-stations not otherwise interconnected, a short-circuit on one division of the network may bring out the lines to other divisions, owing to the "kickback" of power from the converters. On hooking up the lines after such a shut-down, the fault, if it remains, will bring out only the lines to the group of the sub-stations to which it belongs, as converters will not be running on the other divisions to give a "kickback." For this reason there is no need of telephonic communication to get the lines hooked up consecutively on the several divisions from the power house. Each division can go ahead independently, joining up lines as fast as power reaches each point until the faulty line is reached, when the whole division will probably come out but other divisions will not be affected.

Sub-stations with Motor-generator Sets.—The great variety of component units which when mounted on the same shaft constitute a motor-generator set enables such a set to meet very effectively the different requirements called for by railway, power, and lighting service.

The two most important uses for motor-generator sets are for changing frequency and for delivering direct current. When a railway or power system fed by a high-tension current at 25 cycles is to be made to supply a 60-cycle lighting system, a motor-generator set would be used consisting of a 25-cycle synchronous motor and a 60-cycle alternator. The arrangement of the machines must naturally be reversed if the lighting system is to feed the railway system. This use of motor-generators may be illustrated in another way by considering two or more independent power systems of different frequency whose supply is to be rendered interchangeable.

To drive single-phase railway motors it is often necessary to use motor-generator sets with 60-cycle, three-phase synchronous motors and 25-cycle, single-phase generators. By this means the conversion is accomplished without unbalancing the three-phase system.

In designing motor-generator sets the speed must be so chosen that it will correspond to both frequencies, and there must be taken into account whether or not the set is to be reversible. The most frequent use of motor-generators is to convert high-tension alternating current into low-tension direct current. Synchronous motors are preferred for this purpose since they are reversible. The power factor of the transmission line can be easily regulated by changing the field strength, and the motor can be wound without difficulty for 10,000 volts or more. On the other hand, an induction motor is of simpler construction and requires a simpler switching arrangement and no exciter. The synchronous motor may be directly joined to the high-tension line, provided the transmission voltage lies below 15,000. It might also be fed from a transformer bank when the line voltage exceeds the above value. The d.c. generator is of the ordinary type and is built for any required voltage.

As the following table shows, motor-generator sets are used with boosters for voltage regulation. The shunt boosters are principally applied on Edison systems as a means of obtaining different potentials on outgoing feeders; while series boosters are used on railway circuits to compensate for line drop and to maintain approximately constant potential at remote points of distribution.

Motor-generators	Synchronous motor set	240–300 volts 600 volts 1,200 volts 1,500 volts	Shunt boosters
	Induction motor set	240–300 volts 600 volts 1,200 volts 1,500 volts	Series boosters

Since a motor-generator set is employed for the same purpose as a synchronous converter, the question arises which of the two methods is the better in any given case. The points which govern a selection are: (1) Cost; (2) reliability; (3) adaptability; (4) efficiency, and (5) floor space.

In a paper read by E. W. Allen before the Association of Edison Illuminating Companies are discussed the advantages and disadvantages of different apparatus for d.c. sub-stations. An abstract of the paper is given in the following tables and analysis of the figures. The machines taken for comparison are of the very latest type, the synchronous converter being of the split-pole type, while the motor-generator sets are equipped with commutating poles.

The synchronous motor-generator set having been selected in the accompanying table as a basis for the comparisons for the different types of sets in regard to their efficiencies, prices, floor space, and weights, conclusions can be drawn as to the advantage of the use of one or of the other machine in cases at hand.

The values for synchronous converters and induction-motor sets are expressed in percentage of the corresponding values in the synchronous-motor set. The sign preceding a figure denotes whether it should be added to or subtracted from the figure given on the synchronous motor. The analysis of the figures shows that the 25- and 60-cycle synchronous converters are superior in efficiency at all loads to either type of motor-generator set, the difference being particularly marked at light load on the 25-cycle machine. The brush friction and windage constitute a relatively larger proportion of the losses in the 60-cycle converter, and its efficiency at light load only exceeds that of the motor-generator set by a small amount. The induction-motor set is more efficient

TABLE OF COMPARATIVE EFFICIENCIES, PRICES, FLOOR SPACES AND WEIGHTS[1]

Efficiencies

Kw. Capacity	25 Cycles			60 Cycles		
	Syn. mot. gen. set	Ind. mot. gen. set	Syn. convtr.	Syn. mot. gen. set	Ind. mot. gen. set	Syn. convtr.
300 Full load	84	+2.4 %	+ 6.5 %	86¾	−1.4 %	+ 2.2 %
¾ load	82¼	+2.1 %	+ 7.6 %	85	−2.0 %	+ 2.1 %
½ load	77	+4.0 %	+12.3 %	81¾	−2.7 %	+ 1.0 %
500 Full load	85½	+2.3 %	+ 6.1 %	87¾	−8.5 %	+ 1.4 %
¾ load	83¾	+2.1 %	+ 7.7 %	86	−1.1 %	+ 1.2 %
½ load	79½	+3.8 %	+11.0 %	83	−2.7 %	....
1,000 Full load	87½	+0.3 %	+ 4.8 %	87¾	−0.3 %	....
¾ load	86	+0.6 %	+ 5.2 %	86	−0.6 %	....
½ load	82¼	+1.7 %	+ 9.4 %	83	−0.3 %	....
2,000 Full load	88¼	+0.6 %	+ 4.5 %	88½	−0.3 %	....
¾ load	85¾	+0.3 %	+ 5.7 %	86½	−0.3 %	....
½ load	82¾	+2.4 %	+ 9.3 %	83	−0.3 %	....
Price per Kw.						
300	$26.17	+0.6 %	− 4.0 %	$25.75	+0.3 %	− 3.0 %
500	24.70	−1.4 %	−11.7 %	23.20	+0.9 %	− 2.0 %
1,000	20.25	−3.0 %	− 0.5 %	19.45	+0.2 %	....
2,000	19.10	−2.0 %	+10.0 %	18.10	+5.0 %	....
Floor Space (Square Feet)						
300	80	Same as syn. mot. gen. set.	+13.7 %	67	Same as syn. mot. gen. set.	+43.0 %
500	122		+ 8.3 %	110		+36.0 %
1,000	136		+25.0 %	140		....
2,000	440		− 8.4 %	435		....
Weights (Pounds)						
300	50,000	−2.0 %	−36.0 %	48,000	−2.0 %	−16.7 %
500	68,000	−4.4 %	−30.0 %	65,000	−4.6 %	−15.4 %
1,000	98,000	−8.1 %	− 5.0 %	92,000	−3.3 %	....
2,000	215,000	−7.0 %	− 2.3 %	212,000	−1.4 %	....

[1] *Electrical World*, Nov. 14, 1908.

than the synchronous-motor set at 25 cycles, but less efficient at 60 cycles. The difference in efficiency is considered in all cases when the cost of fuel for steam generation is an important factor. If water power is used, this item may lose its significance.

With the exception of the 1,500- and 2,000-kw., 25-cycle units, the synchronous converters are less expensive in both frequencies than the motor-generator sets. In the larger sizes the 25-cycle induction-motor sets are less expensive than either of the synchronons machines. The high cost of the 300-kw., 25-cycle induction-motor set is due to the external starting devices, which constitute a relatively large proportion of the cost of the motor. The 60-cycle synchronous motor-generator sets are less expensive in all sizes than the induction machines listed at this frequency. If induction regulators are used with the older types of the converters for regulating the voltage, the advantage in smaller the cost of the converter is almost entirely eliminated.

The additional floor space required for converters with transformers must be considered in sub-stations where floor space is valuable, such as in large cities. If the transformers can be located on a gallery, a greater capacity in synchronous converters than in motor-generator sets can be installed in a sub-station of a given floor space, inasmuch as induction regulators and series boosters are no longer necessary with the split-pole type of synchronous converters.

It is often necessary to increase the output of sub-stations having limited floor space and head room by the installation of larger units. Instances of this kind may be found in sub-stations located in the basement of office buildings in large cities. The conditions here imposed have been successfully met by vertical synchronous converters. In Chicago, for example, the use of this type permitted the output of a sub-station to be increased 50 per cent. over that possible with any other machine available.

The weight of the motor-generator set is, owing to the high speed, slightly less than that of the synchronous converter, with its accessories, the induction motor weighing less in both frequencies than the synchronous-motor set.

Continuity of service can, with careful attendance, be obtained from all three classes of machines. The excellent protection afforded by the use of speed-limiting devices, reverse-current relays, and circuit breakers practically eliminates the possibility of damage due to runaways or overloads.

Synchronous converters equipped with regulating poles can be used to obtain a wide range in voltage without materially increasing the cost over that required to give a lower range. The excitation of a split-pole converter can be controlled by means of an automatic regulator and the direct-current voltage kept constant even with wide variations in the voltage of the alternating-current supply. The regulator may also be adjusted to hold a constant load on the converter and cause storage batteries or other machines to carry fluctuations in the load beyond a predetermined amount. Advantage can be taken of this point when the energy is purchased from a transmission company whose rate of charge is based on maximum demand. In case induction and synchronous motors are used in the same system, the lagging current of the former can be compensated by overcompounding the synchronous motor, so that the resultant current from the central station will be in phase with the voltage. This ability to furnish a leading current to the line and improve the power factor, and hence the regulation and output, of the generating and transmitting equipment constitutes one of the greatest advantages of synchronous over induction machines. The synchronous motor-generator set is therefore best suited for a system carrying a low power-factor load (see Chap. XVII on synchronous condensers) as extra capacity for improving the power factor can be furnished at a small increase in the first cost cost and at the same time the d.c. voltage may be easily and independently regulated by means of the shunt rheostat or by overcompounding the d.c. generator. Motor-generator sets are well suited for lighting purposes. In Europe they are used almost exclusively and are preferred to converters even in d.c. railway service, as European converters, which have a frequency of 50 cycles, possess the same disadvantages as our 60-cycle machines for railway service.

Storage batteries are installed in the majority of lighting sub-stations, and direct-current starting is to be recommended in such cases for all three classes of machines. If d.c. cannot be obtained from such a source, a failure of the a.c. supply will cause a complete shut-down of the sub-station and require the starting of at least one unit from the a.c. side, after which d.c. will be available for starting the remaining units. All the machines described are suited for starting either from d.c. or a.c. side. The resistance in the secondary circuits of the induction machine limits the current at starting to a value considerably less than that required at full load.

Synchronous motors are started by means of starting compensators. Synchronous converters can be started from one-half or one-third and two-thirds taps or the secondary of the transformers.

Twenty-five-cycle converters are better adapted to railway service than motor-generator sets, as they respond more readily to fluctuations in load and possess a greater overload capacity. Those equipped with regulating poles are superior to all types of machines listed in those qualities which ordinarily govern the selection of low-frequency sub-station apparatus, with the exception of cases where precise voltage regulation is necessary.

Sixty-cycle synchronous converters are more efficient than either type of motor-generator set, but afford a relatively small margin for correcting the power factor of the system, which is the reason why synchronous motor-generator sets are more generally used at the higher frequency. Sixty-cycle induction motor-generator sets are the least efficient of all, and as they possess the added disadvantage of lowering the power factor of the line, are seldom recommended for lighting and power sub-stations.

Where unusual conditions of operation are not encountered, considerations of economy, efficiency, and maintenance dictate a selection of apparatus somewhat as follows:

25 cycles	Railway service, 600 volts.........	Synchronous converter.
	Railway service, 1,200 volts........	Synchronous converter.
	Railway service, 1,500 volts........	Synchronous converter.
	Lighting service, 240–300 volts.....	Synchronous converter.
60 cycles	Railway service, 600 volts.........	Converter or motor-generator set
	Railway service, 1,200 volts.......	Motor-generator set or two 600-volt converters in series.
	Railway service, 1,500–1,800 volts..	Motor-generator set.
	Lighting service, 240–300 volts....	Converter or motor-generator set.

In practice motor-generators are used for feeding the two outer wires in three-wire systems, or for feeding all three wires in three-wire systems. In the former case a single generator will be used coupled to a synchronous motor, and in the latter two generators driven by one induction or synchronous motor are required. For railway service there is used one 600-volt generator with an induction motor. Motor-generators are also used as exciter sets, booster sets, balancer sets in three-wire systems, or as charging sets for storage batteries.

Transformer Sub-stations.—(On transformers see also section on "Synchronous Converter Sub-stations.") A system employing high-tension alternating current possesses not only the advantages due to high tension, but also those resulting from the use of alternating current.

In America as well as in Europe alternating-current railway service is essentially single-phase, although on a few inclined railways, such as the Jungfrau, Gornegrat and Engelberg, and the experimental Berlin-Zossen road, three-phase is used. Single-phase is preferred on account of its simplicity and especially since the new types of single-phase motors have done away with the former lack of good motors with which single-phase current could be used.

In the following paragraphs there will be pointed out the various methods employed in practice for generation and distribution of alternating current.

1. Single-phase current is generated and is fed to the overhead lines directly as such or through transformers.

2. Three-phase current may be generated to supply a motor-generator set from which the required single-phase is obtained.

3. The single-phase railway net may be joined through transformers to one leg of the three-phase transmission line.

4. The three legs of the three-phase line are connected to transformers and are made to supply separate portions of the system.

5. The central station may be located at the center of the system and generate two-phase, each leg supplying energy to half of the system.

6. The three-phase line may be connected to three-phase-two-phase transformers which feed separate sections of the line. Each section is supplied from two adjacent sub-stations.

1. In this case it must be remembered that no synchronous converters, self-starting synchronous motors, or induction motors starting under load can be fed from the line. This system is therefore applicable exclusively to railway service and especially for trunk lines using a current at 15 cycles which is subject to no other kind of load.

Single-phase generators are difficult to regulate. They are heavier and more expensive and operate at lower efficiency. The same considerations apply to single-phase transformers. They nevertheless afford simpler generation and distribution, and therefore the switching arrangements are cheaper. When transformers are required, one in each sub-station will suffice. In that case one terminal of the low-tension side of the transformer is grounded and the other is connected to the overhead line.

2. If three-phase is generated, synchronous converters may be used to feed motors and lighting systems. A single-phase railway net supplied from motor-generators does not subject the primary transmission line to unbalanced loads, which is one of the greatest advantages of this system. The set can be used as a frequency changer, with the motor run by a 60-cycle current and

the generator supplying current at 25 cycles. Another advantage is that it can be set up either in the central station or in the sub-station. In the former case both the primary and secondary distributing nets are single-phase, while in the latter the primary is three-phase and the secondary single-phase.

3. Although the system in which a single-phase railway net is fed from one leg of a three-phase system admits of the use of converters and synchronous motors on all three legs of the circuit, it nevertheless causes considerable unbalanced loading, so that the rating of the starting machines is reduced by from 30 to 50 per cent. The generators themselves will have the same disadvantages as the single-phase machines, and moreover will have only two-thirds the rating which they would have in a balanced three-phase system. In this case both the primary and secondary distribution are single-phase.

4. When the length of the overhead line may be divided into three parts or any multiple of three, then the sections are fed singly from the separate legs of the three-phase circuit. Any two adjacent sub-stations will feed the intermediate section from the same phase, so that two transformers in each sub-station will be necessary to care for the service. The three-phase net is balanced subject to the condition that the sections are equally loaded. As this requirement cannot be met, this method of generation and distribution is open to the same objections noted above under 3.

5. In this system, as in the one described above, the two-phase line is balanced only as long as the two sections are loaded equally.

6. In this last system the three-phase is supplied to each sub-station where a three-phase-two-phase transformer bank (consisting of two transformers) is operated. The secondary delivers two-phase current and the phases are separated so that each feeds a portion of the overhead line. As a rule, adjacent sub-stations feed the intermediate section from the same leg, as in case 4. The system is balanced only when the line is correctly subdivided and when the loads in the various sections are equal.

In considering the relative merits of sub-stations, however, it is found that the two systems differ widely.

Consider a single-phase system with sub-stations for stepping down to a lower supply voltage or for supplying single-phase from three-phase e.m.f. Each of the sub-stations must be equipped with one or more transformer banks, which when compared with the equipment for converter sub-stations cost considerably less. The cost of buildings, switching equipment, and attendance is also a great deal less than for the converter stations. Since the voltage in the supply lines for this kind of system is higher than that ordinarily used with direct current, a greater line drop becomes admissible, whence it follows that the sub-stations may be placed farther apart. This naturally results in fewer such stations and in greatly reduced total cost.

In single-phase systems the distance between sub-stations and the cross section of the overhead line for a given service voltage and drop are interdependent. The relation between these quantities should be such that the greatest possible economy is obtained, that is, the least weight of copper with the least number of sub-stations. In this connection the following points should be taken into account:

1. The line drop should not be sufficient to affect the efficiency of the car service or the lighting of the cars.

2. Reserves must be kept on hand in each sub-station for replacing disabled apparatus. The rating of the transformers must be large enough to enable them to carry not only the maximum overload but also the combined normal load of their own section plus that of the next adjacent section (in case of breakdown of an adjacent transformer station).

3. The overhead line should have neither too small nor too large a cross section. The latter specification is to avoid unnecessarily heavy suspension.

4. A smaller number of sub-stations decreases the number of danger points where the connections to the high-tension line are made.

The development in the last years of the *outdoor sub-station* as against the indoor sub-station, with its attendant problems, is an interesting commentary on the adaptability of engineering progress to new fields. Primarily, as in most such cases, the cause was an economic one, the desire to serve from transmission systems the smaller consumers that could not be reached on account of the excessive first cost of installation. Generally, such outdoor sub-stations may be classified as serving the requirements of (1) small isolated plants, (2) large-capacity loads, and (3) apparatus auxiliary to control stations.

As mentioned, it is along the lines of small isolated loads that the development of the outdoor sub-station first started, and this has offered its greatest field to date. Complete outdoor sub-stations up to a few hundred kva. have enabled service to be given to isolated customers and small communities which could not be profitably served directly off the main transmission lines. It is possible to locate such substations almost anywhere, the structural and foundation requirements varying with conditions. The simplicity of design and construction has meant a small investment and a minimum of operating troubles. High-tension switching is usually limited to air-break types, while lightning protection is afforded either by the electrolytic lightning arrester or the crude horn type. Most such installations have been free from elaborate secondary control, that provided being, if any, of the simplest type established in weatherproof housings.

The application of the outdoor sub-station to large-capacity installations, particularly where the secondary control is split up into various circuits with more or less auxiliary apparatus, confronts the constructor with larger problems. The sub-station may combine not only the functions of a switching station, but also that of supplying a large load to a community, and in general its location would not be affected by the type adopted, provided sufficient space were available.

With sub-stations of this character, involving attendance, it is necessary to provide a suitable building designed for at least part of the installation. What part of the high-tension apparatus may properly be installed out of doors depends largely on the saving thereby effected. The actual outdoor apparatus must be provided for separately, and a combination of the indoor and outdoor sub-station will in many cases prove most advantageous. In some cases it is feasible to place the whole outfit out of doors, including the low-tension control equipment, provided the latter is not extensive, but in most cases it is questionable whether the operating features will permit this extreme. In cases where attendance is required it is advisable to install the secondary equipment indoors.

The manner of installing the outdoor equipment will depend upon the size and importance of the layout. In general, the incoming and outgoing lines must have proper terminal structures for connection and distribution to the banks of transformers. The transformers themselves should be provided with proper foundations, preferably of concrete, so designed that they may be removed on rails. The installation of lightning and control equipment is also subject to special requirements.

A most important detail of the outdoor installation is that of providing proper facilities for repairing damaged apparatus and large stations must provide some shelter where such repairs can properly be made. The space required by the outdoor installation will, of course, be large, and adequate provision must be made for preventing anyone from coming in contact with the equipment.

Many central stations operated in connection with transmission systems have a considerable portion of their buildings given up to high-tension equipment and apparatus incidental to its operation. The outdoor installation of such equipment, provided the extra outdoor space is available, would, in many cases, reduce the building cost materially.

The installation of stations of this character is not limited to any particular section of the country, as they are in successful operation in northern regions where subjected to heavy snow and sleet storms and in southern latitudes where intense heat and heavy rainfall are regularly encountered. Investigation of the merit of these installations reveals certain advantages and disadvantages to be met with in such construction as compared with indoor stations.

Advantages: Lower first cost.
- Extension made at less cost.
- Simplicity of layout.
- Cooling of air-blast and oil-cooled transformers found to be more efficient.

Disadvantages: Operating risk and trouble from moisture.
- Danger to public.
- Repairs and handling of apparatus, especially in bad weather.
- Appearance.

Reduction in cost was the cause for the development of this type of construction; it is the reason for its continuance, and will govern its extensive application, when properly interpreted. For the small isolated sub-station of a few hundred kva. the cost will be from 25 to more than 50 per cent. less than that of the corresponding indoor station. For large installations, particularly where a building is provided for secondary control, the percentage of saving is much smaller; the less the cost of the outdoor part of the installation in proportion to the total, the less the saving due to this construction. In connection with the cost reduction of fire risk is important, and outdoor service reduces this to a minimum.

The simplicity of either complete or partial outdoor installation is particularly attractive.

The most serious disadvantages are those of operation and facility of repairs. Experience indicates that operating difficulties are well overcome, and damage to modern apparatus is reasonably rare.

The most important protective apparatus affected by the outdoor installation is the electrolytic lightning arrester. It has been necessary to protect this equipment by housing, but the development of low-temperature electrolyte makes it now feasible to place the entire equipment out of doors, even under the most severe climatic conditions. The winter season, when the severest temperatures are encountered, is also, as a rule, free from lightning disturbances, although it is important to secure continuous service of protective devices of this nature on account of surges occurring on the lines, the necessity being particularly noted where the switching of the high-tension systems is liable to cause serious line disturbances and consequent breakdowns at weak points.

The outdoor transformer is similar to the indoor type except that the cover and leads are designed to shed moisture and provide the necessary higher factor of safety. Both self-cooled and water-cooled transformers are applicable to this service, with due consideration of its requirements. In the former grades of oil suited to the conditions are, of course, necessary, the only serious difficulty being on account of possible freezing. As a rule the load of the transformers will keep the oil sufficiently warm, but it is important that precautions be taken in order to prevent condensation of moisture. If necessary, this is accomplished by small heating coils installed in the tops of the transformers. The freezing of oil itself in idle units or under severe conditions may also be guarded against by the installation of external heating devices under the casings. The freezing of oil may be serious, particularly if transformers are out of service because on energizing them dissipation of heat, with the oil frozen, is liable to be slow, and inasmuch as the temperature rise in large units is about 20° C., the windings might be injured before the oil is liquefied. Little trouble is experienced, however, as ordinarily oil will not freeze at a temperature above $-15°$ C., and special grades solidify at a considerably lower point.

The use of water-cooled transformers involves similar precautions against moisture, with the addition of suitable protection to the water-circulating system, although the water of itself, by its circulation through the transformers, would not freeze except under severest conditions, provided the external connections are properly protected.

There apparently has not been much experience with air-blast transformers out of doors, and it would seem that there might be a field for them as they do not offer the difficulties in giving climatic protection which are encountered with other types of transformers, provided moisture is kept out.

Special outdoor switches of extremely high voltages are operating satisfactorily in severe sleet and snow storms. In general, it may be stated that the use of the outdoor sub-station is not limited by the lack of apparatus suitable for the severe operating conditions.

Miscellaneous **Sub**-stations.—Another type of sub-station contains only storage batteries which are used to feed the lines. A station of this sort is usually located in an annex to a central station or to a converter sub-station, or it may be housed in a building of its own. This depends upon whether or not the battery is to be used as a reserve or equalizer for the converters, or for feeding the line directly, equalizing the load in the transmission line or compensating the voltage loss. Equalizing and compensating tend toward saving in copper and raising the efficiency of the line. Line batteries are connected to the line directly as floating batteries; that is, the charges and discharges are made to depend upon the suddenness of the load fluctuations. If the fluctuations are not sudden enough or if a more sensitive regulation is required, the action of the battery is regulated by a booster located in the central station or in the battery sub-station.

On the electric zone the New York Central & Hudson River R. R. employs two circuit-breaker houses between each two adjacent sub-stations which are fed from the nearest sub-station. They are used to assist in governing the supply for the third rail. Each house contains six circuit breakers, one for each of the connections to the four third rails, one for the auxiliary feeders, and one for a spare in case of failure of the other five. The circuit breakers open automatically on overload, but they can also be opened or closed from the nearest power sub-station. Pilot lamps in these stations indicate whether the breakers are opened or closed.

Lightning-arrester houses are placed where underground feeders change to overhead lines. They contain the necessary lightning protective devices, reactance coils, and disconnecting switches.

CHAPTER XXVII

TYPICAL SUB-STATIONS

Long Island Railroad.—The following description is an abstract of an article by W. N. Smith, published in the *Street Railway Journal:* The high-tension transmission system of the Long

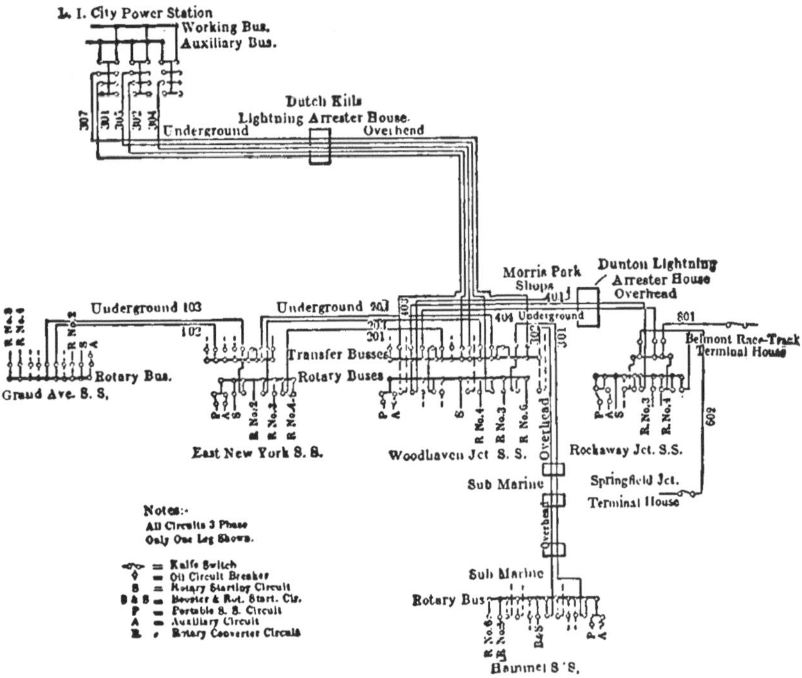

FIG. 277.—Outline diagram of feeder circuits.

Island Railroad is shown above. From the central station in Long Island City five feeders run out in an 18-duct conduit line to Dutch Kills Street, and thence overhead on a line of steel poles to the distributing sub-station at Woodhaven Junction. A branch line of three circuits runs westward from here to East New York sub-station, two circuits running thence to Grand Avenue sub-station, all these being run in underground conduits.

To the east of Woodhaven Junction there are two circuits, run underground to Dunton, where the transmission is changed from underground to overhead, continuing eastward overhead on steel poles to Rockaway Junction sub-station. The branch circuits from Rockaway Junction to the portable sub-station terminal buildings at Belmont Park and Springfield Junction are carried overhead on wooden poles. An additional feeder runs out underground from Woodhaven Junction in the same direction to the repair shop at Morris Park. Southward from Woodhaven two circuits are carried overhead to Hammel sub-station.

The following table gives the present and ultimate equipment of the above-mentioned stations:[1]

Station	Rotary converters, kw.	Transformers, kw.	Boosters, kw.	A.C. feeders	D.C. feeders
Grand Avenue:					
Present installation.....	3—1,000	9—375	......	2	5
Ultimate capacity......	5—1,500	15—550	......	4	11
East New York:					
Present installation.....	3—1,000	9—375	......	5	6
Ultimate capacity......	4—1,500	12—550	......	12	16
Woodhaven Junction:					
Present installation.....	3—1,500	9—550	......	12	10
Ultimate capacity......	6—1,500	18—550	......	18	18
Rockaway Junction:					
Present installation.....	2—1,000	6—375	......	4	6
Ultimate capacity......	4—1,500	12—550	......	11	16
Hammel:					
Present installation.....	2—1,000	6—375	2—162	2	6
Ultimate capacity......	5—1,500	15—550	2—162	5	13
Valley Stream...........	....	....	......	..	..
Two portable sub-stations:					
Each equipped........	1—1,000	3—375	......	1	1

The converters used in the sub-stations are of the Westinghouse type, each provided with a starting motor mounted on an extension of the base of the converter. The 1,000-kw. converters are rated to deliver 1,600 amp. at 625 volts and 1,667 amp. at 600

[1] "Long Island R. R. Sub-stations," *Street Railway Journal*, June 23, 1906.

volts. The three-phase e.m.f. at the alternating end is approximately 370 volts for 625 at the direct-current end. These machines possess eight poles and operate at 375 rev. per min., corresponding to a frequency of 25 cycles per second. The 1,500-kw. converters are rated to deliver 2,400 amp. at 625 volts or 2,500 amp. at 600 volts. They have twelve poles and run at 250 rev. per min. Both types have compound field windings with the shunt winding arranged for self-excitation.

The transformers used for the converters are of the air-blast type. Those for the 1,000-kw. converters are grouped in banks of three 375-kw. transformers to one converter. For the 1,500-kw. machines they are in groups of three, 550 kw. each. The high-tension winding is designed for a normal e.m.f. of 12,000 volts, with taps arranged to enable other voltages to be utilized down to 10,000 volts. The low-tension winding is designed to carry 400 volts normally, with taps which will enable other voltages to be taken off down to 340 volts. The high-tension terminals are at the top of the transformer, and the low-tension at the bottom.

In each station there are four sets of auxiliary transformers which supply energy for the following purposes: (1) To the converter-starting motors. (2) To the motors driving the booster-generators and their exciters. At Hammel station these transformers are made large enough also to drive converter-starting motors at the same time. (3) For driving the transformer blower motors and an induction motor-generator set used to charge the small auxiliary storage battery that supplies energy for the electric switch control system. (4) For house lighting.

At sub-stations 1, 2, 3 and 4, where there are no storage batteries, a group of three transformers is employed to furnish energy for the starting motors of the converters. These are rated at 50 kva. each, the bank being able to start up and synchronize three 1,500-kw. converters simultaneously. They are of the oil-insulated, self-cooling type and reduce the three-phase, 25-cycle current from 12,000 volts to 400 volts. They are placed in a row on the main floor and are connected to the main bus by an automatic oil switch electrically operated from the main control stand.

At sub-station 5, where a storage battery is employed that involves the use of two 162-kw. booster-generators, each driven by a 235-hp. induction motor, there is provided a bank of three

200-kw. air-blast transformers. These are sufficiently large not only for operating the battery booster under maximum conditions of load but also for simultaneously starting one converter without dropping the secondary voltage of the transformers sufficiently to affect the booster regulation. These transformers are set on the main floor over the air ducts for the main transformers and in line with the latter.

The seven 0.5-kva. transformers are provided for the blower motors and the induction motor which operates a small booster-generator used for charging the auxiliary storage battery that furnishes energy for the electrically operated switch-control system. At Woodhaven Junction they are of 10 kw. rating. These transformers are of the oil-insulated type, and the transformation is from 12,000 volts to 400 volts.

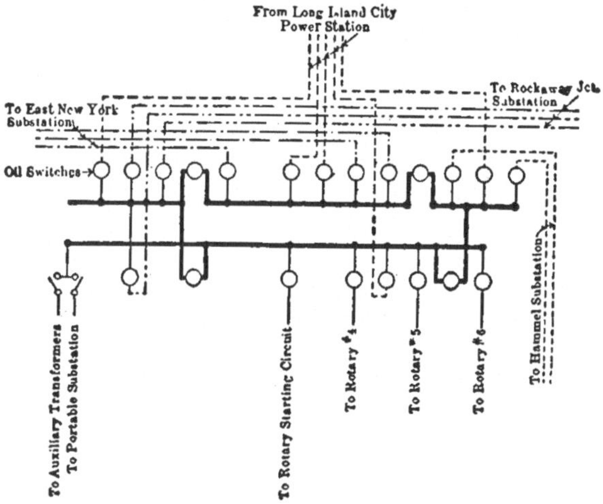

FIG. 278.—Arrangement of connections to high-tension buses at Woodhaven station.

The Woodhaven Junction station and those in East New York and Rockaway Junction distribute alternating current to feeders supplying the outlying sub-stations near the terminals of the railway system. For this purpose they are equipped with two sets of buses, one called the transfer bus and the other the working bus, the former giving flexibility in shifting feeders among all the sub-stations. The converter or working bus in each station receives energy directly through feeders and independently of the transfer bus in that particular station. This enables high-

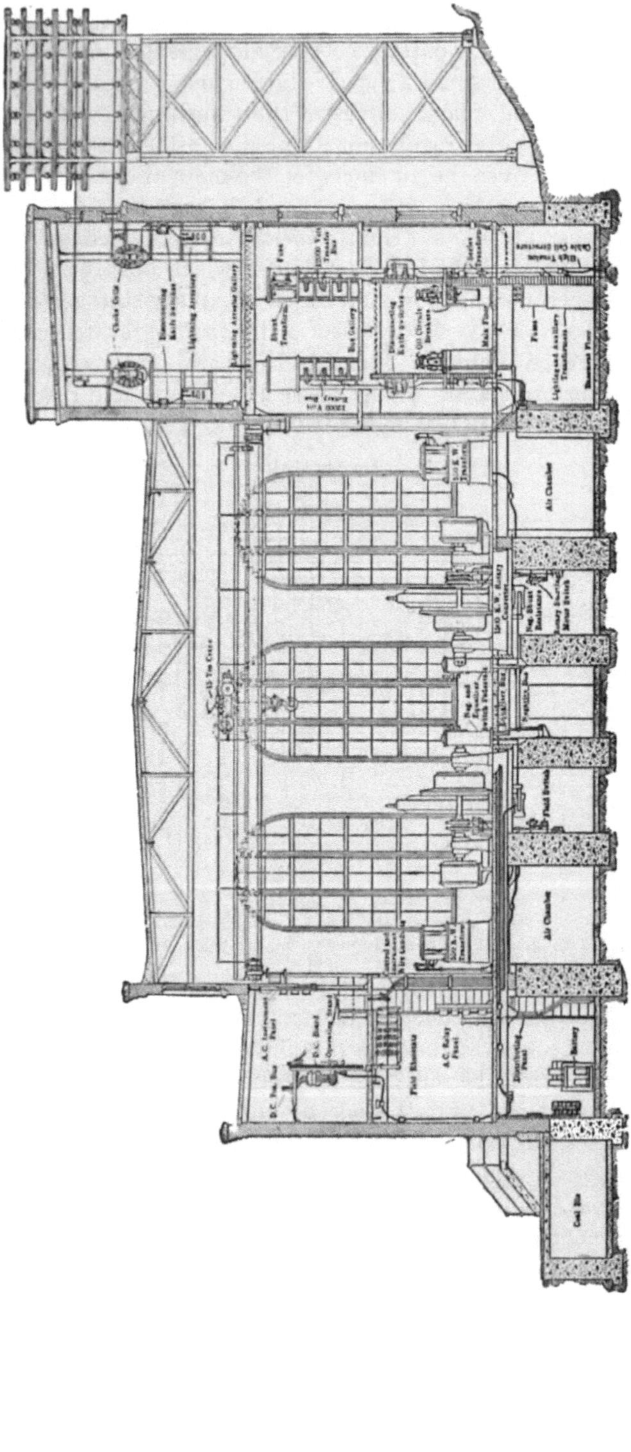

FIG. 279.—Elevation of Woodhaven Junction sub-station, Long Island Railroad.

tension energy to be passed through an intermediate sub-station to one or more beyond, independent of operation of apparatus in the former.

Figure 278 illustrates the high-tension connections at Woodhaven sub-station and shows how the three outgoing branches of the feeder line each receive energy through a separate section of the transfer bus in that station, each section having an independent feeder on the main trunk line coming from the power station. The transfer bus is sectionalized by non-automatic oil switches so that all branches can be run separately or together as desired. By means of tie switches the converter bus can be coupled to either of the three sections of the transfer bus. The other two intermediate sub-stations have a similar equipment, but it is somewhat simpler, as less apparatus is required. In the ultimate installation it is planned to have main feeders run directly from the Long *I*sland *C*ity power station to each of these three principal sub-stations. The feeders now connecting to these transfer buses will then be available as relays.

At sub-stations 1 and 5 the only bus needed is that required for the operation of the converters.

The plan and cross section of the Woodhaven station are shown in Figs. 279 and 280. A side track enters the station and enough space is provided to permit the entrance of a portable sub-station which can be coupled with the apparatus at the station if required. The feeders from the central station run to a tower-like structure in one end of the station where the lightning arresters, reactance coils, and disconnecting switches are located. The arresters are of the Westinghouse low-equivalent type mounted on marble panels which are carried on steel angle-iron framework. The three arresters on the three legs of the high-tension circuits are separated by barriers of asbestos lumber. The arresters are all provided with knife switches so that they can readily be disconnected. There is a reactance coil in series with each main circuit mounted near the top of the steel framework. The arresters are mounted on special porcelain insulators, and the use of wood is entirely dispensed with in the lightning-arrester gallery, thus insuring fireproof construction. The openings in the side through which the cables enter are 18 in. square, enclosed by two glass plates 0.375 in. thick and separated 5 in., having 2.5-in. holes in the centers through which the cables and feeders pass, without touching the glass. Access of rain or

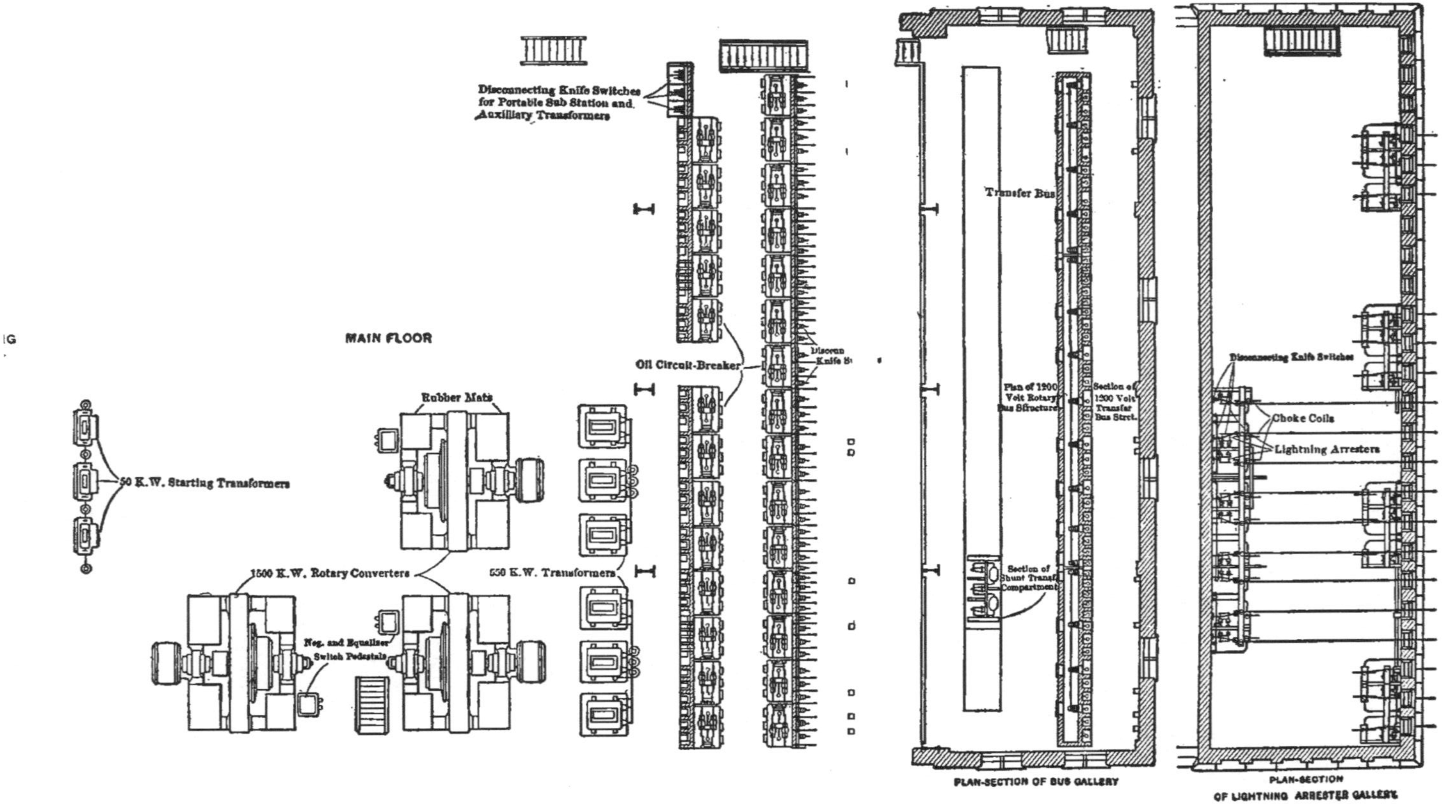

Fig. 280.—Plan of Woodhaven Junction sub-station, Long Island Railroad.

(*Facing page 444.*)

snow through the openings is prevented by a thin brass disk, about 2.5 in. in diameter which is fastened upon each cable between the two glass plates. Standard straight-line insulators are used for supporting the bare wire inside of the building.

From the gallery the cables are led down the main wall to the basement, whence they are run to the respective oil switches on the machine-room floor. These switches are set up in two rows, that nearer the outside wall containing all of those belonging to the distribution system for the other sub-stations, and the other containing the switches for station use. From the feeder oil switches cables are led to the working and transfer buses on the bus gallery. From the latter they run through disconnecting and oil switches to end bells for the underground cables in the basement or to the reactance coils and lightning arresters in the tower for the overhead lines. From the working buses, on the other hand, the cables pass through disconnecting and oil switches to the high-tension delta of the transformers. The positive cables of the machine are led to the positive busbars on the d.c. board on the operating gallery, whence the feeding of the third rail is controlled. The outgoing d.c. feeders are taken out underground in tile ducts. In Fig. 279 there are indicated the mounting of the negative and equalizer buses, the field switch, and the shunt resistance in the foundations of the machines, and also the pedestals with the negative and equalizer switches on the machine-room floor near the machines. On the operating gallery are set the switchboards controlling the a.c. feeders, the bench and instrument boards for the converters, and, as mentioned above, the panels for the d.c. side of the machine and for the d.c. feeders.

The a.c. control apparatus consists of two groups, one taking care of the oil switches for the incoming feeders from power station to transfer bus, for the bus section switches of this set of buses, and for the outgoing high-tension feeders from the transfer bus to the sub-stations. The second group controls the feeder from the power house to the working or converter bus, the switch connecting the transfer and converter buses, and the switches joining the main transformers for the converters with the working bus. The first group has its controlling apparatus on a switchboard consisting of three panels, each of which has provision for mounting six 500-amp. a.c. ammeters, and eight controllers for the oil switches, with eight pairs of signal lamps. The second group is mounted on a separate bank of control benches with instrument panels, which are set up so as not to obstruct the operator's view over the station. Going from left to right, the designation of the control and instrument panels on the bench and overhead framework is as follows:

1. Two converter bus-connecting switches. These connect the converter bus with the transfer bus.
2. Alternating-current feeder direct from power station to converter bus.
3. Blank panel reserved for booster in case of storage-battery installation.
4. Switches connecting the converter bus with transformers supplying converter-starting motors (and booster motors when installed).
5. Blank panel for future converter.

6, 7, 8. Panels for converters installed.

The oil switches for manipulating the 2,000-volt current are three-pole type *C* Westinghouse. Their normal carrying capacity is 600 amp., but they can handle a short-circuit of a maximum kilovolt-ampere rating equivalent to a generator rating of 33,000 kw. All oil switches are automatic except those used for connecting the sections of the transfer bus. The control apparatus by means of which the electrically operated oil switches are worked from the control stand consists of a circuit supplied by a storage battery whose current is conveyed to the two closing coils and one tripping coil of each oil switch. The control circuits are closed either by a controlling switch on the bench, or if control is automatic, by the time-limit relay, which is actuated from a series transformer in each high-tension circuit. The panel with the relays is on the machine-room floor under the operating gallery.

The high-tension busbars each consist of three sets of bars of rolled copper mounted on porcelain pillars and carried in closed compartments placed one above the other in a structure of yellow pressed brick, with alberene stone slabs separating the three tiers. The holes by which the taps enter and leave the compartments are made through alberene stone slabs bushed with heavy porcelain insulating bushings. Where the busbars are sectionalized the compartments in the same tier are completely divided off by stone slabs. The shunt transformers are in separate closed compartments on top of the bus structure. On the back of each of these structures is built a set of vertical septums to separate the cables that enter and leave the structure to tap the busbars. The septums are continuous with those in the upward extension of the back walls of the oil-switch structure.

The air-cooled transformers are set up over air chambers in two rows on both sides of the machine room. Two ventilators supply the chambers, operated by 9.8-hp. motors. The ventilators run at 480 rev. per min. and deliver 18,000 cu. ft. of air per minute at 70° F. against a maintained pressure of 1 oz. per square inch. All disconnecting switches are separated from each side by asbestos barriers. The series transformers for the instruments and relays are mounted in compartments in back of the oil switches.

The arrangements in the other stations are similar to the one described above, but are somewhat simpler, since there are fewer machines and instruments and since no lightning arresters are necessary with the underground lines.

Hammel station differs from the others in that it is provided with a storage battery which is used as a regulator for the converters, though provision is made in the design of the other sub-stations for their ultimate installation. The battery itself comprises 300 elements of the Electric Storage Battery Company's chloride accumulator, each element containing 55 type R plates in regular service. At the temperature of 70° F. they have the following ratings:

Rate, amperes	Time, hours	Rating, ampere-hours
700	8	5,600
1,000	5	5,000
1,500	3	4,500
3,200	1	3,200

The normal rating of the battery is on the basis of one hour, for which time it can be discharged at the rate of 3,200 amp. In case of necessity, however, the battery can discharge at the rate of 6,400 amp. for twenty minutes. For instantaneous fluctuations it can discharge up to a momentary maximum of 9,600 amp.

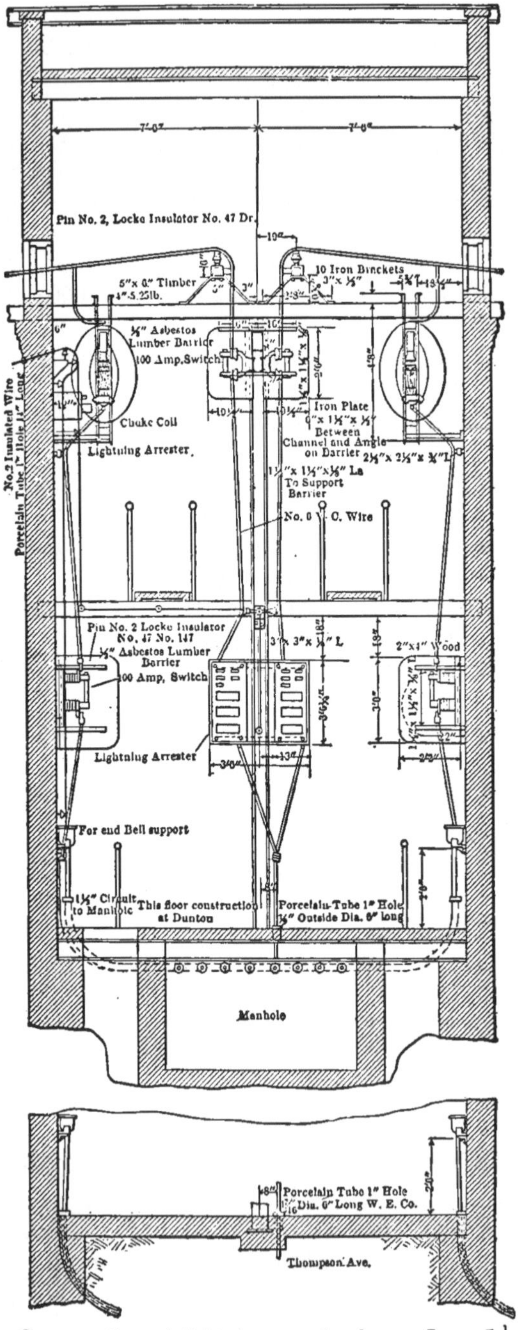

FIG. 281.—Cross section of lightning arrester house, Long Island Railroad.

For charging and discharging of the battery, and hence its proper maintenance as a regulator to maintain a comparatively steady load on the converters, there are used two direct-connected, motor-driven, separately excited boosters. Each consists of one three-phase induction motor for 25 cycles at 400 volts rated at 235 hp., and of a booster-generator rated to deliver 1,200 amp. at 135 volts. The overload capacity of the latter is 1,600 amp. for one hour and 3,200 for two minutes. The transformer equipment which supplies the motor consists of three 200-kva. air-blast transformers. The field of the booster is excited by a small booster exciter-generator driven by a three-phase, 400-volt, 25-cycle induction motor. The strength of the field and the polarity of the booster-exciter field coils are regulated by a carbon regulator manufactured by the Electric Storage Battery Company. By the aid of this regulator the polarity and field strength of the exciter change instantaneously with the fluctuations in the main circuit, which in turn produces changes in the excitation of the booster. This latter regulates charging and discharging of the battery. In this way the fluctuations of the load on the converters may be adjusted within a wide range (from 5 per cent. to 50 per cent.)

Whenever an underground feeder changes to an overhead line, lightning-arrester houses are installed. Figure 277 shows the location of two of these houses, one at Dutch Kills Street and the other at Dunton, and three others on the line going to Hammel sub-station. In Fig. 281 the cross section of the Dutch Kills lightning-arrester house is shown. It is a brick structure and contains room for eight outgoing overhead circuits which leave the house four on a side. The arrester house is 33.15 ft. in length, 17.5 ft. wide, and 30.5 ft. high inside. The steel beams supporting the apparatus extend to the outside of the building, forming a series of racks for the support of transmission cables which are dead-ended upon them. The arresters are all provided with knife switches. Resistance coils are built in, in series with each main circuit, and another knife switch is between the coil and the cable bell. The disposition of the material is such as to economize space and at the same time make each circuit capable of ready access without incurring risk from other apparatus. The incoming cables are carried through the floors by means of ducts reaching to the last manhole in the conduit line and are arranged along the wall running through the switches and through the reactance coils to the various outlets. The arresters are mounted on either side of the steel framework in the center of the building, and the ground connections all run to a single ground lead consisting of a 5.5.-sq. ft. copper plate buried in the ground between layers of crushed coke. The outlets and arresters are the same as in the station previously described.

Coney Island and Brooklyn Railroad Company Sub-station.—Figures 282, 283, 284, 285, and 286 show the arrangements designed by the author for the sub-station in De Kalb Avenue.

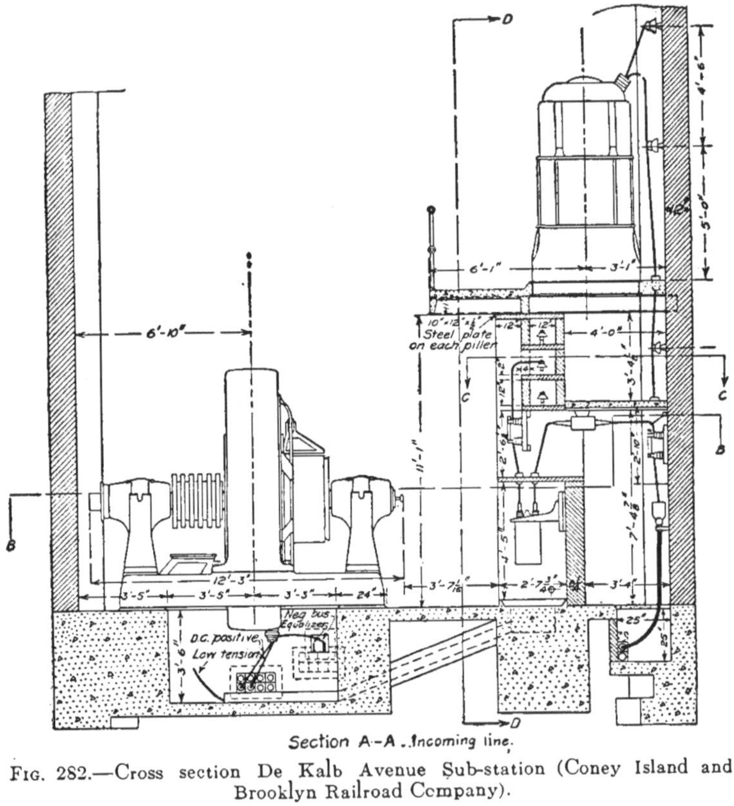

FIG. 282.—Cross section De Kalb Avenue Sub-station (Coney Island and Brooklyn Railroad Company).

The building is typical for city service where real estate is expensive. It is built on a lot 25 ft. wide and is at present equipped with two 1,000-kw. converters. If necessary, the rear of the building can be extended so as to accommodate two additional units. The narrowness of the building necessitates a

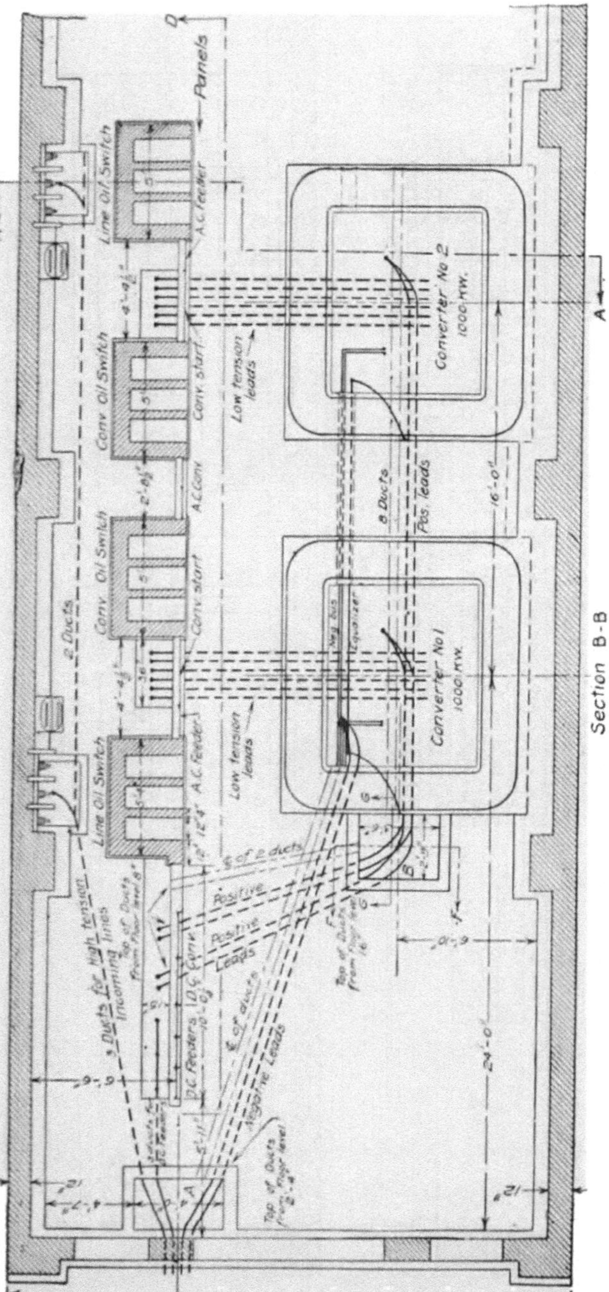

FIG. 283.—Plan of De Kalb Avenue Sub-station (Coney Island and Brooklyn Railroad Company).

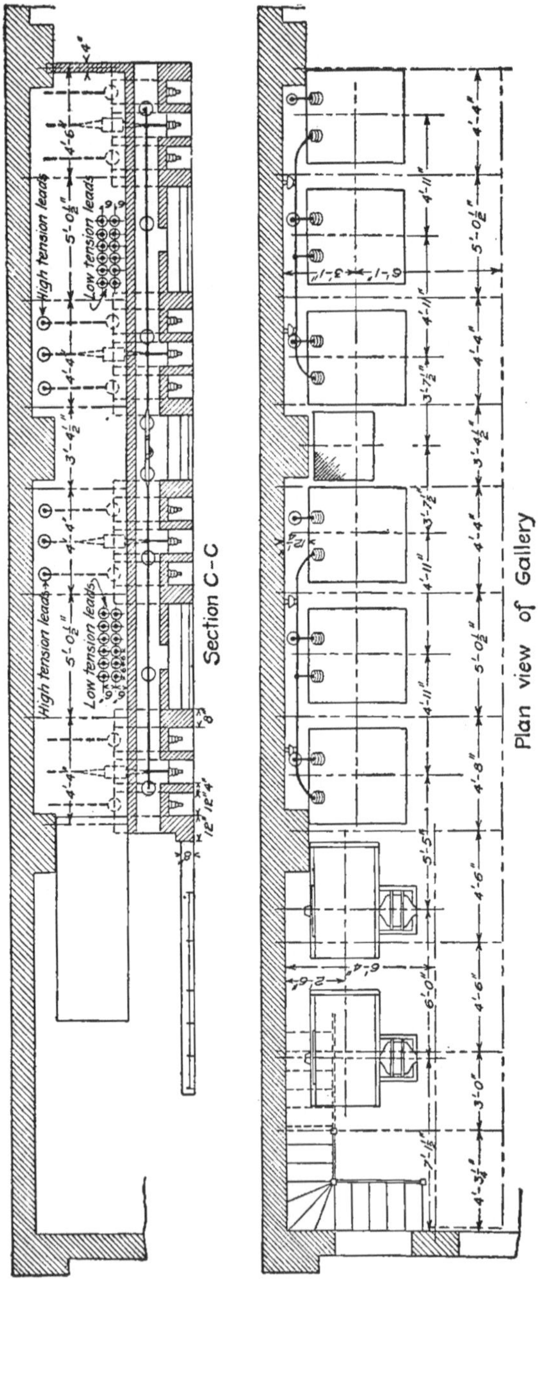

FIG. 284.—Plan of bus structure and of gallery floor of De Kalb Avenue sub-station.

special gallery arrangement for the apparatus. Four three-pole oil switches (two for incoming line and two for the machines) are set up in separate cell structures, the end walls of which are continued upward so as to carry the gallery with the air chamber and busbar compartments. The inner partitions of the structures also run up to the gallery, but serve only as barriers for the separation of the busbar connections. The high-tension three-

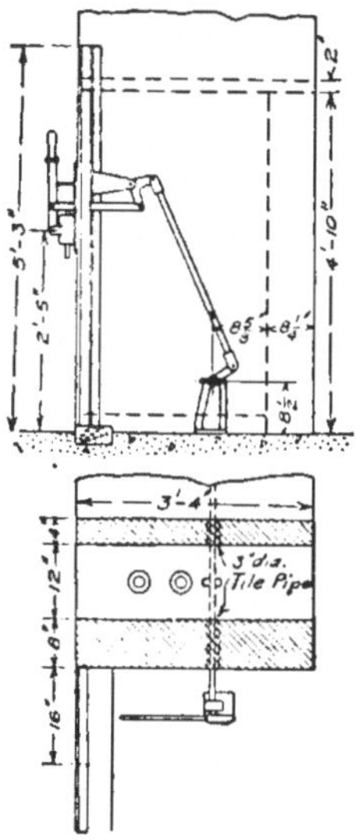

FIG. 285.—Detail of oil-switch pipe mechanism.

conductor cables are led in ducts to the wall back of the oil switches, where they diverge and are joined to the disconnecting switches. From here they run on to the ceiling of the air chamber and to the oil switches. All oil switches are type K4 (G. E. Co.), three single-pole making one triple-pole. They are operated by means of a pipe mechanism and shafting. Each oil-switch set has its own operating board placed at its side and the operating shaft passes through openings in the cell walls. From the oil switches the cables are led through disconnecting switches to the 11,000-volt buses, and from here they pass through another series of disconnecting and oil switches up to the air chamber and the transformers on the gallery. The transformers are air-cooled, 11,000- to 430-volt and 375-kw. rating each. Their high-tension side is star-connected.

The low-tension cables lead from the transformers to the starting panels which are located opposite each machine, whence they run in ducts through the machine foundations to the converters. The positive cables and the negative and equalizer buses are contained in ducts and openings in the converter foundations, the former running from the manhole *B* to their d.c. panels, while the negative cables lead from the same machine, side to the main manhole *A*, to which the outgoing d.c. feeders are also led. On the gallery there are, besides the transformers, two blower sets, each driven by a 4-hp. induction motor at 350 volts and 750 rev. per min. Note the concrete blocks under the individual cells which serve as foundations for the walls supporting the gallery. All a.c. panels are set up between the oil-switch cells. The d.c. board, on the other hand, is in line with the cells at the end of the row. The busbars are sectionalized, and the bus compartment has rather large openings at the section points for operating the disconnecting switches.

Since all machines and apparatus are made in standard forms, it is often found convenient to standardize their arrangement in the station. This makes each station a concrete unit which may be dealt with as such. The equipment for a traction system sub-station might thus be given by a specification formula such, for instance, as follows:

> One M-kw. type Q converter, 25-cycle.
> Three $\frac{M}{3}$-kw. type R transformers.
> One high-tension panelboard, type S.
> One low-tension panelboard, type T.
> One lightning arrester, three-phase, type W
> Oil switches and cables.
> Make up in brick compartments.
> Take one station for each M miles of track.

Early sub-stations concentrated all switching equipment at one point to avoid running about, and large sub-stations are still laid out in the same way, but modern small sub-station practice tends toward the distribution of the switchboard apparatus to the points most convenient for the connections, since the few steps from one switch to another are probably more beneficial than harmful, in that they give a hasty or nervous man a second or so to think between operations.

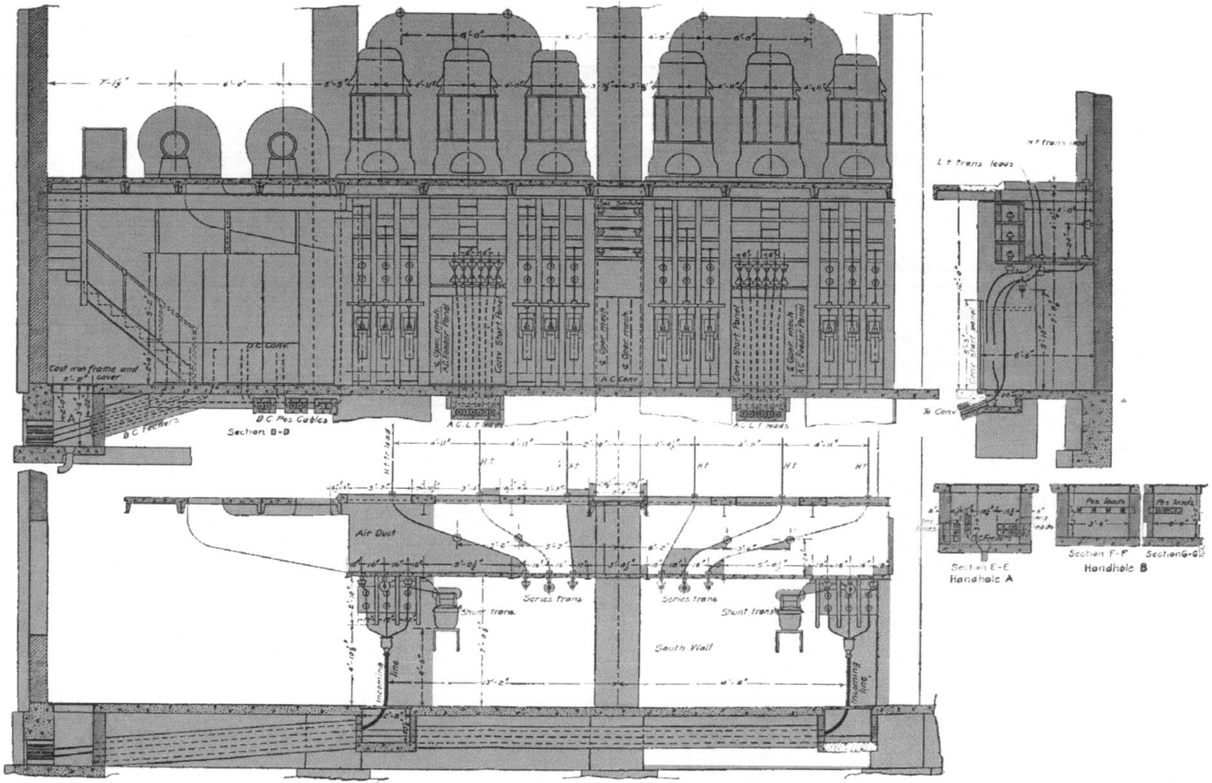

FIG. 286.—Elevation of galleries of De Kalb Avenue sub-station.

(Facing page 454)

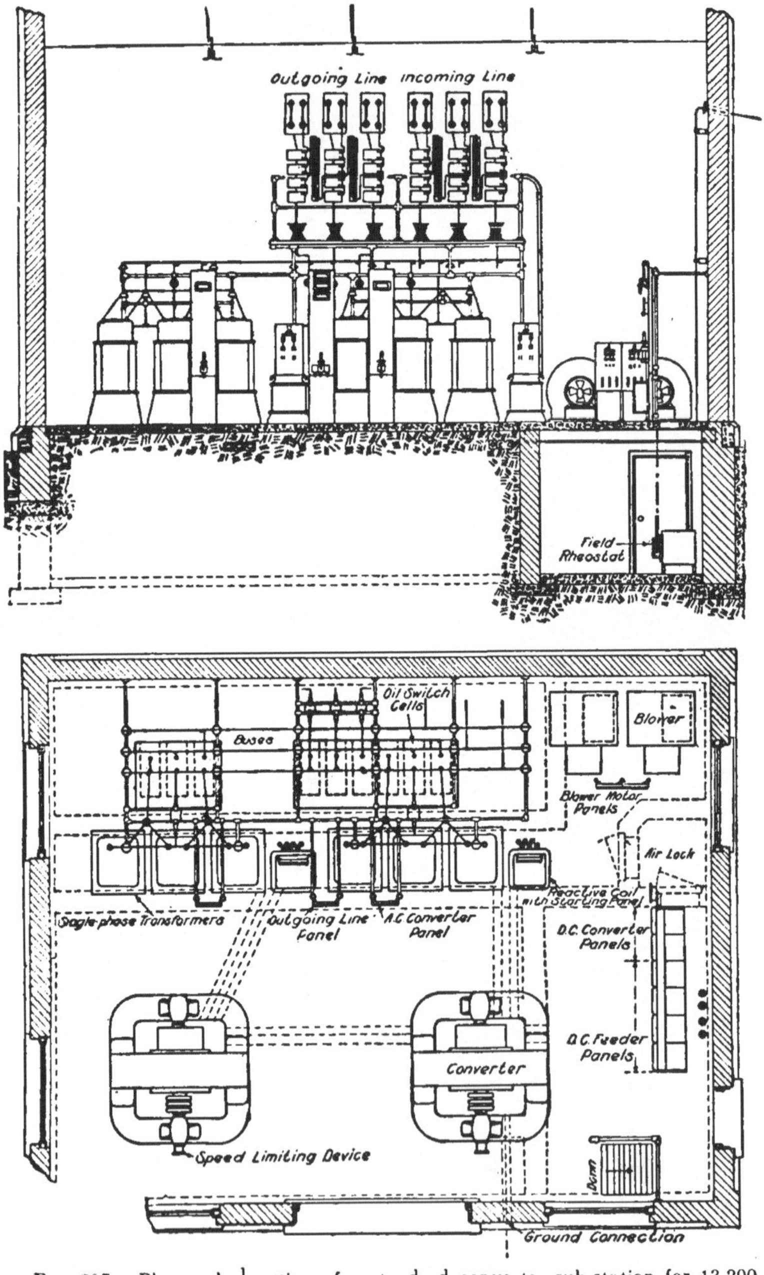

FIG. 287.—Plan and elevation of a standard converter sub-station for 13,200 volts, with 300-kw. machines and single-phase air-blast transformer.

In the following figures, 287 to 294, there are shown a number of arrangements for traction system sub-stations for various high-tension values and kilowatt ratings of the machines, and for different types of transformers and oil switches. These plans were prepared by the General Electric Company and are intended to give the normal arrangements for the particular kind of service for which they are designed. It is assumed in all cases that there is sufficient floor space available.

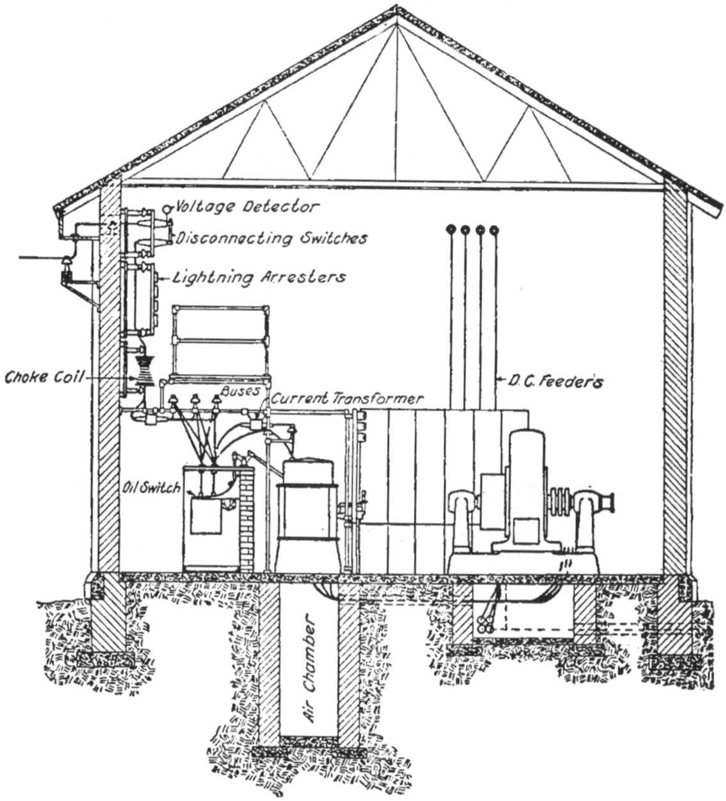

FIG. 288.—Cross section of station shown in Fig. 287.

In most of these stations the high-tension panels are located in front of the transformers and opposite their respective converters, the low-tension a.c. starting switches are just alongside, on top of the reactive coils, and the field break-up and equalizer switches (those next used in the normal course of starting) are mounted

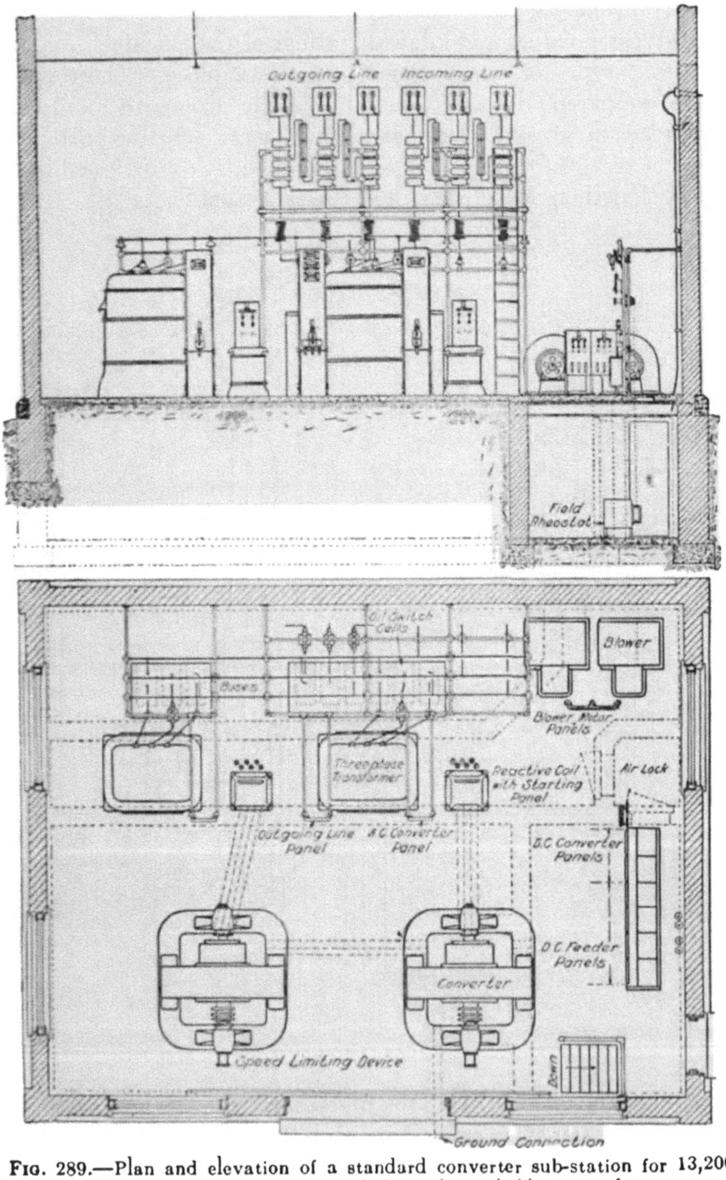

Fig. 289.—Plan and elevation of a standard converter sub-station for 13,200 volts, with 300-kw. machines and three-phase air-blast transformer.

on the machine frames just across the operating passageway. Care should be taken to get the field break-up, equalizing, and series shunt switches on the side of the converter frame toward

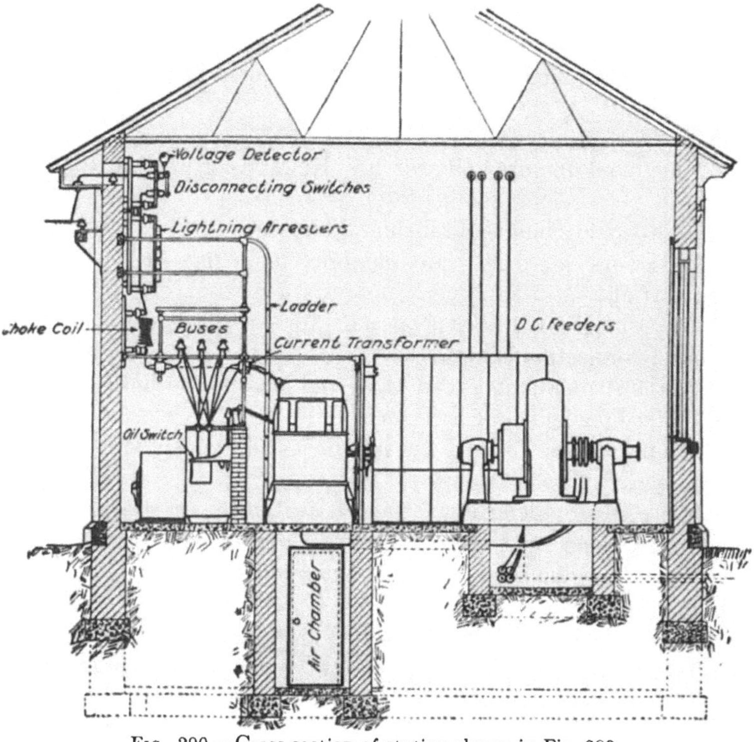

FIG. 290.—Cross section of station shown in Fig. 289.

the switchboard, so that the position of the equalizer switch, etc., can be seen when paralleling machines. The d.c. voltmeter should also be visible from the position of any equalizer switch. The blowers are located at the switchboard end, so that if shut down by the shutting down of one transformer bank, the absence of their hum will be noticeable.

Figures 287 and 288. Installed at present:

Two 300-kw., three-phase synchronous converters, 25-cycle.

Six 110-kw., single-phase air-blast transformers, $^{13,200}/_{2,300}$ volts, Δ connected.

Two blower sets (motor-operated).

Two reactance coils with starting panels.

Three alternating-current high-tension panels: Two machines, one outgoing feeder.

Two alternating-current low-tension panels: Two induction motors for blowers.

Six direct-current low-tension: Two machines, four feeders.

Two lightning arresters: Three-phase for Δ circuit, multigap, multiplex.

Six reactance coils.

Nine hand-operated oil switches, K2 or K4 type, S. P. S. T., in cells.

High-tension buses and insulator supports.

Brickwork for cells, air chamber, machine and basement foundation.

Tile ducts for low-tension a.c., for equalizer, negative, and ground connection cables.

Wall outlets for overhead high- and low-tension lines.

Gallery for lightning arresters.

Frame support for high-tension buses and panels.

Cables.

Space is provided for another oil switch set for a second outgoing feeder or for eventual use of oil switches for the incoming feeder. The building can be extended to the left without interrupting the service.

Figures 289 and 290. At present installed:

Two three-phase, 25-cycle, 300-kw. synchronous converters.

Two 330-kw., three-phase air-blast transformers, $^{13,200}/_{370}$ volts, Δ connected.

Two blower sets (motor-operated).

Two reactance coils with starting panel.

Three alternating-current high-tension panels: Two machines, one outgoing feeder.

Two alternating-current low-tension panels: Two induction motors for blowers.

Six direct-current panels: Two machines, four feeders.

Two lightning arresters: Three-phase for Δ circuits, multigap, multiplex.

Six reactance coils.

Nine hand-operated oil switches, K2 and K4 type, S. P. S. T., in cells.

High-tension buses and insulator supports.

Brickwork for cells, air chambers, machine foundations, and basement.

Tile ducts for low-tension alternating current, for equalizer, negative, and ground connection cables.

Wall outlets for overhead high-tension and low-tension lines.

Gallery for lightning arresters.

Frame support for high-tension buses and panels.

Cables.

Space is provided for another oil switch set for a second outgoing feeder or for eventual use of oil switches for the incoming feeder. The building can be extended to the left without interruption of service.

For a similar station with oil-cooled transformers the equipment is as follows:

Two three-phase, 25-cycle, 300-kw. synchronous converters.

Six single-phase, 110-kw. oil-cooled transformers, $^{13,200}/_{2,300}$ volts, Δ connected,

Two reactance coils with starting panel.

Three alternating-current high-tension panels: Two machines, one feeder.

Six direct-current low-tension panels: Two machines, four feeders.

Lightning arresters: Three-phase for Δ circuit, multigap, multiplex.

Six reactance coils.

Nine hand-operated oil switches, K2 or K4 type, S. P. S. T., in cells.

High-tension buses and insulator supports.

Brickwork for cells and machine foundations.

Tile ducts for low-tension direct-current equalizer, negative, and ground-connection cables.

Wall outlets for overhead high-tension and low-tension lines.

Gallery for lightning arresters.

Frame support for high-tension buses and panels.

Cables.

Figures 291 and 292.

Three six-phase, 25-cycle, 500-kw. synchronous converters.

Nine single-phase, 185-kw. air-blast transformers, $^{33,000}/_{2,300}$ volts, Δ connected.

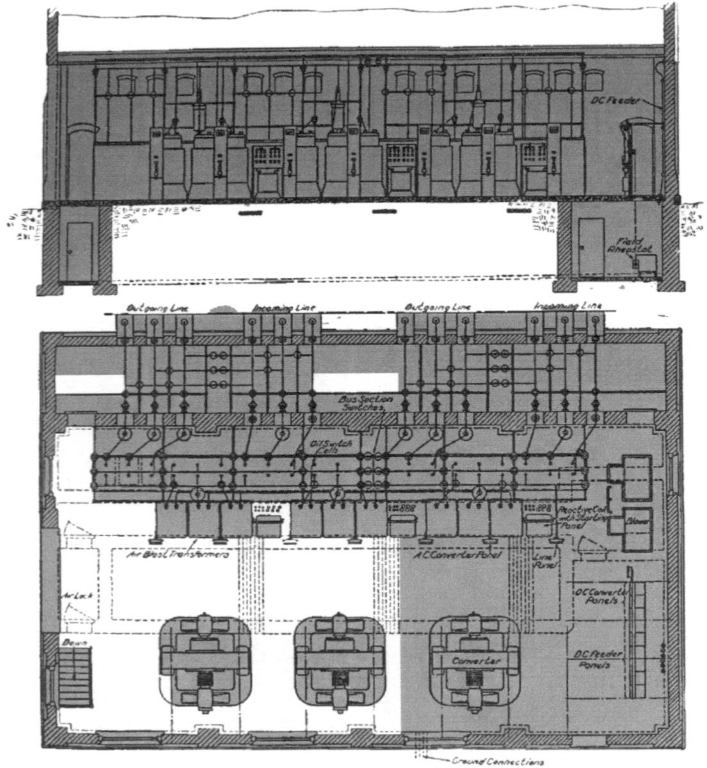

Fig. 291—Plan and elevation of a standard converter sub-station for 33,000 volts, with 500-kw. machines and single-phase air-blast transformers.

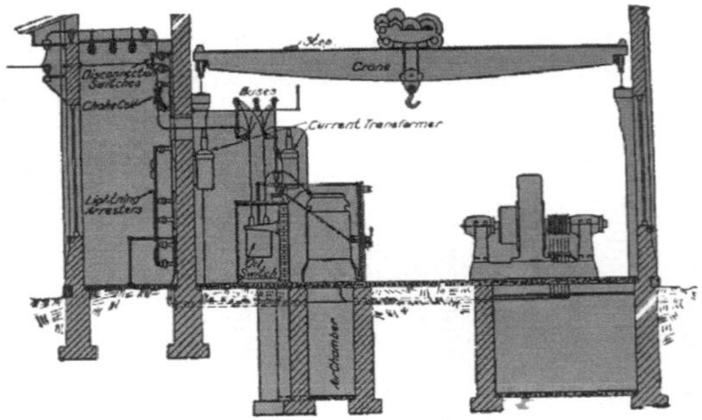

Fig. 292.—Cross section of station shown in Fig. 291.

(Facing page 461.)

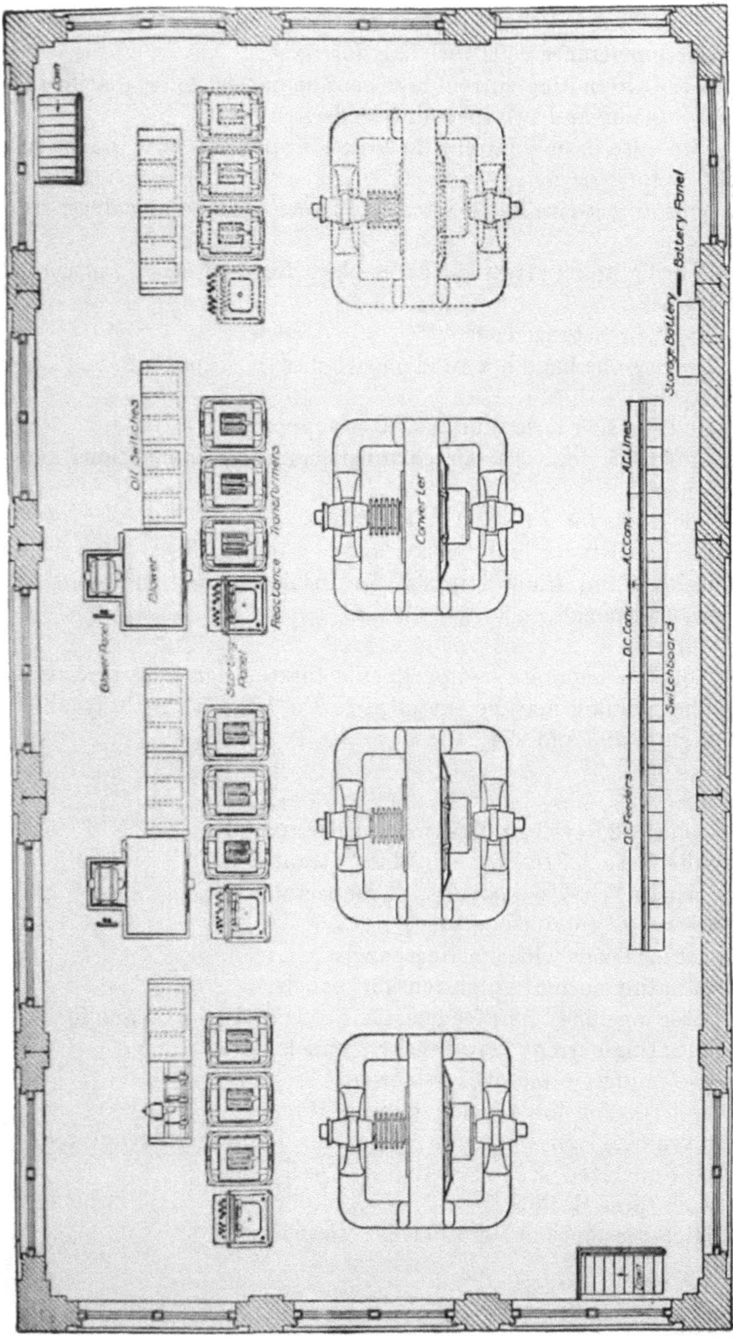

FIG. 293.—Plan of a standard converter sub-station for 13,200 volts with 1,000-kw. machines and single-phase air-blast transformers.

Two blower sets (motor-operated).

Three reactance coils with starting panels.

Seven alternating-current high-tension panels: Three machines, two outgoing and two incoming feeders.

Two alternating-current low-tension panels: Two induction motors for blowers.

Nine direct-current low-tension panels: Three machines, six feeders.

Four lightning arresters: Three-phase for △ circuits, multigap, multiplex.

Twelve reactance coils.

Twenty-one hand-operated oil switches, K6 type, S. P. S. T., in cells.

High-tension buses and insulator supports.

Brickwork for cells, air chambers, machine foundations and basement.

Tile ducts for low-tension a.c. cables.

Wall outlets for high-tension and low-tension lines.

Railing and frame support for insulators and high-tension buses and panels.

Cables.

Note the separate compartments for the lightning arresters.

The building may be extended to the left with little trouble.

Figures 293 and 294.

	Present Equipment	Room for Future Addition
Six-phase, 25-cycle, 1,000-kw. converters..	3	1
Single-phase, 375-kw. air-blast transformers, $^{13,200}/_{2,300}$ volts, △ connected.	9	3
Blower sets (motor-operated)............	2	1
Reactance coils with starting panels......	3	1
Alternating-current high-tension panels, three machines, four feeders............	7 and 1 blank for	1
Alternating-current low-tension panels, two induction motors for blowers......	2	1
Direct-current low-tension panels, three machines, twelve feeders...............	15 and 1 blank for	1
Motor-operated H3 oil switches in cells, each three S. P. S. T..................	7	

High-tension bus and insular supports.

Static dischargers for the underground transmission line.

Brickwork for cells, compartments, air chambers, machine foundations, and basement.

Tile ducts for high-tension and low-tension feeders.

Frame supports for switchboard and insulators.

Cables.

Storage battery for motors on H3 switch (with panels).

Space and foundations are provided for installing a fourth converter with accessories. The building may be extended in either direction.

Portable Sub-stations.—One of the chief reasons why portable sub-stations have won favor with some railroads, besides their regular use in handling centralized loads in time of special traffic, is that of utility during construction and extension periods before the time of completion of the regular station. In building a new line or an extension it is often very difficult to determine in advance the most advantageous locations for the sub-stations. A portable sub-station can be tried at several points and under different conditions of load distribution, and thus the question may be settled by actual data instead of by calculations which may not be very accurate. The same takes place when soliciting a customer for industrial load from a high-tension circuit; it is difficult to estimate the power requirements. Again, the normal load on a sub-station can be carried by one rotary converter unit, and frequently no reserve equipment is provided in the sub-station for use in case of accident or for supplying abnormal current demands. To meet these conditions a temporary sub-station must be installed. The portable sub-station therefore insures continuity of power supply with a minimum investment in permanent sub-stations. On a line which is heavily loaded only on special days or seasons the portable outfit saves a large investment in copper and sub-station equipment and also provides additional power at any point where the traffic may be abnormally heavy temporarily.

Portable sub-station apparatus consists generally of a standard rotary converter equipment interchangeable with that used in the permanent sub-stations. Special conditions may make motor-generators more desirable in certain cases. The car bodies may be constructed of wood or steel. The chief objection to the wooden car is the possibility of fire either from the outside or from electrical disturbances in the interior.

Under the description of the Long Island Railroad system two such stations were mentioned which were used during the racing season at Belmont Park and at the Metropolitan Race Track. Each of these stations is equipped with a 1,000-kw. converter, identical with those used in the regular stations, and with three

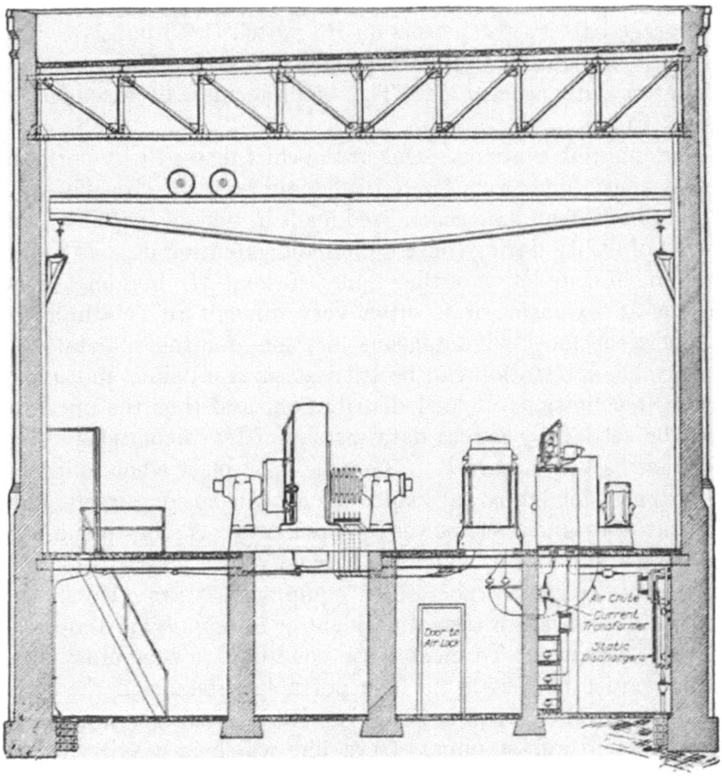

FIG 294.—Cross section of station shown in Fig. 293.

375-kw. air-blast transformers with the accompanying ventilator equipment and auxiliary apparatus. Figure 295 shows the arrangement of this apparatus in the car. The floor of the car is of very strong steel construction while the sides and roof are made as light as possible. The converter is set up in the end of the car, and this part can easily be taken apart so that the machine may be dismounted or removed if necessary. At the other end there are the three symmetrically arranged transformers. They are easily removable and can be lifted through the roof

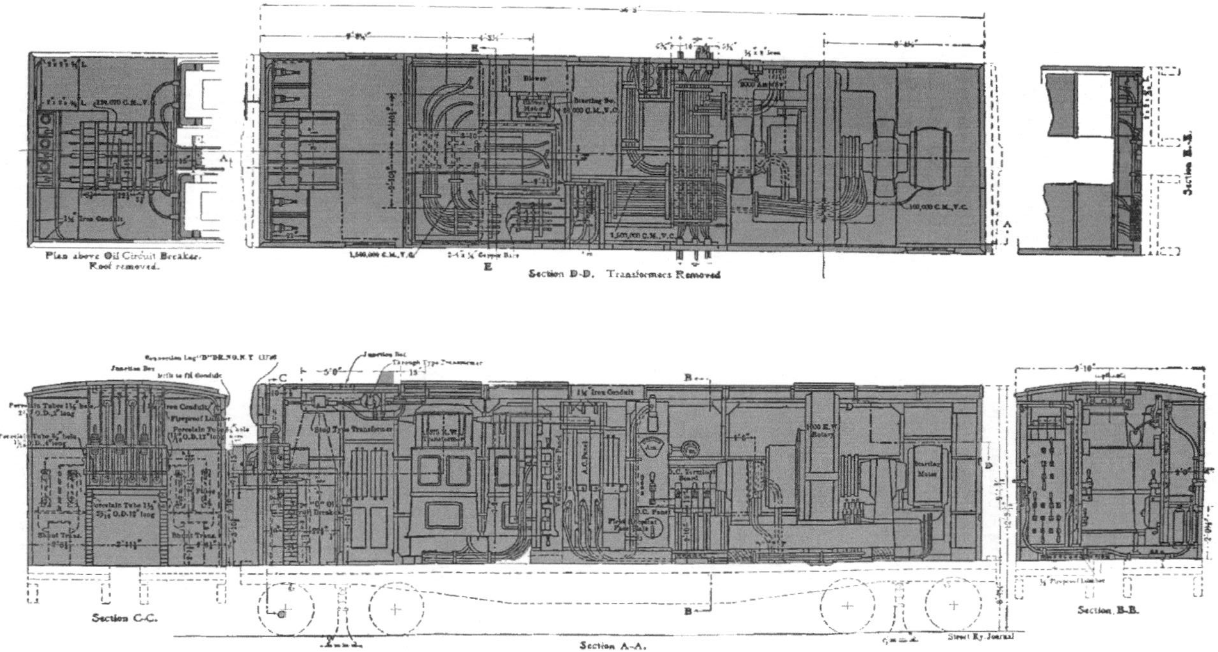

FIG. 295.—Location of electrical apparatus in portable sub-station of the Long Island Railroad.

(Facing page 464.)

of the car by the traveling cranes of any of the sub-stations. They are mounted on a raised chamber which is supplied with air from the blower. The latter delivers 4,500 cubic feet of air per minute at a pressure of 1 oz. It is driven by a three-phase, 3-hp., 400-volt induction motor which is supplied from the a.c. board from the low-tension side of the transformers.

The high-tension lines come into the car through three inlets on the high-tension side. They are first led to an oil switch (type *C* Westinghouse, 600 amp.) enclosed in a yellow pressed-brick cell, whence they run along the roof of the car to the delta connection on the high-tension side of the transformers. The space between the oil-switch cells and the transformers is accessible through doors in the side of the car. There are three switchboards in the car. The first is provided with several switches making possible four combinations of voltage from the low-tension side of the transformers, and the second and third are the a.c. and the d.c. boards. The converter is started by an induction motor. The induction-motor section also contains the switchboards and is accessible through doors. The d.c. feeders leave the car on one side near the d.c. side of the machine. When in service these cars are housed in specially constructed sheds where they are connected up with the high-tension line. The necessary lightning arresters are therefore installed in the towers of these sheds, the arrangement being similar to that described above for lightning-arrester houses.

Direct current for operating the oil-switch solenoids is drawn directly from the third rail, and the solenoids must therefore be wound for 500 volts. Note the method of mounting the shunt transformers on either side of the oil switch and of the series transformer on the roof of the car. All cables in the car are laid under the machines and transformers. The control and operation of the machines are exactly the same as in all of the other sub-stations of the system.

In order to secure a rigid base for the machines when in service, the car is lifted off its trucks and springs. The car alone weighs 49,000 lb., and the weight of the equipment is 142,400 lb.

Figures 296 and 297 show a portable sub-station equipped with General Electric machines and apparatus, used by the Cincinnati and Columbus Traction Company. The converter is a 400-kw., three-phase, 25-cycle machine and delivers a 600-volt direct current. A three-phase air-blast transformer, $^{16,500}/_{3,300}$-volt-

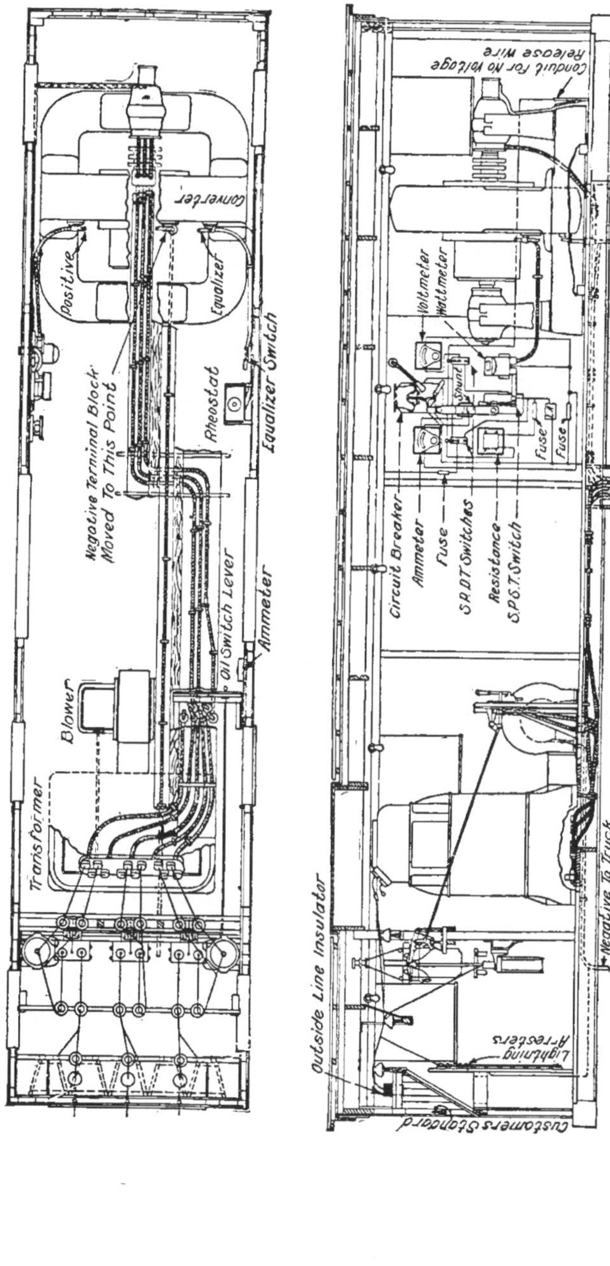

FIG. 296.—Plan and elevation of a portable sub-station with 400-kw. converter and a three-phase, 25-cycle, 16,500/3 300 – 370 volt air-blast transformer.

370-volt, delivers the necessary low-tension current for the converter. The latter is started on the a.c. side through two double-throw switches, which at starting connect the machine to a low-voltage and when running to a 370-volt circuit. The machine and transformers are placed in opposite ends of the car in order to balance the weight as much as possible. A number of wooden blocks are fastened to the floor of the car and hold the cast-iron frame of the machine in place. The high-tension lines are brought in through inlets protected against rain, etc., and are

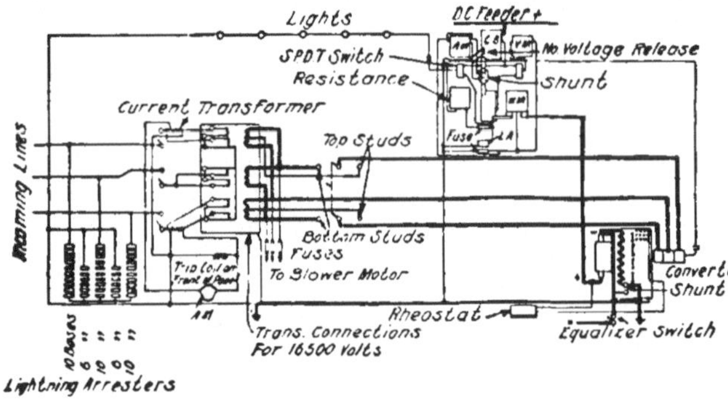

FIG. 297.—Wiring diagram of a portable sub-station.

led to the three single-pole oil switches which are mounted on wooden supports on one side of the car. These supports also carry the operating mechanism, which is actuated from the switchboard through a linkage system. The starting and control panels are placed side by side, while all instruments for both the d.c. and a.c. sides are mounted on individual bases directly on the walls of the car. The negative machine terminals are directly connected with the steel work of the car. A disconnecting switch is provided for connecting the negative side with the equalizer bus in case the car is used as a reserve station or is run in parallel with a stationary sub-station. The positive cables run from the circuit breaker through the side of the car to a terminal stud on the outside to facilitate connection to the third rail. Lightning arresters are provided near the inlet of the high-tension cables. The use of lightning arresters in wooden cars is not recommended, on account of the fire risk. However, for voltages up to 13,200 the lightning arresters are small and can readily be housed in asbestos lumber compartments, or the car can be protected by lining that section of it with asbestos. There are a number of openings in the roof over the transformers so that these may be accessible for repairs or removal. In the middle of the car there are two doors, and besides this there are a number of windows in the sides for illumination.

The approximate dimensions for cars with machines of different kilowatt ratings are given in the following table. The first three are those recommended by the General Electric Company, and the fourth is that used by the Long Island Railroad, where Westinghouse machines are employed:

Rating of converter	Length		Width		Height		Weight complete
Kw. Volts	Ft.	In.	Ft.	In.	Ft.	In.	Lb.
300 and 13,200	36		7	6	8		76,500
500 and 33,000	40		8	9	9		97,000
500 and 60,000	50		8	6	9	3	150,000
(Motor-generator)							
1,000 and 12,000	36	8	9	10	9		200,000

Transformer Stations.—A distribution system for single-phase railway service with single-phase transmission is indicated in Fig. 298. The connections between the intermediate and terminal sub-stations are also shown. In the intermediate station the incoming and outgoing lines are connected to 22,000-volt transfer buses, which supply the auxiliary buses in the station. Both terminals on the high-tension side of the transformers are connected to the auxiliary bus, and one terminal on the low-tension side is grounded while the other is connected to the 2,200-volt bus. The trolley wire is sectionalized in front of both sub-stations and the immediate section is fed by both of the stations. The low-tension feeders have single-pole automatic oil switches, and the high-tension have double-pole switches. With the exception of the outgoing feeders the end station is identical with the intermediate stations. All incoming and outgoing feeders are protected by lightning arresters.

Figure 299 shows the distribution for a single-phase railway system, with three-phase transmission, double track. The high-tension side is analogous to that shown in Fig. 298, with the exception that there are banks of two transformers connected

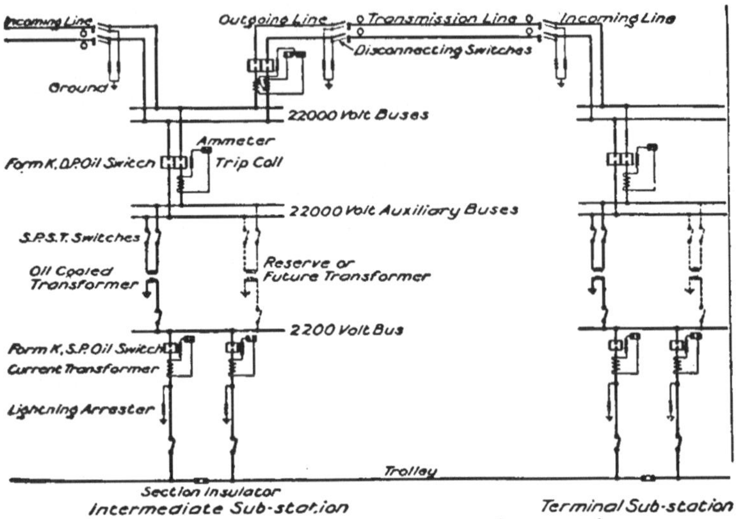

FIG. 298.—Distribution system for single-phase railway, single-phase transmission.

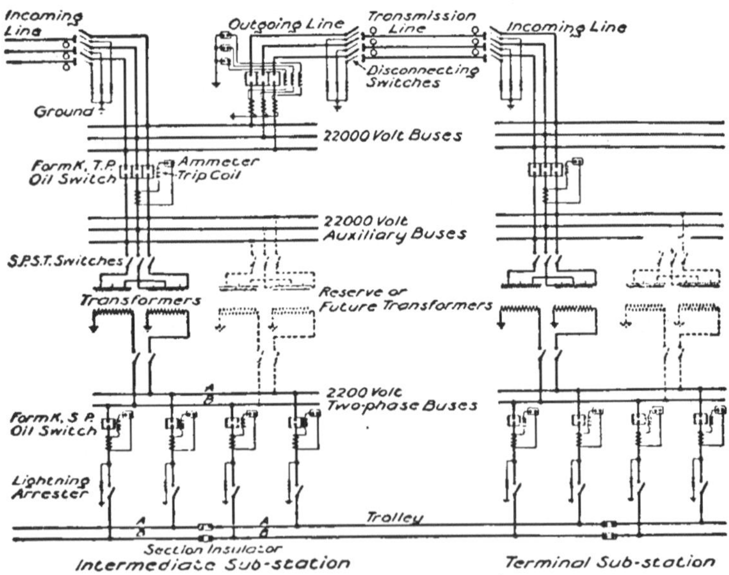

FIG. 299.—Distribution system for single-phase railway, three-phase transmission, double track.

three-phase-two-phase to supply the low-tension buses, instead of one, as before. The two phases supply different sections of the two trolley lines, a given phase always feeding the same trolley. If only one trolley is used, its sections are supplied alternately by both phases.

Rochester Division of the Erie Railroad.—An abstract of a paper by W. N. Smith, in the *Street Railway Journal.*[1] This line was the first in this country to operate electric cars on a single-phase system over the tracks of an operating steam railroad. Further, it was the first to use 11,000 volts working pressure on a trolley and the first instance of a single-phase traction system receiving power from a 60,000-volt transmission line. The energy is supplied from the plant of the Ontario Power Company at Niagara Falls at 60,000 volts pressure, which is stepped down at the Avon sub-station to 11,000 volts working pressure. Figures 300, 301, and 302 are the plans and cross sections and wiring diagram of this sub-station.

The building is of brick resting on solid concrete foundations, the roof and floors being of reinforced concrete. Thefloors are supported upon steel beams, but the roof beams are of reinforced concrete like the slabs which they support. The building is absolutely fireproof, the doors and windows being of kalomein construction and fitted with wire glass. It is 39 ft. 8 in. by 44 ft. outside and 29 ft. 10 in. high from the top of the foundations to the top of the parapet. In the basement are located one of the transformer oil tanks and the oil pump. The main floor is divided into three rooms, the main transformer room being 43 ft. by 17 ft. and extending the full height of the structure to allow room for the high-tension busbars which are carried over the transformers. The remaining space on the main floor is divided into a high-tension room (through which the 60,000-volt wires enter and which is the location of the high-tension circuit breakers, 16 ft. 8 in. by 19 ft. 8 in.) and the operating room, which is 19 ft. 8 in. by 24 ft., where all the 11,000-volt switching apparatus and the measuring instruments are located. Directly over the operating chamber is a mezzanine floor, reached by an iron staircase, in which are located 11,000-volt lightning arresters, the 60,000-volt reactance coils, and the 60,000-volt series transformers.

[1] W. N. Smith: "Single-phase Electric Motive Power on the Rochester Division of the Erie Railroad," *Street Railway Journal*, Oct. 12, 1907.

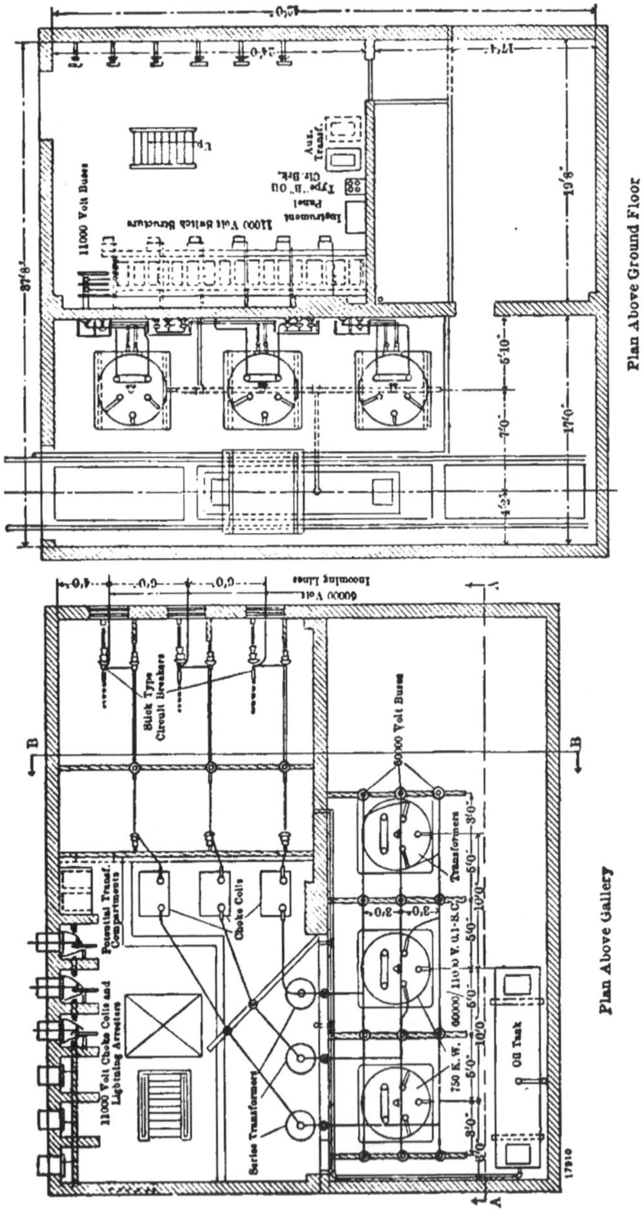

FIG. 300.—Plan of Avon sub-station.

The transmission line terminates at the lightning-arrester yard in the rear of the sub-station. The arrangement of the 60,000-volt lightning arrester consists of three horn gaps arranged one behind the other on each of the three conductors, the first gap being 4.75 in. across, the second 5 in., and the third 6 in. A concrete column is in series with the first gap, an electrolytic arrester in series with the second, and a 5-ft. fuse of No. 18 copper wire in series with the third, that is, between horn and

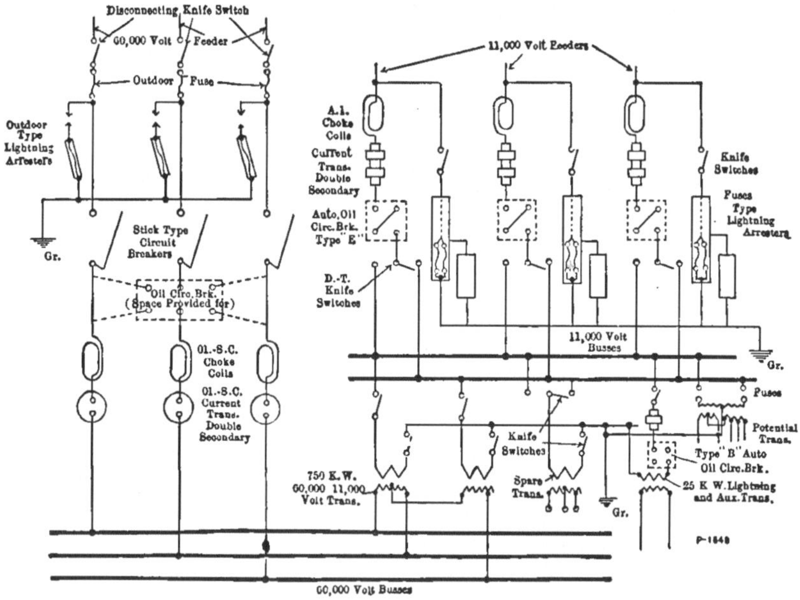

FIG. 302.—Diagram of connections of the Avon sub-station.

ground. Both horns of each gap are 1.5-in. round iron. Between the line and the first arrester there is a hook-type knife switch, and between the last arrester and the lead going into the sub-station there is a No. 18 copper-wire fuse in each conductor, placed horizontally. These fuses are enclosed in wooden tubes about 5 ft. long, wrapped with torpedo twine. The entire arrangement of lightning-arrester gaps, fuses, and switches is mounted upon eighteen chestnut poles and a suitable elevated platform railed off and fitted with a gate to keep out trespassers.

The theory on which these arresters are installed is that for ordinary slight static disturbances in the line the arrester having the lower striking e.m.f. will discharge, and since it has in series

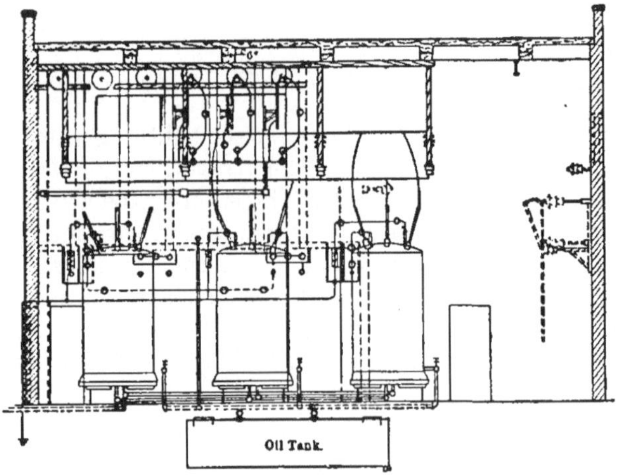

Section A-A

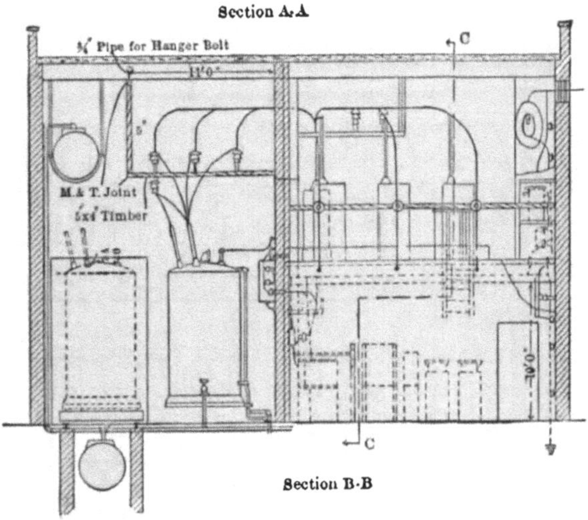

Section B-B

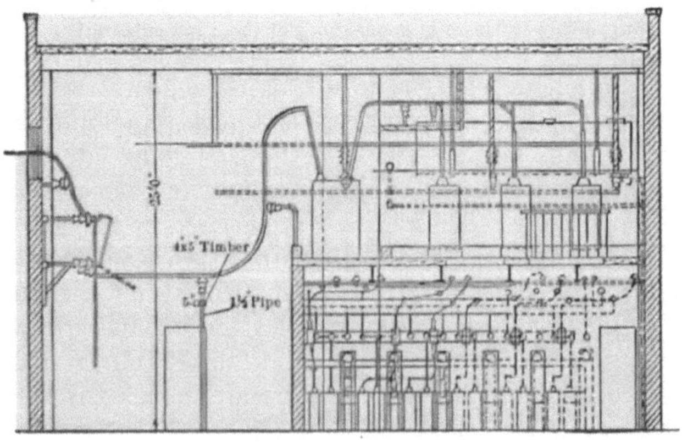

Section C-C

FIG. 301.—Plan and elevation of Avon sub-station.

(*Facing page* 472.)

with it a comparatively high resistance, the resultant disturbance due to the current which follows the discharge will be comparatively slight. A more severe static disturbance causes the arrester having the lowest gap and the arrester having the next higher gap to discharge simultaneously, thus affording two paths to ground, the combined resistance and inductance of which are much lower than that of the first path. When both arresters discharge, the disturbing effect of the system, due to the generated current which follows, is much more severe than that in the first instance. For a direct lightning stroke on the line the three arresters would discharge simultaneously, the long, thin copper wire in the case of the arrester with the highest air gap blowing and interrupting the arc upon it, and the disturbance of the circuit finally ending upon the other two arresters.

The three high-tension conductors enter the sub-station through glass disks held in 36-in. tile set in the upper portion of the rear wall of the sub-station. Within the sub-station the wires first pass to three 60,000-volt stick-type circuit breakers mounted directly inside of the rear wall. Thence they run over bare copper conductors to the three oil-insulated reactance coils situated on the mezzanine floor, thence through three oil-insulated series transformers, also on the mezzanine floor, and finally through a wide opening in the division wall to the 60,000-volt busbar in the transformer room. The busbars are mounted upon porcelain insulators on wooden cross-arms at a convenient height over the transformers.

The transformers are of the Westinghouse oil-insulated, water-cooled type, each of 750 kw. For the present installation two only are used, the third and middle one being the spare. The high-tension connections are such that in case of one transformer failing while in service its connection can quickly be taken off the busbars and put on the spare transformer. The transformer windings are fitted with taps enabling the three-phase-two-phase "Scott connection" to be used. The low-tension windings can be so connected that 11,000 or 22,000 volts can be obtained, the latter to be used in case another sub-station for a 40- or 50-mile extension is added. The low-tension windings also have six taps, allowing small variations in the secondary voltage. One end of each low-tension winding is directly grounded to the boiler-iron case, which in turn is directly connected to the track return circuit by means of a No. 4 copper-stranded cable. The transformer cases are made of boiler iron and are set on a square cast-iron base which is mounted on three pairs of wheels running upon an iron sub-base set in the concrete floor of the room. A track runs lengthwise across the room directly in front of the transformers. A transfer truck runs upon this, and on the top of the truck there is another set of small wheels which line up with those on which the transformer cases are set.

Two cylindrical boiler-iron oil tanks are provided. One is located in the basement directly under the transformer room so that the oil from the transformer can readily be drained into it, and the other is suspended from the concrete roof beams at the top of the transformer room, close to the side wall of the building. This is intended to act as a reservoir for distributing oil back into the transformers. The oil is pumped from the lower to the upper tank by means of a steam pump supplied from the boiler room in the adjacent division roundhouse, where steam is always available.

The water circulation is by gravity, the supply coming from the railroad company's water-tank system. There are three separate water-cooled coils in each transformer case, each one controlled by its own valve.

The low-tension busbars run along the division wall of the operating room, and directly beneath them are three type E Westinghouse automatic oil switches, one on each of the two trolley feeders, the third or middle one being the spare. One pole of each of the three oil switches is connected to the center pole of a double-throw hook-type switch, by means of which it is thrown upon either busbar. The other pole of the oil switch runs directly to the feeder. The outgoing lead from the middle or spare oil switch can instantly be thrown upon either one of the feeders should the oil switch controlling that feeder be temporarily disabled. The outgoing 11,000-volt feeders run up to the mezzanine floor directly over the operating room, where they emerge from the building through perforated glass disks set in 18-in. round tiles. Before emerging there are tapped to them two Westinghouse low-equivalent lightning arresters set in brick compartments and reinforced by two electrolytic lightning arresters of the 11,000-volt type.

A set of call bells is provided so that when the oil switch is open a bell is rung in the car inspection shop adjoining. Also, if the temperature of any transformer runs above normal, a bell circuit connected to the thermometers in the top of the transformer tank is similarly made to operate. The station itself does not require the continuous presence of an attendant. The working force is so organized that the car repair men are always available for manipulating the sub-station oil switches.

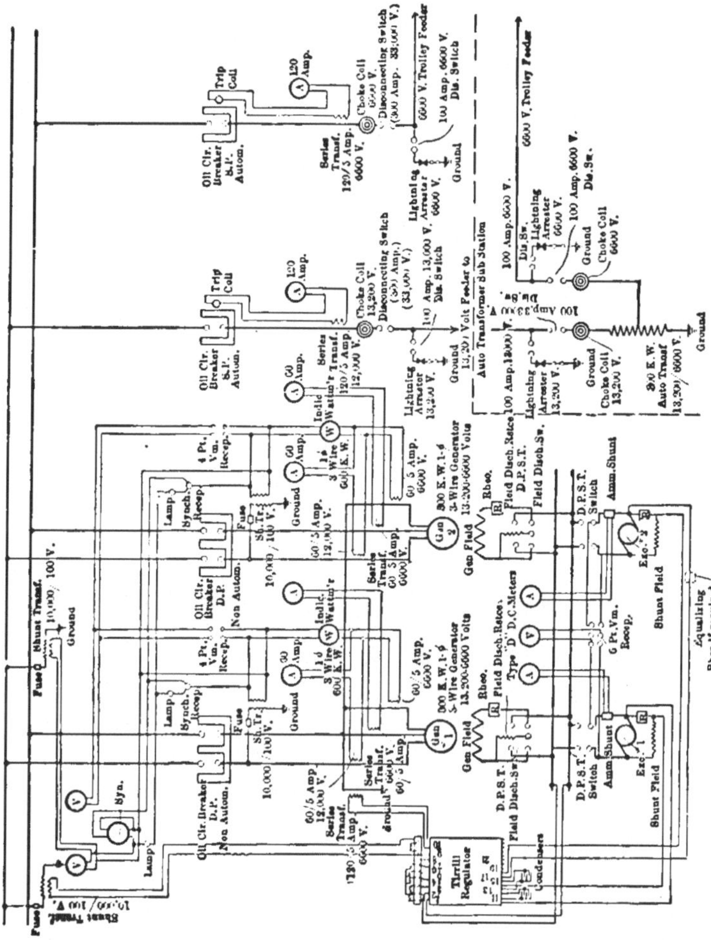

FIG. 303.—Wiring diagram of power house and sub-station of the Windsor, Essex and Lake Shore Rapid Railway

On the mezzanine floor there is room for an oil switch eventually to be installed for the 60,000-volt incoming line. There is also space for a fourth transformer and for a number of type E oil switches.

Windsor, Essex and Lake Shore Rapid Railway.—In Fig. 303 there are shown the wiring diagrams of the central and sub-stations of the above system, which employs 6,600 volts pressure in the overhead trolley.

The central station is equipped with two generators of 500 kw., 25-cycle, single-phase, three-wire, flywheel type. The windings of these generators are such that from the terminals there may be obtained current at 13,200 volts pressure and also at 6,600 volts. The two windings may be used in series for obtaining these two voltages or may be connected in parallel for obtaining the full rating of the machine at 6,600 volts. One of the three terminals of each machine is grounded. The exciting current for each generator is furnished by a 30-kw., 125-volt belted generator, the field current of which is varied by a Tirrill regulator to obtain smooth voltage regulation for the units.

Energy is supplied to two buses in the station, one at 13,200 volts which feeds the sub-station, and one at 6,600 volts which feeds directly a section of the trolley line. The 13,200-volt single-wire transmission line feeds a 300-kw. auto-transformer in the sub-station, which in turn supplies the other section of the trolley wire at 6,600 volts. The station is equipped with the necessary oil switches for the supply to the buses and the two outgoing feeders. The sub-station contains only the auto-transformer and the necessary lightning arresters and reactance coils. The switchboard for regulating and controlling the output of the generating station comprises five panels, one exciter, two machines, one 6,600-volt feeder, and one 13,200-volt feeder. Enclosed in concrete cells back of the switchboard are two machine and three feeder switches. These are distant-controlled oil switches, type E. The high-tension wiring within the station is composed of lead-covered cables enclosed in fiber conduits. The trolley wire is divided into two sections, the 18-mile section fed from the power house and the 12-mile section from the sub-station.

The Spokane Inland Railroad Company.—The Inland Empire System of this railroad operates with energy purchased from the Washington Water Power Company. The output of this company's plant is delivered at 60 cycles, three-phase and 4,000 volts. In the purchase of the energy the charges are based upon the maximum demand during each month, and for this reason it became very desirable to employ some means to flatten the railway load curve, and it was necessary also on account of the motor characteristics to change the frequency from 60 to 25 cycles before feeding the energy to the railway transmission line. The problem was solved by combining phase-changing induction motor-generator sets on the same shafts with direct-current railway machines which utilize a large storage battery as a flywheel. (See also Chap. V on storage batteries for a.c. systems.)

The station equipment consists of four main units each made up of three machines. In each set a 1,000-hp. induction motor takes three-phase, 60-cycle current at 4,000 volts from the incoming Washington Water Power lines. The machine is mounted on one end of a shaft in the middle of which is a 1,000-kw., single-phase, 2,200-volt, 25-cycle generator and at the other end of which is a d.c. machine rated at 1,100 amp. and 550 volts. These three machines of Westinghouse manufacture operate at 550 rev. per min. When the load on the generator is light the d.c. machine runs as a generator and charges a 275-cell storage battery, and when the load is heavy the d.c. machine runs as a motor, taking current from the battery and assisting the three-phase induction motor in driving the single-phase generator. The single-phase generator has a somewhat larger power than the three-phase induction motor, the idea being that the three-phase motor in connection with the d.c. machine will take nearly as full power from the line under a large variation of the single-phase load. At times when the single-phase load is excessive the d.c. machine, taking current from the battery, makes up the deficiency. By this method a more uniform load from the three-phase line is had. The periods of light load are thus diminished and the peak of heavy load materially flattened. These machines are mounted on a common bed plate, each machine having its own pair of bearings and all being connected by fixed couplings.

The station equipment also includes two battery boosters of 960 amp. rating and three 50-kw. exciter sets for the single-phase generators. These exciters are driven by 75-hp. Westinghouse motors taking three-phase current at 125 volts pressure.

The feed lines from the Washington Water Power plant enter the station on a gallery and pass to a hand-operated disconnecting oil switch and then through reactance coils, instrument transformers, and down through the gallery floor to a busbar set. From this power bus leads pass to the two Westinghouse type *C* oil switches, having remote control and located on the gallery floor. These switches admit of flexibility in feeding two sets of 4,000-volt busbars at the back of the switchboard on the main floor. The motors of the phase-changing sets can be operated from either set of busbars. The oil switches for the motors are type F Westinghouse, from which energy is fed through starting resistance in the secondaries and rheostatic controllers to the induction motors.

The battery across which the d.c. machines are connected is made up of 275 chloride cells, type R-33, having a discharge rate of 2,880 amp., furnished by the Electric Storage Battery Company.

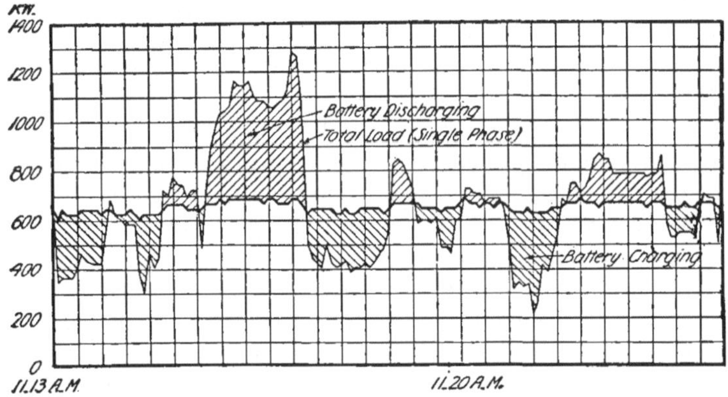

FIG. 304.—Power curve showing smoothing effect of storage battery in the Spokane and Island Railway System.

Figure 304 gives an idea of the effect which the battery has on the supply line load. The single-phase railway load is very irregular, while the three-phase supply line load is comparatively smooth. Regulation of the battery booster is controlled by a carbon regulator operated by changes of current in the three-phase supply line. Small changes in the current in this line cause the battery to charge from or discharge to the d.c. machine and thus keep the power-supply load curve flat.

A 30-panel switchboard from which all the equipment and the battery are controlled is located in front of the gallery on the main floor. On the panels for the 750-hp. d.c. machines starting switches are provided for taking current from the battery for starting, thus not putting the large units on to the three-phase, 4,000-volt supply line until the machine is up to speed. The single-phase, 25-cycle generators are controlled by Tirrill regulators. For synchronizing, a synchroscope having an illuminated dial with an illuminated pointer is mounted on a pedestal in front of the switchboard.

The step-up transformers for raising the single-phase, 25-cycle current from the generator pressure of 2,200 volts to line pressure of 45,000 volts are mounted on cars made of structural iron shapes which stand in brick compartments. These compartments are provided with rails set in the concrete floor, and in front of the row of compartments is a track running to a cross track by which the transformers can be taken to the end of the transformer room. The transformer car runs from the transfer car to the short track leading under the crane at the end of the station. From each one of the five 1,250-kw., 2,200- to 45,000-volt, single-phase, step-up, oil-insulated, water-cooled transformers two leads pass to the gallery floor above to 60,000-volt type *L* Westinghouse oil switches and then through reactance coils and hand-operated disconnecting switches to the transmission line.

The local trolley section is fed from the phase-changing station through a 6,600-volt panel on the station switchboard supplied by three 375-kw., 2,200- to 6,600-volt transformers.

When this system is completed there will be fifteen transformer sub-stations located 10 miles apart. The equipment of each includes three 375-kw. oil-insulated transformers connected in parallel and fed through one hand-operated 60,000-volt oil switch provided with automatic release. Lightning arresters are placed on both high- and low-tension sides of the transformers, and 6,600-volt oil switches serve to disconnect the transformers from the trolley.

The Southern Sierras Power Company. *San Bernardino Terminal Station.*—This system carried the idea of outdoor stations to its logical extreme and, with the exception of secondary transformer stations, where a local distribution is controlled through a switchboard, in which case a house of just sufficient size is used, no housing or protection of any sort from wind or weather has been provided.

The transmission system of the Southern Sierras Power Company includes six transformer stations on the high-tension 140,000-volt side, and there are eight stations which receive their current supply from the intermediate 33,000-volt circuits. The transmission line is 238 miles in length at 140,000 volts, while the present capacity of the outdoor sub-stations reaches 12,000 kw.

The transmission line terminates at the San Bernardino station at two double-break horn-gap air-break switches mounted

FIG. 6.—San Bernardino terminal station of Southern Sierras Power Company.

on towers. The switches are motor-operated and controlled from the switchboard in the near auxiliary steam plant. There are three sets of A-frame structures, two of which are on either side of a hollow square. (See Fig. 305.) The third is opposite to one of the main switch sets and is connected to it by a bus line, and in this structure are mounted the disconnecting switches from which the circuit is carried to a set of electrolytic lightning arresters. The bus from the other main switch is supported by a two-pole structural. steel frame set in line with the A-frames first mentioned. Between the end A-frame structures is strung a cross bus from which connections are made to both upper buses, and these A-frames carry disconnecting switches for connections to step-down transformers. But one set of these transformers is in place. There are three transformers in this set of 4,000 kva. capacity each. They are wound for 87,000 volts primary to be star-connected to the 140,000-volt circuits. The secondaries are wound for 36,000 and 6,000 volts, the latter tap permitting the turbo-generators of the auxiliary steam plant to feed direct to these transformers. These transformers owing, to their high voltage windings and their large capacity, are very heavy and bulky and are provided with three sets of water-cooling copper coils each. The cases are mounted on low trucks which rest on concrete piers 5 by 10 ft. A sub-grade transfer track passes in front of the transformers and extends to the door opening in the rear of the steam power house, where a transformer may be taken for repair. A 33,000-volt bus line above the transformer and over the transfer track is carried on two end A-frame structures which straddle the track. From this bus connections are made to three outgoing circuits and to the reactive coils. Two of the outgoing circuits are equipped with 33,000-volt outdoor oil switches mounted on steel towers. The third circuit has an air-break horn switch, and a similar switch is interposed in a circuit which feeds a secondary set of transformers. This set comprises three water-cooled transformers of 2,000 kva. each, mounted on wheels so as to be movable to the transfer track described. These are wound for 36,000 to 6,000 volts and are intended to connect the steam-power circuits with the outgoing lines and incidentally with the main transmission through the larger transformers. An angle-steel framework over these transformers carries on one side the 36,000-volt disconnecting switches and on the other those for the 6,600-volt circuit.

The reactive coils are placed in a one-story reinforced-concrete building 42 by 22 ft. There are two sets of these coils in use. Each set, consisting of three coils, is placed parallel with one side of the building. By placing reactance in the form of an open-core inductance coil across each phase of the circuit the capacity effect of the rransmission line becomes neutralized and the voltage is brought to a point within the limit of ordinary operation, especially to allow synchronizing within limits and without putting a dangerous strain on the generators, transformers, etc. In using high voltages on long transmission lines the capacity of the circuit would serve, if the line were open at the receiving end, under certain conditions, to give a higher voltage than on the initial end and in any case a higher voltage than the operating voltage which the machine on the receiving end may be designed for. The purpose of these reactive coils is therefore purely one of line regulation. In operating the reactancc coils they are arranged so that the two sets may be placed in series with each other or, by a change in connections, may be placed simgly, cutting one set out, or may be placed in parallel, depending on the amount of reactance necessary for the condition at hand. These various changes are made automatically by three sets of three-pole, 36,000-volt oil switches. The coils themselves consist of twelve flat coils of one turn per layer and having sixty turns of ribbon wire, all very heavily insulated. The coils are mounted on hard-wood frames heavily varnished.

Sub-station of Amherst Power Company at Chicopee, Mass. — This system adopted the so-called semi-outdoor sub-station, in which all the 66,000-volt equipment, including the step-down transformers, was placed outdoors, as were also the aluminum lightning arresters on the 13,000-volt outgoing lines, while 13,200-volt buses and switches, the control switchboard, the 2,300-volt distributing switchboard, and the railway apparatus were placed inside.

The choice between indoor and outdoor apparatus in different parts of the equipment was based on the following considerations:

1. The 66,000-volt apparatus was the most bulky part of the equipment, and therefore the greatest saving in building space could be accomplished by putting it outdoors.

2. The more complicated switching necessary with the numerous 13,200-volt and 2,300-volt outgoing circuits and synchronons-motor circuits would have been much more difficult and expensive to handle in an outdoor installation. Hence the switches were placed inside.

3. The control switchboard, with its motors and instruments, the storage battery, and the railway sets are all unsuited for outdoor operation. The relative location of the remainder of the apparatus, that is, whether it would be placed indoors or outdoors, was determined principally by comparison of estimated first cost, since no well-founded objection to the use of outdoor aluminum lightning arresters, oil switches, transformers, or buses exists.

The saving in the first cost of the building may be arrived at very closely on the basis that two proposed buildings erected on the same site for the same purpose and of the same general design and material will, within reasonable limits, have first cost in direct proportion to their cubic contents. A building of the dimensions necessary to accommodate indoors all the apparatus, including the 66,000-volt equipment, allowing for the same switching scheme and providing a space for future expansion equal to that which is incorporated in the present station and outdoor foundations, would cost on the above basis 202 : 86 = 2.35-fold of the cost of the present building. After allowing for the extra cost of outdoor apparatus and the cost of outdoor foundations and steel structures, the saving in this case was an amount equal to 11 per cent. of the cost of the complete station and equipment as built.

Attention may be called to same special features of the outdoor equipment. The oil in the outdoor transformers freezes at $-15° C$. It is therefore necessary to keep the temperature of the oil in the spare transformer above this point if it is to be ready for immediate service. This will be accomplished by connecting the spare transformer to the bus on the low-tension side from Jan. 1 to March 10 of each year according to the location of the station. During the remainder of the year the temperature of the oil in the spare transformer does not go down to its freezing point when left without artificial heat.

The transformer and oil switches are connected by a complete piping system to a two-compartment oil tank buried in the ground, so that oil can be drained from any transformer or oil switch into its respective compartment of the oil tank, and from there can be pumped through a portable oil filter press back into the transformers or switches. Alarm thermometers on the transformers are connected to bells in the station. There is installed in the cover of each transformer a small heating unit taking 200 watts at 110 volts. The object of this is to keep the air in the top of the transformers, and especially the spare transformer when not excited, a little above temperature of the air outside, so that in the process of breathing, which is sure to be present to some extent in any large transformer subjected to varying temperatures, the air taken in, even though heavily laden with moisture, will have that moisture vaporized by coming in contact with the warm air inside, rather than being further condensed by coming in contact with cooler air inside, which latter condition would occur if there were no artificial heat inside the cover of the transformer.

When it becomes necessary to disassemble a transformer it is rolled on to a transfer truck upon which it is run into a compartment at one end of the station where a hoist is available of sufficient capacity to lift and remove the core. Switch hooks used for operating the disconnecting switches outdoors are fitted with a copper cone at about the middle of the handle, which cone is grounded when the hook is to be used. All the outdoor aluminum lightning arrester cones are painted white to overcome a difficulty experienced in warm climates where the absorption of the sun's heat by dark-colored cones has raised the temperature of the electrolyte of the arrester to a point where it was damaged. Two coats of white paint on these cones have been sufficient to keep the temperature of the arresters below the danger point in these installations. This same treatment may be applied with similar results to the transformers where excessive temperatures are reached on account of the sun's direct rays.

Each outdoor station is surrounded by an iron picket fence and all the equipment is set on concrete foundations extending below frost line and about 15 in. above the final level of the ground.

APPENDIX

TABLE I.—CARRYING CAPACITY OF BARE AND INSULATED WIRE

Amperes

Size, B. & S. gage	National Electrical Code		Insulated wires in moldings, Kennelly's rule	Bare wire in still air, temp. rise 50° F.	Size, cir. mils	National Electrical Code		Insulated wires in moldings, Kennelly's rule	Bare wire in still air, temp. rise 50° F.
	A, rubber	B, weatherproof				A, rubber	B, weatherproof		
Solid									
18	3	5	4.5	6.0	300,000	270	400	281	373
17	...	...	5.4	7.2	400,000	330	500	349	463
16	6	8	6.4	8.5	500,000	390	590	413	549
15	...	...	7.6	10.2	600,000	450	680	474	631
14	12	16	9.1	12.1	700,000	500	760	532	708
13	...	...	10.8	14.4	800,000	550	840	587	781
12	17	23	12.9	17.1	900,000	600	920	641	852
11	...	...	15.3	20.4	1,000,000	650	1,000	694	922
10	24	32	18.3	24.3	1,100,000	690	1,080	746	991
9	...	...	21.6	28.7	1,200,000	730	1,150	797	1,058
Strand					1,300,000	770	1,220	846	1,123
8	33	46	31.3	41.5	1,400,000	810	1,290	894	1,187
7	...	...	37.3	49.5	1,500,000	850	1,360	942	1,250
6	46	65	44.3	58.8	1,600,000	890	1,430	989	1,312
5	54	77	52.5	69.7	1,700,000	930	1,490	1,035	1,373
4	65	92	62.7	83.3	1,800,000	970	1,550	1,080	1,433
3	76	110	74.4	98.8	1,900,000	1,010	1,610	1,125	1,492
2	90	113	88.6	117.6	2,000,000	1,050	1,670	1,169	1,550
1	107	156	105.4	140.0	....	....	.....	.....	.....
0	127	185	127.8	169.8	....	....	.....	.....	.....
00	150	220	151.7	201.5	....	....	.....	.....	.....
000	177	262	180.8	240.2	....	....	.....	.....	.....
0000	210	312	215.2	286.0	....	....	.....	.....	.....

The values in the fourth column are such that twice the current given will cause a rise in temperature of 36° F.

Table II.—Recommended Current-carrying Capacities for Cables and Watts Lost per Foot

For each of four equally loaded paper-insulated, lead-covered cables installed in adjacent ducts in the usual type of conduit system where the initial temperature does not exceed 70° F., the maximum safe temperature for continuous operation being taken at 150° F.

(Copyright by Standard Underground Cable Co., 1906)

Size, B. & S. gage	Safe current in amp.	Watts lost per foot at 150° F.	Size, cir. mils	Safe current in amp.	Watts lost per foot at 150° F.
14	18	0.97	300,000	323	4.22
13	21	1.03	400,000	390	4.61
12	24	1.09	500,000	450	4.91
11	29	1.15	600,000	505	5.16
10	33	1.25	700,000	558	5.36
9	38	1.39	800,000	607	5.56
8	45	1.53	900,000	650	5.71
7	53	1.67	1,000,000	695	5.86
6	64	1.85	1,100,000	740	6.01
5	76	2.08	1,200,000	780	6.13
4	91	2.31	1,300,000	820	6.25
3	108	2.54	1,400,000	857	6.37
2	125	2.77	1,500,000	895	6.49
1	146	3.00	1,600,000	933	6.61
0	168	3.23	1,700,000	970	6.73
00	195	3.46	1,800,000	1,010	6.85
000	225	3.69	1,900,000	1,045	6.97
0000	260	3.92	2,000,000	1,085	7.09

TABLE III.—RECOMMENDED POWER-CARRYING CAPACITY IN KILOWATTS DELIVERED

Three-conductor, Three-phase Cables
(Copyright, 1906, by Standard Underground Cable Co.)

Size, B. & S. gage and cir. mils	1,100 V.	2,200 V.	3,300 V.	4,000 V.	6,600 V.	11,000 V.	13,200 V.	22,000 V.
	Kilowatts							
6	92.	183.0	275	333	549	915	1,098	1,831
5	104.	217.0	326	395	652	1,087	1,304	2,174
4	130.	260.0	390	473	781	1,301	1,562	2,603
3	154.	309.0	463	562	927	1.544	1.854	3.089
2	179.	358.0	536	650	1,073	1,788	2,145	3,575
1	209.	418.0	626	759	1,253	2,088	2,506	4,176
0	240.	481.0	721	874	1,442	2,402	2,884	4,805
00	279.	558.0	836	1,014	1,674	2,788	3,347	5,577
000	322.	644.0	965	1,172	1,931	3,217	3,862	6,435
0000	372.0	744.0	1,115	1,352	2,231	3,717	4,462	7,435
250,000	413.0	827.0	1,240	1,503	2,480	4,132	4,960	8,264

TABLE III.—RECOMMENDED POWER-CARRYING CAPACITY IN KILOWATTS DELIVERED.—(*Continued.*)

Single-conductor Cables, A.C. or D.C.
(Copyright, 1906, by Standard Underground Cable Co.)

Size, B. & S. gage and cir. mils	125 V.	250 V.	500 V.	1,100 V.	2,200 V.	3,300 V.	6,600 V.	11,000 V.
	Kilowatts							
6	8.0	16.0	32	70	141	211	422	704
5	9.5	19.0	38	84	167	251	502	836
4	11.4	22.8	45	100	200	300	601	1,001
3	13.5	27.0	54	119	238	356	713	1,188
2	15.6	31.2	62	138	275	413	825	1,375
1	18.3	36.5	73	161	321	482	964	1,606
0	21.0	42.0	84	185	370	554	1,104	1,848
00	24.4	48.8	97	215	429	644	1,287	2,145
000	28.1	56.3	113	248	495	743	1,485	2,475
0000	32.5	65.0	130	286	572	858	1,716	2,860
300,000	40.5	80.8	162	355	711	1,066	2,132	3,553
400,000	48.8	97.5	195	429	858	1,287	2,574	4,290
500,000	56.3	112.5	225	495	990	1,485	2,970	4,950
600,000	63.1	126.3	253	556	1,111	1,667	3,333	5,555
700,000	69.8	139.5	279	614	1,228	1,841	3,683	6,138
800,000	75.9	151.8	304	668	1,335	2,003	4,006	6,677
900,000	81.3	162.5	325	715	1,430	2,145	4,290	7,150
1,000,000	86.9	173.8	348	764	1,529	2,294	4,587	7,645
1,100,000	92.5	185.0	370	814	1,628	2,442	4,884	8,140
1,200,000	97.5	195.0	390	858	1,716	2,574	5,148	8,580
1,400,000	107.1	214.3	429	943	1,885	2,828	5,656	9,427
1,500,000	111.9	223.8	448	985	1,969	2,954	5,907	9,845
1,600,000	116.6	233.3	467	1,026	2,053	3,079	6,158	10,263
1,700,000	121.3	242.5	485	1,067	2,134	3,201	6,402	10,670
1,800,000	126.3	252.5	505	1,111	2,222	3,333	6,666	11,110
2,000,000	135.3	271.3	543	1,194	2,387	3,581	7,161	11,935

These tables are based on the "Recommended Current-carrying Capacity of Cables." A power-factor = 1 was used in the calculation, and hence the values found in the last table are correct for direct currents. For alternating currents the kilowatts given in both tables must be multiplied by the power factor of the delivered load.

TABLE IV.—EQUIVALENT CONDUCTOR AREAS
Of Single Conductors of Any Size from 0000 to No. 15 in a Stated Number of Smaller Conductors

Size, B. & S. gage	In 2 conductors	In 4 conductors	In 8 conductors	In 16 conductors	In 32 conductors	In 64 conductors	In 2 conductors, one each of:
0000	No. 0	No. 3	No. 6	No. 9	No. 12	No. 15	Nos. 00 and 1
000	No. 1	No. 4	No. 7	No. 10	No. 13	No. 16	Nos. 0 and 2
00	No. 2	No. 5	No. 8	No. 11	No. 14	No. 17	Nos. 1 and 3
0	No. 3	No. 6	No. 9	No. 12	No. 15	No. 18	Nos. 2 and 4
1	No. 4	No. 7	No. 10	No. 13	No. 16	......	Nos. 3 and 5
2	No. 5	No. 8	No. 11	No. 14	No. 17	......	Nos. 4 and 6
3	No. 6	No. 9	No. 12	No. 15	No. 18	......	Nos. 5 and 7
4	No. 7	No. 10	No. 13	No. 16	......	......	Nos. 6 and 8
5	No. 8	No. 11	No. 14	No. 17	......	......	Nos. 7 and 9
6	No. 9	No. 12	No. 15	No. 18	......	......	Nos. 8 and 10
7	No. 10	No. 13	No. 16	......	......	......	Nos. 9 and 11
8	No. 11	No. 14	No. 17	......	......	......	Nos. 10 and 12
9	No. 12	No. 15	No. 18	......	......	......	Nos. 11 and 13
10	No. 13	No. 16	......	......	......	......	Nos. 12 and 14
11	No. 14	No. 17	......	......	......	......	Nos. 13 and 15
12	No. 15	No. 18	......	......	......	......	Nos. 14 and 16
13	No. 16	......	......	......	......	......	Nos. 15 and 17
14	No. 17	......	......	......	......	......	Nos. 16 and 18
15	No. 18	....	....	....	....	....	....

For the same temperature rise more current can be carried by using divided circuits, and the greater the number of divided circuits for the same equivalent cross section, the greater the amount of current that can be carried. (See tables of carrying capacities.)

TABLE V.—SIZE AND WEIGHT OF STANDARD COPPER AND ALUMINUM WIRES

Size, B. & S.	Circular mils	Diameter, inches	Lb. per mile[1]	
			Copper	Aluminum
....	1,000,000	1.152	16,140	4,870
....	900,000	1.092	14,530	4,380
....	800,000	1.035	12,910	3,890
....	700,000	0.963	11,300	3,410
....	600,000	0.891	9,690	2,920
....	500,000	0.819	8,070	2,430
....	450,000	0.770	7,260	2,190
....	400,000	0.728	6,460	1,947
....	350,000	0.679	5,650	1,703
....	300,000	0.630	4,840	1,460
....	250,000	0.590	4,040	1,217
0000	211,600	0.530	3,420	1,030
000	167,800	0.470	2,710	817
00	133,100	0.420	2,150	648
0	105,500	0.375	1,703	513
1	83,690	0.330	1,351	407
2	66,370	0.291	1,071	323
4	41,740	0.231	674	203
6	26,250	0.183	420	128

[1] Increased 1 per cent. over weight of solid wire on account of stranding.

TABLE VI.—WATTS PER FOOT LOST IN SINGLE-CONDUCTOR CABLES AT DIFFERENT MAXIMUM TEMPERATURES WITH DIFFERENT CURRENTS

(Copyright, 1906, by Standard Underground Cable Co.)

Size, B. & S. gage and cir. mils	Current in amperes					
6	66	81	93	104	114	123
5	74	91	105	117	128	138
4	84	102	117	131	144	153
3	93	114	132	148	161	175
2	105	128	148	166	181	196
1	118	148	166	186	203	220
0	132	162	187	209	228	247
00	149	181	210	235	256	277
000	166	204	235	263	288	311
0000	186	229	264	295	323	350
300,000	222	273	315	352	385	416
400,000	248	315	363	406	445	480
500,000	288	352	406	455	498	537
600,000	315	385	445	497	545	587
700,000	341	416	480	538	588	635
800,000	364	446	514	575	628	679
900,000	386	473	545	610	666	720
1,000,000	407	498	575	642	703	758
1,100,000	426	522	602	674	736	796
1,200,000	446	546	630	705	772	833
1,300,000	462	568	655	732	802	866
1,400,000	480	590	681	761	834	900
1,500,000	496	610	704	788	862	931
1,600,000	512	629	726	812	889	960
1,700,000	529	649	750	837	916	990
1,800,000	543	667	770	862	943	1,018
1,900,000	557	686	792	886	970	1,048
2,000,000	573	705	813	910	995	1,075
Temp. of cond. in degrees F.	**Watts lost per foot**					
100	1.81	2.71	3.62	4.52	5.43	6.33
125	1.91	2.87	3.82	4.78	5.73	6.69
150	2.00	3.00	4.00	5.00	6.00	7.00

The watts lost per foot means the amount of electric power lost in heating the conductor, and is equal to the product of the resistance per foot of cable times the square of the current in amperes.

For two-conductor cables the watts corresponding to the different currents must be multiplied by two, and to obtain the currents corresponding to the watts in the table multiply the currents given in the table by 0.707.

For three-conductor cables the watts corresponding to the currents in the table must be multiplied by 3, and to obtain the currents corresponding to the watts in the table multiply the currents given in the table by 0.577.

TABLE VII.—PER CENT. POWER LOSS PER MILE PER THOUSAND KILOWATTS DELIVERED, THREE-PHASE SYSTEM, COPPER WIRE, UNITY POWER FACTOR

(H. Pender, *Proceedings A. I. E. E.*, June, 1908)

Size, B. & S.	Cir. mils	Pounds per mile 3-stranded wires		Kilovolts delivered									
		Copper	Aluminum	2	6	10	20	30	40	50	60	80	100
	500,000	24,200	7,290	2.7	0.31	0.11	0.027	0.012	0.0069	0.0044	0.0031	0.0017	0.0011
	450,000	21,700	6,500	3.1	0.34	0.12	0.031	0.014	0.0076	0.0049	0.0034	0.0019	0.0012
	400,000	19,300	5,830	3.4	0.38	0.14	0.034	0.015	0.0086	0.0055	0.0038	0.0021	0.0014
	350,000	16,900	5,100	3.9	0.44	0.16	0.039	0.017	0.0098	0.0063	0.0044	0.0025	0.0016
	300,000	14,500	4,370	4.6	0.51	0.18	0.046	0.020	0.011	0.0073	0.0051	0.0029	0.0018
	250,000	12,100	3,650	5.5	0.61	0.22	0.055	0.024	0.014	0.0088	0.0061	0.0034	0.0022
0000	211,600	10,200	3,080	6.5	0.72	0.26	0.065	0.029	0.016	0.010	0.0072	0.0041	0.0026
000	167,800	8,130	2,450	8.2	0.91	0.33	0.082	0.036	0.020	0.013	0.0091	0.0051	0.0033
00	133,100	6,430	1,940	10.0	1.1	0.41	0.10	0.046	0.026	0.016	0.011	0.0064	0.0041
0	105,500	5,100	1,540	13.0	1.4	0.52	0.13	0.058	0.033	0.021	0.014	0.0081	0.0052
1	83,690	4,040	1,220	16.0	1.8	0.65	0.16	0.072	0.041	0.026	0.018	0.010	0.0065
2	66,370	3,220	969	21.0	2.3	0.83	0.21	0.092	0.052	0.033	0.023	0.013	0.0083
3	52,630	2,540	768	26.0	2.9	1.0	0.26	0.12	0.065	0.042	0.029	0.016	0.010
4	41,740	2,010	609	33.0	3.7	1.3	0.33	0.15	0.082	0.053	0.037	0.021	0.013
5	33,100	2,002	.....	41.0	4.6	1.7	0.41	0.18	0.10	0.066	0.046	0.026	0.017
6	26,250	1,587	.....	52.0	5.8	2.1	0.52	0.23	0.13	0.084	0.058	0.033	0.021

Table VII, giving the per cent. power loss per mile per 1,000 kw. delivered at unity power factor over a three-phase copper line, will be found useful in determining the size of wire required to transmit a given amount of power under given conditions.

Note.—The foregoing table is calculated for copper wire and unity power factor. The size of wire to use under other conditions can be readily found by determining from the table the size of copper wire to use in case the power factor were unity and then for the actual conditions selecting a wire as follows:

For copper wire and
- 90 per cent. power factor, take a wire of 25 per cent. greater cross section (on B. & S. gage a wire larger by one number).
- 80 per cent. power factor, take a wire of 60 per cent. greater cross section (on B. & S. gage a wire larger by two numbers).

For aluminum wire and
- 100 per cent. power factor, take a wire of 60 per cent. greater cross section (on B. & S. gage a wire larger by two numbers).
- 90 per cent. power factor, take a wire of twice the cross section (on B. & S. gage a wire larger by three numbers).
- 80 per cent. power factor, take a wire of two and one-half times the cross section (on B. & S. gage a wire larger by four numbers).

For any voltage not given in the table,
- the size wire to use can be readily determined by remembering that the cross section for a given per cent. power loss varies inversely as the square of the voltage, e.g., for 11 kilovolts delivered multiply the cross section corresponding to 10 kilovolts by 10/11 – 0.83.

Example: 20,000 kilowatts is to be transmitted 30 miles over a three-phase system with a total loss in the line of 10 per cent. of the power delivered, the delivered pressure to be 60,000 volts; to find the size of wire required.

The power loss per mile per 1,000 kw. delivered is

$$\frac{10}{30 \times 20} = 0.017 \text{ per cent.}$$

From the table the nearest size copper wire corresponding to 60 kilovolts and a loss of 0.017 per cent. is No. 1 B. & S. (this will give a total loss slightly greater than 10 per cent.). If the power factor of the load at the end of the line is 90 per cent., No. 0 copper should be used (see note below table); if the power factor is 80 per cent., No. 00 wire should be used.

If aluminum is to be used, the size of wire under the various conditions would be for unity power factor, No. 00; for 90 per cent. power factor, No. 000; and for 80 per cent. power factor, No. 0000.

TABLE VIII.—OVERLOAD CARRYING CAPACITIES OF CABLES

(By Ralph W. Atkinson and H. W. Fisher, *Proceedings A. I. E. E.*, Feb., 1913

Per cent.	No. 6 B. & S. gage			No. 4 B. & S. gage			No. 2 B. & S. gage			No. 0 B. & S. gage		
	80%	50%	0%	80%	50%	0%	80%	50%	0%	80%	50%	0%
	Time in minutes			Time in minutes			Time in minutes			Time in minutes		
200	0.6	1.3	1.8	0.8	1.6	2.0	1.0	2.2	2.9	1.5	3.1	4.5
175	0.9	2.0	2.7	1.2	2.5	3.0	1.5	3.2	4.5	2.3	5.0	6.5
150	1.6	3.2	4.5	2.0	4.0	5.5	2.7	5.5	7.0	4.0	8.5	10.0
125	3.5	7.0	8.5	4.5	9.0	11.0	6.0	12.0	15.0	9.0	16.0	20.0

Per cent.	No. 000 B. & S. gage			300,000 cir. mil			500,000 cir. mil			700,000 cir. mil		
	80%	50%	0%	80%	50%	0%	80%	50%	0%	80%	50%	0%
	Time in minutes			Time in minutes			Time in minutes			Time in minutes		
200	2.1	4.5	6.0	3.2	7.5	9.5	5.0	10.0	13.0	6.0	12.0	15.0
175	3.2	6.5	9.0	5.0	11.0	13.0	7.0	14.0	17.0	8.5	16.0	22.0
150	5.5	12.0	15.0	9.0	16.0	20.0	12.0	21.0	26.0	14.0	26.0	33.0
125	13.0	23.0	28.0	18.0	31.0	39.0	23.0	40.0	45.0	28.0	50.0	60.0

Per cent.	1,000,000 cir. mil			1,200,000 cir. mil			1,500,000 cir. mil			2,000,000 cir. mil		
	80%	50%	0%	80%	50%	0%	80%	50%	0%	80%	50%	0%
	Time in minutes			Time in minutes			Time in minutes			Time in minutes		
200	7.5	15.0	19.0	8.5	17.0	22.0	11 0	21.0	26.0	12.0	24.0	30.0
175	11.0	22.0	27.0	13.0	25.0	31.0	16.0	30.0	36.0	18.0	34.0	40.0
150	18.0	33.0	41.0	20.0	37.0	45.0	25.0	46.0	55.0	28.0	50.0	65.0
125	35.0	60.0	73.0	40.0	70.0	85.0	50.0	80.0	95.0	55.0	95.0	110.0

The data given are the lengths of time during which the loads given in the left-hand column can be carried, when the loads given in the first horizontal line of each table have been continuously applied previous to the application of the overload. Load is expressed in percentage of continuous carrying capacity, that is normal load. It must be noted that a considerable variation in the time frequently makes only a few degrees difference in the temperature.

To determine the time during which a three-conductor cable may carry a given per cent. overload use the table as for a single conductor having twice the cross-section area. For a two-conductor cable use the table as though for single-conductor cable of 50 per cent. greater cross-section.

TABLE IX.—TIME REQUIRED TO ATTAIN ABOUT 90 PER CENT. OF FINAL TEMPERATURE
(By Ralph W. Atkinson and H. W. Fisher, *Proceedings A. I. E. E.*, Feb., 1913)

Size, B. & S. gage and cir. mils	Hours	Size, B. & S. gage and cir. mils	Hours
6	¾	500,000	2
1	1	700,000	2½
000	1¼	1,000,000	3
300,000	1½	1,500,000	3¾
		2,000,000	4¼

It will require 50 per cent. longer time to attain 97 per cent. of final temperature and will require 50 per cent. less time to reach about 68 per cent. of final temperature.

The time given in this table may also be taken as the time interval which must elapse between overloads. If repeated oftener, the allowable duration of overload is decreased. If repeated at intervals of one-half this time, the allowable duration of overload is decreased 35 per cent.

Example of the Use of Tables VIII and IX.—It is required to determine the time an 800,000-cir. mil cable can carry 1,200 amp. when it carries continuously 400 amp., its continuous capacity being 650 amp. Express the loads as 185 per cent. and 62 per cent. of normal rating. From the tables it will be found that this load may be carried for about 11 minutes at intervals of 2½ hours.

TABLE X.—REACTANCE PER MILE OF A NO. 0000 B. & S. WIRE = X_1
(H. Pender, *Electrical World*, June 10, 1909)

Distance apart of wires in feet	15 cycles	25 cycles	40 cycles	60 cycles	125 cycles
1	0.128	0.213	0.340	0.510	1.063
2	0.149	0.248	0.396	0.594	1.328
3	0.161	0.268	0.429	0.644	1.341
4	0.170	0.283	0.452	0.678	1.413
5	0.176	0.294	0.470	0.705	1.470
6	0.182	0.303	0.485	0.728	1.516
7	0.187	0.311	0.498	0.746	1.555
8	0.191	0.318	0.508	0.763	1.589
9	0.194	0.324	0.518	0.777	1.618
10	0.197	0.329	0.526	0.790	1.645
15	0.210	0.350	0.559	0.839	1.748
20	0.218	0.364	0.582	0.874	1.820
25	0.225	0.375	0.601	0.901	1.877

TABLE XI.—RESISTANCE PER MILE OF COPPER AND ALUMINUM STRANDED WIRES AND REACTANCE INCREMENT[1] X_2

(H. Pender, *Electrical World*, June 10, 1909)

Size, cir. mils and B. & S. gage	Ohms per mile at 20° C.[1]		Difference[2] in reactance per mile of any size wire and that of a No. 0000 B. & S. wire = X_2				
	Copper	Aluminum	15 cy.	25 cy.	40 cy.	60 cy.	125 cy.
1,000,000	0.0566	0.0894	−.024	−.039	−.063	−.094	−.196
900,000	0.0629	0.0993	−.022	−.037	−.059	−.088	−.183
800,000	0.0707	0.1118	−.020	−.034	−.054	−.081	−.168
700,000	0.0808	0.1278	−.018	−.030	−.048	−.073	−.152
600,000	0.0943	0.1490	−.016	−.026	−.042	−.063	−.132
500,000	0.1131	0.1788	−.013	−.022	−.035	−.052	−.109
450,000	0.1257	0.1987	−.011	−.019	−.031	−.046	−.095
400,000	0.1414	0.224	−.010	−.016	−.026	−.039	−.080
350,000	0.1616	0.255	−.008	−.013	−.020	−.031	−.064
300,000	0.1886	0.298	−.005	−.009	−.014	−.021	−.044
250,000	0.226	0.358	−.003	−.004	−.007	−.010	−.021
0000	0.267	0.423	.000	.000	.000	.000	.000
000	0.337	0.533	+.004	+.006	+.009	+.014	+.029
00	0.425	0.672	+.007	+.012	+.019	+.028	+.059
0	0.536	0.848	+.011	+.018	+.028	+.042	+.088
1	0.676	1.068	+.014	+.023	+.038	+.056	+.117
2	0.852	1.347	+.019	+.029	+.047	+.070	+.147
4	1.355	2.14	+.025	+.041	+.066	+.098	+.205
6	2.15	3.41	+.032	+.053	+.084	+.127	+.264

[1] Stranded wire, copper 98 per cent., aluminum 62 per cent. conductivity, resistance increased 1 per cent. on account of stranding; temperature coefficient, 0.42 per cent. per degree centigrade.

[2] The total reactance of a wire for any spacing and frequency is $x = x_1 + x_2$, where x_1 = the reactance of a No. 0000 wire under the same conditions.

TABLE XII.—CAPACITY SUSCEPTANCE PER MILE OF TWO PARALLEL STRANDED AERIAL WIRES, FREQUENCY ONE CYCLE-SECOND

(H. Pender, *Electrical World*, June 10, 1909

Susceptance between wires = b_1

Size, cir. mils and B. & S.	1	2	3	4	6	8	10	12
1,000,000	9.3×10^{-8}	7.5×10^{-8}	6.8×10^{-8}	6.3×10^{-8}	5.8×10^{-8}	5.5×10^{-8}	5.3×10^{-8}	5.1×10^{-8}
500,000	8.3×10^{-8}	6.9×10^{-8}	6.3×10^{-8}	5.9×10^{-8}	5.4×10^{-8}	5.2×10^{-8}	4.9×10^{-8}	4.8×10^{-8}
250,000	7.5×10^{-8}	6.3×10^{-8}	5.8×10^{-8}	5.5×10^{-8}	5.1×10^{-8}	4.9×10^{-8}	4.7×10^{-8}	4.5×10^{-8}
0000	7.4×10^{-8}	6.2×10^{-8}	5.7×10^{-8}	5.4×10^{-8}	5.0×10^{-8}	4.8×10^{-8}	4.6×10^{-8}	4.5×10^{-8}
1	6.6×10^{-8}	5.6×10^{-8}	5.2×10^{-8}	5.0×10^{-8}	4.6×10^{-8}	4.4×10^{-8}	4.3×10^{-8}	4.1×10^{-8}
6	5.8×10^{-8}	5.0×10^{-8}	4.7×10^{-8}	4.5×10^{-8}	4.2×10^{-8}	4.0×10^{-8}	3.9×10^{-8}	3.8×10^{-8}

Note.—The charging current per mile of single-phase line (2 miles of wire) is equal to $bnE \times 10^3$ for a single-phase line; the charging current per wire per mile of line (3 miles of wire) is equal to $1.16 \times 10^3\ nbE$, where n is the cycles per second, E the kilovolts between wires.

Table XIII.—Rubber-covered Wire

Insulation for Voltages Between 0 and 600, as Recommended by National Board of Fire Underwriters

B. & S. Gage	Thickness
18 to 16	1/32 inch
15 to 8	3/64 inch
7 to 2	1/16 inch
1 to 0000	5/64 inch
Cir. mils	
250,000 to 500,000	3/32 inch
500,000 to 1,000,000	7/64 inch
Over 1,000,000	1/8 inch

Table XIV.—Insulation for Voltages Between 600 and 3,500

B. & S. Gage	Thickness
14 to 1	3/32 inch
0 to 0000	3/32 inch, covered by tap or braid
Cir. mils	
250,000 to 500,000	3/32 inch, covered by tap or braid
Over 500,000	1/8 inch, covered by tap or braid

TABLE XV.—VARNISHED-CAMBRIC-INSULATED CABLES (GENERAL ELECTRIC COMPANY)
Single Conductor

Size, B. & S. gage and cir. mils		1000 V. working			3,000 V. working			5,000 V. working			7,000 V. working			10,000 V. working			15,000 V. working		
		Thick. insul., ins.	Thick. lead, ins.	Dia. in ins.	Thick. insul., ins.	Thick lead, ins.	Dia. in ins.	Thick. insul., ins.	Thick. lead, ins.	Dia. in ins.	Thick. insul., ins.	Thick. lead, ins.	Dia. in ins.	Thick. insul., ins.	Thick lead, ins.	Dia. in ins.	Thick. insul., ins.	Thick. lead, ins.	Dia. in ins.
Solid	6	1/16	3/64	0.40	3/32	3/64	0.46	9/64	3/64	0.55	3/16	1/16	0.68	1/4	1/16	0.80	11/32	3/32	1.05
	4	1/16	3/64	0.44	3/32	3/64	0.50	9/64	1/16	0.63	3/16	1/16	0.72	1/4	1/16	0.84	11/32	3/32	1.10
Strand	6	1/16	3/64	0.42	3/32	3/64	0.48	9/64	3/64	0.57	3/16	1/16	0.70	1/4	1/16	0.82	11/32	3/32	1.08
	4	1/16	3/64	0.47	3/32	3/64	0.53	9/64	1/16	0.65	3/16	1/16	0.75	1/4	1/16	0.87	11/32	3/32	1.12
	2	1/16	1/16	0.56	3/32	1/16	0.62	9/64	1/16	0.71	3/16	1/16	0.81	1/4	5/64	0.96	11/32	3/32	1.18
	1	5/64	1/16	0.63	3/32	1/16	0.66	5/32	1/16	0.79	3/16	5/64	0.88	1/4	3/32	1.04	3/8	3/32	1.29
	0	5/64	1/16	0.67	3/32	1/16	0.70	5/32	5/64	0.86	3/16	3/32	0.95	1/4	3/32	1.08	3/8	3/32	1.33
	00	5/64	1/16	0.72	3/32	1/16	0.75	5/32	3/32	0.94	3/16	3/32	1.00	1/4	3/32	1.12	3/8	3/32	1.37
	000	5/64	1/16	0.77	3/32	5/64	0.83	5/32	3/32	0.99	3/16	3/32	1.05	1/4	3/32	1.17	3/8	3/32	1.42
	0000	3/32	5/64	0.89	7/64	5/64	0.92	5/32	3/32	1.04	3/16	3/32	1.11	1/4	3/32	1.23	3/8	3/32	1.48
250,000		3/32	3/32	0.97	7/64	3/32	1.00	5/32	3/32	1.09	3/16	3/32	1.15	1/4	3/32	1.28	3/8	3/32	1.53
300,000		3/32	3/32	1.03	1/8	3/32	1.09	5/32	3/32	1.15	13/64	3/32	1.24	1/4	3/32	1.38	3/8	7/64	1.63
400,000		3/32	3/32	1.12	1/8	3/32	1.16	5/32	3/32	1.24	13/64	3/32	1.36	1/4	3/32	1.48	3/8	1/8	1.73
500,000		3/32	3/32	1.21	1/8	3/32	1.27	5/32	3/32	1.33	7/32	3/32	1.46	1/4	7/64	1.57	3/8	1/8	1.82
750,000		3/64	3/32	1.42	1/8	3/32	1.45	11/64	7/64	1.58	7/32	7/64	1.67	17/64	7/64	1.80	25/64	1/8	2.05
1,000,000		7/64	3/32	1.58	1/8	3/32	1.61	11/64	7/64	1.73	7/32	1	1.85	17/64	1/8	1.96	13/32	1/8	2.23
1,250,000		1/8	7/64	1.78	1/8	7/64	1.77	....	...	....	....	...	....	....	....	....	....	....	....
1,500,000		1/8	1/8	1.93	1/8	1/8	1.93	....	...	....	....	...	....	....	....	....	....	....	....
2,000,000		1/8	1/8	2.15	1/8	1/8	2.15	....	...	....	....	...	....	....	....	....	....	....	....

The columns headed "Dia. in Ins." give over-all diameter of the finished cable, which is practically the same for either braided or leaded.

TABLE XVI.—VARNISHED-CAMBRIC-INSULATED CABLES (GENERAL ELECTRIC COMPANY)
Triple Conductors
(All dimensions given in inches)

Size, B. & S. gage and cir. mils	1,000 V. working			3,000 V. working			5,000 V. working			7,000 V. working			10,000 V. working			15,000 V. working		
	Thick. insul.	Thick. lead.	Dia.	Thick. insul.	Thick. lead	Dia.	Thick. insul.	Thick. lead	Dia.	Thick. insul.	Thick. lead	Dia.	Thick. insul.	Thick. lead	Dia.	Thick. insul.	Thick. lead	Dia.
6	1/16-1/32	1/16	0.87	5/64-3/64	5/64	1.03	3/32-5/64	3/32	1.20	1/8-1/8	3/32	1.42	5/32-5/32	3/32	1.62	3/16-3/16	3/32	1.82
4	1/16-1/32	1/16	0.97	5/64-3/64	5/64	1.14	3/32-5/64	3/32	1.30	1/8-1/8	3/32	1.53	5/32-5/32	3/32	1.73	3/16-3/16	3/32	1.92
2	1/16-1/32	5/64	1.17	5/64-3/64	5/64	1.27	3/32-5/64	3/32	1.43	1/8-1/8	3/32	1.66	5/32-5/32	3/32	1.86	3/16-3/16	1/8	2.11
1	1/16-1/32	5/64	1.25	5/64-3/64	3/32	1.38	3/32-3/32	3/32	1.55	1/8-1/8	1/8	1.81	11/16-11/64	1/8	2.10	13/64-13/64	1/8	2.30
0	1/16-1/32	5/64	1.34	5/64-1/16	3/32	1.50	3/32-3/32	3/32	1.64	1/8-1/8	1/8	1.89	11/64-11/64	1/8	2.19	13/64-13/64	1/8	2.39
00	1/16-1/32	3/32	1.47	5/64-1/16	3/32	1.60	3/32-3/32	3/32	1.74	1/8-1/8	1/8	1.99	11/64-11/64	1/8	2.29	13/64-13/64	1/8	2.48
000	1/16-1/32	3/32	1.58	3/32-1/16	3/32	1.78	3/32-3/32	1/8	1.91	1/8-1/8	1/8	2.10	11/64-11/64	1/8	2.40	13/64-13/64	1/8	2.59
0000	5/64-1/32	1/8	1.84	3/32-1/16	1/8	1.97	3/32-3/32	1/8	2.03	1/8-1/8	1/8	2.23	11/64-11/64	1/8	2.52	13/64-13/64	1/8	2.72
250,000	3/32-3/64	1/8	1.97	3/32-1/16	1/8	2.07	3/32-3/32	1/8	2.13	1/8-1/8	1/8	2.33	11/64-11/64	1/8	2.63	13/64-13/64	1/8	2.82

Under "Thickness of Insulation," the first column is the thickness of insulation on each conductor and the second column is the thickness over all. The column "Diameter" is the over-all diameter of the finished cable and is approximately the same for either braided or leaded.

Table XVII.—Sparking Distances

The following table gives the sparking distances between sharp points corresponding to different alternating-current voltages when the ratio between maximum and mean effective voltages is $\sqrt{2} = 1.41$. The values given were derived from experiments made by the Standard Underground Cable Co.

(Copyright, 1906, by Standard Underground Cable Co.)

Volts	Spark distance, A or B	Volts	Spark distance A	Spark distance B	Volts	Spark distance A	Spark distance B
1,000	0.028	21,000	1.092	1.097	44,000	2.370	2.506
2,000	0.098	22,000	1.143	1.150	45,000	2.432	2.580
3,000	0.159	23,000	1.195	1.206	46,000	2.495	2.60
4,000	0.216	24,000	1.247	1.260	47,000	2.560	
5,000	0.270	25,000	1.300	1.314	48,000	2.625	
6,000	0.324	26,000	1.353	1.373	49,000	2.692	
7,000	0.378	27,000	1.405	1.427	50,000	2.760	
8,000	0.432	28,000	1.460	1.485			
9,000	0.487	29,000	1.512	1.540			
10,000	0.540	30,000	1.566	1.600			
11,000	0.595	31,000	1.620	1.655			
12,000	0.644	32,000	1.675	1.712			
13,000	0.695	33,000	1.728	1.772			
14,000	0.746	34,000	1.785	1.833			
15,000	0.797	35,000	1.840	1.895			
16,000	0.845	36,000	1.900	1.958			
17,000	0.897	37,000	1.945	2.020			
18,000	0.945	38,000	2.012	2.085			
19,000	0.995	39,000	2.062	2.153			
20,000	1.042	40,000	2.127	2.220			
......	.....	41,000	2.190	2.290			
......	.....	42,000	2.247	2.360			
......	.....	43,000	2.308	2.434			
......	.....	......	.....	.....			
......	.....	......	.....	.....			

American Institute of Electrical Engineers' table

Volts	Spark distance
10,000	0.47
20,000	1.04
25,000	1.30
30,000	1.625
40,000	2.45
45,000	2.95
50,000	3.55

Column "A" gives spark distance with 10-in. concave metallic shields, the plane of whose edges was 1 in. back of the needle points.

Column "B" gives the spark distance without shields.

INDEX

T